T0134475

Advances in Intelligent Systems and Computing

Volume 1068

The series "Advances in Intelligent Systems and Computing" contains publications on theory, applications, and design methods of Intelligent Systems and Intelligent Computing. Virtually all disciplines such as engineering, natural sciences, computer and information science, ICT, economics, business, e-commerce, environment, healthcare, life science are covered. The list of topics spans all the areas of modern intelligent systems and computing such as: computational intelligence, soft computing including neural networks, fuzzy systems, evolutionary computing and the fusion of these paradigms, social intelligence, ambient intelligence, computational neuroscience, artificial life, virtual worlds and society, cognitive science and systems, Perception and Vision, DNA and immune based systems, self-organizing and adaptive systems, e-Learning and teaching, human-centered and human-centric computing, recommender systems, intelligent control, robotics and mechatronics including human-machine teaming, knowledge-based paradigms, learning paradigms, machine ethics, intelligent data analysis, knowledge management, intelligent agents, intelligent decision making and support, intelligent network security, trust management, interactive entertainment, Web intelligence and multimedia.

The publications within "Advances in Intelligent Systems and Computing" are primarily proceedings of important conferences, symposia and congresses. They cover significant recent developments in the field, both of a foundational and applicable character. An important characteristic feature of the series is the short publication time and world-wide distribution. This permits a rapid and broad dissemination of research results.

** Indexing: The books of this series are submitted to ISI Proceedings, EI-Compendex, DBLP, SCOPUS, Google Scholar and Springerlink **

More information about this series at http://www.springer.com/series/11156

António Pedro Costa · Luís Paulo Reis ·
António Moreira
Editors

Computer Supported Qualitative Research

New Trends on Qualitative Research
(WCQR2019)

 Springer

Editors
António Pedro Costa
Department of Education and Psychology,
Research Center on Didactics
and Technology in Education
of Trainers (CIDTFF)
Ludomedia and University of Aveiro
Aveiro, Portugal

Luís Paulo Reis
Faculty of Engineering, Artificial
Intelligence and Computer Science
Laboratory (LIACC)
University of Porto
Porto, Portugal

António Moreira
Department of Education and Psychology,
Research Center on Didactics
and Technology in Education
of Trainers (CIDTFF)
University of Aveiro
Aveiro, Portugal

ISSN 2194-5357   ISSN 2194-5365   (electronic)
Advances in Intelligent Systems and Computing
ISBN 978-3-030-31786-7   ISBN 978-3-030-31787-4   (eBook)
https://doi.org/10.1007/978-3-030-31787-4

Preface

This book contains a selection of the articles accepted for presentation and discussion at the fourth World Conference on Qualitative Research (WCQR2019), held in Porto, Portugal, October 16–18, 2019. WCQR2019 was organized by Lusófona University of Porto (ULP) and Ludomedia. The conference organization also had the collaboration and/or sponsoring of several universities, research institutes and companies, including Aveiro University/CIDTFF, National Centre for Research Methods (NCRM), Asian Qualitative Research Association (AQRA), Interdisciplinary Research Centre for Education and Development (CeiED), Ibero-American Congress on Qualitative Research (CIAIQ), Microio, webQDA, ATLAS.ti and Optimal Workshop.

WCQR2019 builds upon several successful events, including WCQR2018, held in Lisbon, Portugal, CIAIQ/ISQR2017, held in Salamanca, Spain, and CIAIQ/ISQR2016, held in Porto, Portugal. The conference's focus was on Qualitative Research with emphasis on methodological aspects and their relationship with research questions, theories and results. This book is mainly focused on the use of Computer-Assisted Qualitative Data Analysis Software (CAQDAS) for assisting researchers in using correct methodological approaches for Qualitative Research projects.

WCQR2019 featured four main application fields (Education, Health, Social Sciences, Engineering and Technology) and seven main subjects: Rationale and Paradigms of Qualitative Research (theoretical studies, critical reflection about epistemological, ontological and axiological dimensions); Systematization of approaches with Qualitative Studies (literature review, integrating results, aggregation studies, meta-analysis, meta-analysis of qualitative meta-synthesis, meta-ethnography); Qualitative and Mixed Methods Research (emphasis on research processes that build on mixed methodologies but with priority to qualitative approaches); Data Analysis Types (content analysis, discourse analysis, thematic analysis, narrative analysis, etc.); Innovative processes of Qualitative Data Analysis (design analysis, articulation and triangulation of different sources of data—images, audio, video); Qualitative Research in Web Context (eResearch, virtual ethnography, interaction analysis, Internet latent corpora, etc.); and

Qualitative Analysis with the Support of Specific Software (usability studies, user experience, the impact of software on the quality of research and analysis).

In total, after a careful review process with at least three independent reviewers for each paper, a total of 21 high-quality papers from WCQR were selected for publication, totalling over 66 authors from eight countries, including Brazil, Lithuania, Ireland, New Zealand, Kenya, Poland, Portugal and Spain. The volume also features three invited papers authored by Keynote Speakers from the WCQR conference.

We would also like to take this opportunity to express a special word of acknowledgement to the members of the WCQR2019 organization—Catarina Brandão, Conceição Ferreira, Elisabete Pinto da Costa and Sónia Mendes – for their hard and fine work on the scientific management, local logistics, publicity, publication and financial issues. We also express our gratitude to all the members of the WCQR Program Committees and to the additional reviewers, as they were crucial for ensuring the high scientific quality of the event. We would also like to acknowledge all the authors and delegates whose research work and participation made this event a success. Finally, we acknowledge and thank all Springer staff for their help in the production of this volume.

October 2019   António Pedro Costa
                                                                                      Luís Paulo Reis
                                                                                      António Moreira

Organization

Scientific Committee

Ana Amélia Carvalho	University of Coimbra, Portugal
Ana António	CeiED—Interdisciplinary Research Centre for Education and Development, Portugal
Ana Isabel Rodrigues	Polytechnic Institute of Beja, Portugal
Ana María Pinto Llorente	University of Salamanca, Spain
Ana Paula Sousa Santos	University of the Azores, Portugal
António Carrizo Moreira	University of Aveiro, Portugal
António Dias de Figueiredo	University of Coimbra, Portugal
Branislav Radeljic	University of East London, UK
Brigitte Smit	University of South Africa, South Africa
Carl Vogel	Trinity College Dublin, Ireland
Carla Galego	CeiED—Interdisciplinary Research Centre for Education and Development, Portugal
Catarina Brandão	Faculty of Psychology and Educational Sciences, University of Porto, Portugal
Cecília Guerra	University of Aveiro, Portugal
Celina Pinto Leão	University of Minho, Portugal
Dely Elliot	University of Glasgow, UK
Effie Kritikos	Northeastern Illinois University, USA
Elena Rozhdestvenskaya	National Research University Higher School of Economics, Russia
Elizabeth Pope	University of West Georgia, USA
Emanuela Girei	Sheffield University Management School, UK
Fátima Marques	ESEL—Lisbon School of Nursing, Portugal
Fernando Albuquerque Costa	University of Lisbon, Portugal
Gilberto Tadeu Reis da Silva	Federal University of Bahia, Brazil

Helder José Alves Rocha Pereira	University of the Azores, Portugal
Helena Presado	ESEL—Lisbon School of Nursing, Portugal
Helga Rafael	ESEL—Lisbon School of Nursing, Portugal
Henrika Jormfeldt	Halmstad University, Sweden
Inês Amaral	Miguel Torga Higher Institute, Portugal
Isabel Cabrita	University of Aveiro, Portugal
Isabel Huet	University of West London, UK
Isabel Pinho	University of Aveiro, Portugal
Izabela Ślęzak	University of Lodz, Poland
Jaime Ribeiro	Polytechnic of Leiria, Portugal
Jakub Niedbalski	University of Lodz, Poland
João Amado	University of Coimbra, Portugal
Jose Siles-González	University of Alicante, Spain
Judita Kasperiuniene	Vytautas Magnus University, Lithuania
Kathleen Gilbert	Indiana University, USA
Katrin Niglas	Tallinn University, Estonia
Lakshmi Balachandran Nair	Utrecht University, Netherlands
Lia Oliveira	University of Minho, Portugal
Lubomir Popov	Bowling Green State University, USA
Lucila Castanheira Nascimento	University of São Paulo, Brazil
Luís Paulo Reis	University of Porto, Portugal
Luís Sousa	Atlântica University, Portugal
Mabel Segu	Deusto University, Spain
Marcel Ausloos	University of Leicester, UK
Marcos Teixeira de Abreu Soares Onofre	University of Lisbon, Portugal
Maria José Brites	Lusófona University of Porto, Portugal
Martin Tolich	University of Otago, New Zealand
Martina Gallarza	University of Valencia, Spain
Miguel Serra	ESEL—Lisbon School of Nursing, Portugal
Mirliana Pereira	University of Chile, Chile
Miroslav Rajter	University of Zagreb, Croatia
Neringa Kalpokaite	IE University, Spain
Óscar Ferreira	ESEL—Lisbon School of Nursing, Portugal
Patricia López	TEC—Costa Rica Institute of Technology, Costa Rica
Philia Issari	Centre for Qualitative Research in Psychology and Psychosocial—University of Athens, Greece
Rosa Godoy	University of São Paulo, Brazil
Rui Neves	University of Aveiro, Portugal
Rui Vieira	University of Aveiro, Portugal
Sandra Saúde	Polytechnic Institute of Beja, Portugal

Contents

Empowering Qualitative Research Methods in Education with Artificial Intelligence

Luca Longo[(⊠)] [iD]

School of Computer Science, Technological University Dublin,
Dublin, Republic of Ireland
luca.longo@dit.ie

Abstract. Artificial Intelligence is one of the fastest growing disciplines, disrupting many sectors. Originally mainly for computer scientists and engineers, it has been expanding its horizons and empowering many other disciplines contributing to the development of many novel applications in many sectors. These include medicine and health care, business and finance, psychology and neuroscience, physics and biology to mention a few. However, one of the disciplines in which artificial intelligence has not been fully explored and exploited yet is education. In this discipline, many research methods are employed by scholars, lecturers and practitioners to investigate the impact of different instructional approaches on learning and to understand the ways skills and knowledge are acquired by learners. One of these is qualitative research, a scientific method grounded in observations that manipulates and analyses non-numerical data. It focuses on seeking answers to why and how a particular observed phenomenon occurs rather than on its occurrences. This study aims to explore and discuss the impact of artificial intelligence on qualitative research methods. In particular, it focuses on how artificial intelligence have empowered qualitative research methods so far, and how it can be used in education for enhancing teaching and learning.

Keywords: Artificial intelligence · Qualitative research · Methods · Data analysis · Education · Teaching · Learning · Behaviourism · Constructivism · Cognitivism · Automated reasoning · Knowledge representation · Machine learning · Planning · Perception · Natural language processing

1 Introduction

Artificial Intelligence has been disrupting many sectors and disciplines, offering tools to conduct research and support innovation that were not available a couple of decades ago. This has been facilitated by the technological progress and the availability of instruments and technologies for collecting, storing and analysing data of many forms. Artificial Intelligence has contributed to many

© Springer Nature Switzerland AG 2020
A. P. Costa et al. (Eds.): WCQR 2019, AISC 1068, pp. 1–21, 2020.
https://doi.org/10.1007/978-3-030-31787-4_1

fields and disciplines, accelerating scientific discovery [32]. It has been informing natural language research [50], the social sciences [83], medicine [38,106],
finance [3], psychology and human behaviour research [105], neuroscience [40]
just to mention a few. However, there is a discipline in which the potential of
artificial intelligence has not been fully explored and explored yet: education.
The key goal of education is to design approaches, methodologies, methods and
techniques for enabling learning and facilitating the acquisition of knowledge
and skills. Many educational methods are used within education with teaching
the core one. Teaching can occur in many forms and it is the aim of pedagogy
to investigate different methods and how they can affect and enhance learning
[18]. In contrast to other disciplines where quantitative research methods are frequently employed, education and pedagogy heavily rely on qualitative methods
to extract patterns from non-numerical data and allow inferences.

As a researcher in artificial intelligence and computer science as well as a
third-level pedagogist, I strongly support the use of technology in general, and
AI-based tools and techniques in particular to support teaching and promote
learning. This article is intended for the broad audience of pedagogists, psychologists, social scientists and educational practitioners, on average not having the
formal and technical skills usually belonging to engineers and computer scientists. Its aim is to informally define artificial intelligence and its main goals, so
as to provide readers with the basic and core components of this fast growing
discipline. Subsequently, the learning theories mainly used within education by
pedagogists are introduced and the main qualitative research methods generally used by them are briefly described. A discussion follows, showing how the
theoretical and practical advances in artificial intelligence have empowered qualitative research data analysis methods so far. Eventually, it proposes ways of
how these advances can advance pedagogy in particular, through specific contributions to each educational learning theories (Fig. 2).

2   Background

Artificial intelligence, education and qualitative research methods have long his-
tories and for those scholars working mainly in one of them, some notions and
concepts belonging to the others might not sound very familiar. For this purpose, the content of this section is aimed at providing readers with the relevant
background information and at describing the main core notions and concepts
behind each of these fields of research.

2.1   Artificial Intelligence

Artificial Intelligence (AI) can be traced back to 1943 when McCulloch and
Pitts proposed a model of artificial neurons inspired by the physiological and
functional properties of a neuron in the brain, and building on the propositional
logic of Russell and Whitehead as well as the theory of computation brought
forward by Alan Turing [80]. From this, artificial intelligence has tremendously

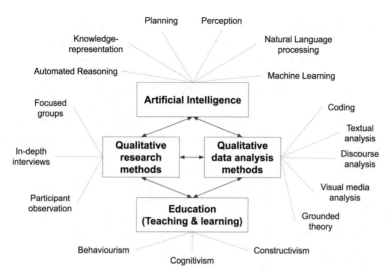

Fig. 1. Exploration of the impact of artificial intelligence for empowering qualitative research, its data analysis methods in education, teaching and learning

evolved and significantly extended its boundaries. It has seen early achievements and great expectations but it has also gone through difficult times [9]. Early research was mainly devoted to knowledge-based systems and the development of expert systems enabling the commercialisation of tools for supporting various tasks in industrial settings. After a period known as the AI winter, in middle 80s, neural networks returned to the scene, and with the re-design of the back-propagation algorithm, they posed the basis of modern machine learning [58]. In recent years, AI has gone through a revolution concerning the methodologies employed in research and the actual content being used for research. It is now more common to exploit existing theories for novel applications rather than proposing new ones. Late 90s have seen the emergence of intelligence agents to solve the main sub-problems of AI and then, in the early 2000s, with the availability of large datasets and computational power, it become norm to tackle these problems with a stronger emphasis on data and less on the theoretical approach to apply. Defining AI is not a trivial task since many definitions have been proposed by many scholars in the last 70 years. However, a set of major goals have emerged within AI: knowledge representation, automated reasoning, automated planning, machine perception, machine learning, natural language processing (Fig. 1).

Knowledge-representation focuses on the formal representation of information about the world in a form that can be actually used by computers for solving tasks. It is a branch of AI that is connected to psychology because it builds upon the capabilities of humans to represent knowledge and information as well as to solve problems employing it [104]. Examples includes ontologies [76], frames and rules [124], semantics nets [25]. *Automated reasoning* is another branch of AI concerning logic and reasoning, strictly connected to the goals of knowledge representation. It is aimed at designing and building intelligent agents with automated reasoning capabilities. Example includes reasoning under uncertainty [98], defeasible and non-monotonic reasoning [70] as well as argumentation [67]. *Automated planning* is a sub-field of AI devoted to the design and development of automated scheduling strategies or sequences of actions [31]. This usually concerns the development of intelligent agents that aim to find optimal solutions in a complex multi-dimensional space and it is often connected to decision theories. Applications include planning and scheduling for autonomous vehicles or mean of transportations [90], reconfigurable production systems [11,103] and in general distributed systems [120,121]. *Machine perception* is that sub-field of AI that concerns the construction of artificial systems able to interpret data that resembles the human way through the use of senses and the context around them [134]. With the technological progress and the explosion of sensor-based technologies, early systems that were focusing on analysing data, mainly collected through keyboards and mouse [72], have now been empowered with a wider range of sensory input that is close to the way humans perceive, including computer vision [28], machine hearing [75], and machine touch [13]. *Machine learning* is probably the fastest-growing sub-field of AI concerned with the development of statistical models and algorithms that are able to perform specific tasks without the use of explicit rules, instructions and directives [6]. Machine learning have seen a tremendous acceleration of research outputs and have significantly informed and contributed to the advance of the other sub-branches of artificial intelligence [47]. Some of the plethora of applications includes classification of facial expressions [4] or cancers [22], forecasting in stock markets [43] and image processing [123] just to mention a few. Eventually, *natural language processing* is aimed at designing and developing artificial agents that are capable of interpreting and processing natural language. Example of problems includes speech recognition [42], natural language generation [108] and machine translation [51].

2.2   Qualitative Research Methods

A number of research methodologies and methods are employed by scholars in science. One of these is qualitative research methods [135]. These are scientific methods of observation that, in contrast to quantitative research methods, make use of non-numerical data such as textual information, images, audio and video. It is not grounded on the occurrences of things or their frequencies or measures. Rather it focuses on investigating why and how a phenomenon takes place, by understanding the meaning of concepts, definitions and features of things. The core element is the descriptive observations made by scholars and its main strength lies in its ability to provide complex textual descriptions of how people experience things. Qualitative research has a very long history and its literature is very fast that would be impossible to review all of its methods. However, a number of classes of qualitative research methods have been identified. These are referred to as semi-structured methods and include participant observations, in-depth interviews and focus groups [99].

Participant observation is a method for collecting data on naturally occurring behaviours within their usual contexts [126]. The goal of this method is to get a richer familiarity with a group of people while interacting in their cultural environments, usually over an extended period of time [89]. For example, within education, observation can be conducted through analysis of documents produced by learners in the classroom or other environments [7]. *In-depth interviews* are more relevant on an individual basis, when the researcher is interested in the history of a learner, her/his own perspectives and experiences [60]. Considerable effort is spent by the researcher to grasp the essence of the phenomenon under investigation and to use literary devices to transmit it as rich as possible [143]. *Focus group* is a method that focuses on eliciting data on the cultural norms of a group of people. The main goal is to shape broad overviews and summary of certain issues concerning to a group or represented subgroups [137]. The purpose is to promote a comfortable atmosphere of disclosure of information by people who can share their beliefs, ideas, attitudes and experiences about a topic [142].

The format of the questions usually administered by a qualitative researcher is open-ended. Thus the main type and shape of the data produced by the application of qualitative research methods are field notes, audio, sometimes video, recordings, transcripts and images. These methods seek to explore hypotheses about a phenomenon rather then confirming them. Data collection instruments are flexible as well as those approaches and techniques used for eliciting and categorising responses to questions. The main goal is to describe variation, explain relationships, investigate individual experiences and group norms. Some aspects of research studies are adjustable, as the addition or exclusions of questions or their rewording. The responses by participants might impact the subsequent questions. In nature, the design of a qualitative research study is iterative: it evolves and can be updated according to data and what is learned [63].

2.3   Qualitative Data Analysis Methods

A frequently used method to analyse qualitative data is qualitative *coding*. It refers to that process of assigning descriptive labels to pieces of data with the aim of helping researchers in the development of their theories and hypotheses [2]. It is a process of organising qualitative information in a systematic order employing techniques such as linkage, grouping, aggregation to support meaning extraction and formulation. Unfortunately, this process is often time-consuming as it demands scholars to analyse data with great precision in order to identify interesting patterns, assign descriptive labels and categories according to commonly shared characteristics [116]. However, with the technological progress and the plethora of technologies that can be employed for data gathering, the availability of data is exponentially growing. Therefore, for researchers it would be impossible to go through thousands of records containing qualitative information. This would not support the identification of inconsistencies within data and the development of robust hypotheses.

Coding enables *content analysis*, a strict and systematic process of summarising and reporting written data composed by a set of procedures for the rigorous investigation, analysis and verification of the contents [79] for making replicable and valid inferences [54] of any written communicative materials, such as documents, interviews transcriptions, speeches, intended for people other than the research analyst. It often uses categorisation as a means to reduce large quantities of data. In a nutshell, content analysis involves coding, categorising the unit of analysis such as words and sentences in to meaningful categories and descriptive labels, comparing and linking them as well as concluding by drawing theoretical conclusions and hypotheses. In detail, content analysis starts with the definition of a research question and the population from which the unit of analysis of text are to be sampled. A strategy for defining the sample of interest is set according to the notions of representativeness, access and generalisability. It follows a definition of the context of the generation of the document which includes, for instance, the analysis of who was involved in data collection/generation/transcription, origins of documents, corroboration and authenticity of data. Then the unit of analysis has to be decided as well as the codes to be used in it enabling the construction of the categories for grouping key features of the written data. Subsequently, coding and categorisation of data has to be performed as well as its analysis that includes activities such as extrapolation of trends, patterns and differences, standards and indices as well as linguistic representations. These often produce numerical tabular data that can be used for statistical analysis by employing quantitative research methods. Eventually, summarisation follows with the researcher making speculative inferences and generating theories and hypotheses [18].

Beside content analysis other approaches exist that do not fragment the text as in coding. The assumption here is that words can carry many meanings thus their nuances are context dependent and their separation is not always the best approach. Example of contexts include conversations, narratives and autobiographies, specific types of discourses. In *discourse analysis*, words and sentences are semantically linked to each other and their meaning is influenced by those before and after [30]. Qualitative research analysis can be conducted also on visual media such as images and videos. They are a form of text or discourse thus coding or discourse analysis methodologies can be employed. Visual images can be analysed by reading their meanings and reflectively disclosing the researcher point of view, perspective, values and background. Interpreting an image is a subjective process, thus a more formal approach to minimise the bias and guarantee objectivity must be used as for the coding process. The stage of theory generation offered by qualitative research methods is linked to *Grounded Theory*. This is a methodology for developing theories that are grounded in data that is systematically gathered and analysed [14]. Theories are derived inductively from the analysis of an underlying study and a reflection and interpretation is conducted on the phenomena under investigation. It can be referred to as a set of procedures for explicating the relationship among pieces of data and a set of descriptive categories from which a plausible explanation of the investigated phenomena can be generated. The first step of this methodology is theoretical sampling in which data is collected iteratively until theoretical saturation is reached, that means when the researcher has enough information to describing the underlying phenomena under investigation. Coding is subsequently applied to collected data to disassembling and reassembling it. It first breaks collected data down into manageable chunks, parts such as lines, paragraphs or sections to facilitate meaning and pattern extraction. Examples include open, axial and selective coding techniques. It then employs categorisation to reassemble these parts and produce concepts thus a new understanding that investigates differences and similarities across a number of different cases. Initially, confusion is high, but as the iterative process continues, themes emerge and the analysis becomes more structured and organised. Through the application of constant comparison, core categories are identified, accounting for most of the collected pieces of data and their relationship that far. Eventually, saturation is reached when no new categories are formed, that means when no new insight, categories as well as relationships and properties can be produced. Formally, theoretical saturation is connected to the notion of theoretical completeness that means it is reached when a theory can be shaped and successfully able to explain the data satisfactorily and fully [18].

3   Extending Qualitative Data Analysis Research Methods with Artificial Intelligence

Qualitative data analysis is key to generate preliminary research hypothesis that can be further tested with quantitative research methods. Nowadays, new and less intrusive data gathering technologies can significantly automatise and speed the acquisition process of qualitative data. Their advantage is that they allow to collect higher dimensional information and larger sample sizes. However, the drawback is that this new multi-dimensional and larger amount of data cannot be inspected qualitatively and individually by qualitative researchers in its entirety. Therefore, there is the need of intelligent solutions to minimise the time and labour demanded to qualitative data analysts to interpret qualitative data, but at the same time to take full advantage of the richness of information and depth of knowledge that can be extracted from it. Artificial intelligence, with its theoretical and practical advances, as well as its novel intelligent applications can offer many solutions to this problem. This section is devoted to the introduction and description of novel methods for qualitative data analysis offered by artificial intelligence and its advances. It is intended to provide qualitative research practitioners with notions and techniques that can be employed to extend their traditional methods for data analysis. To achieve this goal, the reminder of this section reviews intelligent applications, created within the larger field of artificial intelligence, that work on qualitative data. These applications will by organised and grouped by considering the main goals of artificial intelligence (Sect. 2.1, and the type and form of qualitative data being considered (Sect. 2.2).

Representing qualitative knowledge into a formal form that either humans and computer can interpret is one of the goals of artificial intelligence. [122] and [52] proposed a method for generating semantic networks from text. These types of networks are aimed at representing a knowledge-base that embeds semantic relations between concepts in a graph [125] and thus can help the qualitative researcher to explore knowledge and extract meanings. Similarly, ontologies play an important role in representing and organising knowledge. They are graphs but usually contains fewer formal semantics than semantic networks. They are aimed at representing entities, events, ideas along with their interdependent relations and properties, by using a system of categories [127]. For example, [145] introduced a method for deriving domain ontologies from the concept maps semiautomatically generated from textual information for educational purposes. A similar work proposed a method for the automatic acquisition of taxonomies, special concept hierarchies from a text corpus in order to formulate rules and relations in an abstract but concise way and foster knowledge search and reuse [17].

Knowledge representation is strictly connected to reasoning, another important sub-field of artificial intelligence. Knowledge-graphs, ontologies, taxonomies can help organise text into interlinked concepts and categories, providing qualitative researchers with a formal tool to build knowledge and extract rules that can be used in a reasoning system. This is a formal tool that generates conclusions from available knowledge inductively or deductively using an underlying logic [74]. Information in reasoning systems can be formalised using the notion of rule, a monological structure that links premises, built upon evidence and/or the knowledge or beliefs of the researcher, to a claim or conclusion. These can be connected to each other also into a dialogical structure, through the notion of conflict and their interaction can reveal the most rationale conclusion/s [67]. Reasoning systems can help qualitative researchers to reason over available knowledge characterised by uncertainty, partiality and conflictuality and can reveal special cases in which previous beliefs and intuitions are no longer valid [70]. An example of application of formal reasoning can be found in [109], where authors attempted to represent and formally model the ill-defined construct of mental workload [112] or for the prediction of mortality in elderly using fragmented qualitative knowledge related to biomarkers [110,111].

Knowledge representation, reasoning systems and their tools can enable qualitative researchers in planning actions or decisions [49]. Automated planning supports the realisation of strategies and sequences of action from a complex space of possible solutions. It is strictly related to decision theory and given a description of the possible initial states of the underlying domain under consideration, the possible desired goals and actions, automated planning allows to synthesise a plan that is guaranteed to generate a state which contains the desired goals [90]. For example, [53] introduced a system for automatically synthesizing curricula and dynamically constructing learning paths even from disjoint learning objects with the goal of providing learners with a personalised learning solution that accounts for their preferences, profile needs and abilities. Similarly, [29] focused on planning learning routes employing a constraint programming approach that required, among other things, to model the profile of learners, to learn concepts and to understand tasks and attain concepts at different competence levels.

Each act, word and gesture of an interviewed participant is important for a qualitative researcher. Machine perception, a subfield of AI, can provide methods for classification and recognition of symbolic human behaviours such as gestures [84,88] and verbal behaviour such as speech [35]. The latter not only can be employed to transcribe digital interviews automatically, thus significantly reducing human labour [12,46], but it can be used for emotion recognition in speeches [102,138] and dialogs [59], usually tasks for humans. Emotions can be also recognised by automatic analysis of images [26] or hybrid multimodal approaches including images and speeches [10].

A recent article [15] explored the use of machine learning to support qualitative coding through the use of visual tools. In general machine learning can be divided into supervised and unsupervised. The former requires labelled data and it is useful for prediction and forecasting, while the latter is suitable for the exploration of high-dimensional data, its clustering and patterns extraction. These can be related to different grounded theories [41]. According to [87], the former could be related to Glaserian grounded theory [129] because of the availability of ground truth, the labels, equivalent to the Glaser's coding families [33]. The latter can be related to the Straussian grounded theory [130] because it is unconstrained to prior theories. Both can be used in different orders and can extend traditional qualitative research methods. For example, on one hand, [1,140] used topic modelling for automatic, unsupervised discovery of hidden semantic structures and clusters of similar words, in a text body. Similarly, [36] proposed an unsupervised sentiment extraction approach from textual data. Work on opinion and sentiment analysis is extremely vast [64,65,97,132]. On the other hand, [85] used supervised machine learning to identify the most representative sentences that affected success on Kickstarter, a crowdfunding website where artists and entrepreneurs seek funding, by analysing million phrases and other variables commonly present on these sites. Similarly, [101] and [23] employed supervised machine learning respectively for breast cancer classification via histology images and classification of normal/abnormal brain MRI images. A hybrid approach by [114] firstly used Latent Dirichlet Allocation and Formal Concept Analysis as two unsupervised techniques for respectively extracting topics and deriving a concept hierarchy. The outputs of these approaches were individually used as the input of a machine learning classifier for the classification of short-text documents. Similarly, the whole class of deep learning-based classification techniques can perform automatic feature extraction and learn representations from pictures in a supervised [91] or unsupervised fashion [131]. Many other applications exist that segments images [66] or perform sentiment classification from text [133].

A method to help researchers to discover patterns from qualitative data has been proposed in [136]. A number of semistructured qualitative interviews on the experience of participants in using distance education were conducted. Audio recordings were transformed into textual transcripts and an automatic natural language processing tool identified relevant first and second order categories. A graph was subsequently developed to associate first and second order categories, and subsequently employed for theme discovery. Natural language processing has been shown to be a potential solution for coding problems [37] and a combination of computational and manual techniques can preserve the strengths of traditional content analysis, but at the same time offering systematic rigour and contextual sensitivity [61]. Yan et al. proposed using natural language processing jointly with machine learning to train a classification model of initially labelled codes and subsequently used to automatically generate codes. A human-in-the-loop approach subsequently involves human to supervise and to correct predicted codes and use the outcome of this process to create more robust and accurate models [62]. This method allows for a faster categorisation of the unit of a qualitative linguistic corpus. [21] compared two methods for developing rules for extracting coded text. The first included a manual approach in which an expert, in natural language processing, developed rules to extract the coded segments. The second employed machine learning to train models capable of predicting codes. Results showed how the former approach worked better with smaller sample sizes. However, the latter approach achieved better accuracy with larger sample sizes.

4   Enhancing Education, Teaching and Learning with Artificial Intelligence for Qualitative Data Analysis

The discipline of education has a long history and has always attracted many scholars, practitioners and scientists all interested in understanding how us, as humans, learn and how to facilitate the acquisition of knowledge and skills. Education is often directed and involves teachers, lecturers, professors and many other educators who orchestrate classrooms, instructional material, delivery methods and teaching activities. These notions are the key components of pedagogy, the theory and practice of education that aims at understanding learning and supports the growth of learners in various educational contexts with an emphasis of the individual interactions and group dynamics. Various learning theories have emerged in the years, each focusing on different aspects on how knowledge is acquired, maintained and processed during learning. These can be summarised in three main categories: behaviourism, constructivism, cognitivism [39, 44, 117].

Behaviourism, grounded in psychology, is concerned with observable behaviour of learners. It is a reductionist methodological approach to learning assuming that all types of behaviours, regardless of their complexity, can be reduced to a simple association between external stimulus and response in an environment without considering the internal mental states of learners [141]. *Constructivism*, grounded in the earlier work of Piaget [139] and Vygotsky [48], is focused on how learners construct their knowledge from their experiences. It is grounded on the notion of prior knowledge, affected by the environment and social connections of learners. Constructivists assume that learning is an active constructive process with the learner as the core element of information processing who creates own internal subjective representation, due to their own prior knowledge, by acquiring knowledge in objective reality [128]. *Cognitivism*, grounded in cognitive psychology, focuses on cognition, largely neglected within behaviourism [77]. It argues that the way humans think actually affect their behaviour thus this cannot be regarded as behaviour in and of itself. The core notion behind cognitivism is that the mind is seen as a closed box which should be opened and investigated. Cognitivists essentially regard learners as complex information processors who assimilate and expand knowledge through cognitive development absorbing stimuli within different environments and processing them to produce knowledge and support the formation of skills [45,95]. The learning theories have been used and compared many times [24] within the field of instructional design [81,94]. With the technological progress, a new theory has recently emerged: connectivism [34]. It essentially focused on the Internet technologies and how these have created a new class of opportunities for learners to share information across themselves through the World Wide Web.

Artificial intelligence, as previously mentioned, with its theoretical and practical advances, as well as its novel intelligent applications can offer many solutions to the problem of qualitative data analysis and extraction of meanings. Unfortunately, their use in educational contexts is often under explored. This is mainly caused by the background of instructions, lecturers and educational practitioners which is very often less technical than that of computer scientists, engineers and other practitioners working within artificial intelligence. The former usually tend to employ traditional qualitative research methods for data gathering and analysis, while the latter usually prefer quantitative research methods. However, in science, making full use of the advantages of these two methods is known to enrich data analysis, enhance meaning and knowledge extraction as well as supporting hypothesis generation, testing and confirmation [8,20]. This section is devoted to the description of how the methods, developed so far within artificial intelligence that deal with qualitative data, as described in Sect. 3, can empower education, teaching and learning. In particular, as a third-level instructor, I am interested in exploring the possibilities of how these methods can support my work in the classroom. Thus I focus on describing a set of potential applications of these methods for enhancing teaching and learning in the classroom with a reference to the underlying pedagogical theory, as described in Sect. 4.

The main goal of behaviourism is to observe the behaviours of learners over time. Arranging the setting of a typical third-level classroom in a way that an instructor can observe the behaviour of the learners against the learning strategy employed is not a trivial task. Learners are more inclined to change their behaviours and engage in the classroom if they experience positive feelings during learning as well as from the approval from their fellow learners. They tend not to engage in behaviours that are unpleasant but they are rather inclined to form habits from engaging and pleasant behaviours. AI solutions for face and emotion recognition can be employed by instructors in real-time in the classroom and can serve as significant sources of data for improving instructional design and observing behaviours over time. For example, [27] implemented a neural network for recognising emotions based on facial expressions. Recognising emotions in real-time in the classroom can help an instructor understand the mental states of learners and adjust content delivery accordingly or as a means to temporally break the activities. The solution can be implemented with minimal effort as only a few cameras, placed in different corners of a room, are required for capturing faces and recognising emotions. The bottleneck is represented by training the neural network that can take many days. However, with novel transfer learning solutions, within the sub-field of machine learning, pre-trained networks specialised in specific tasks, in this case face recognition and then emotion recognition, are available and can be easily deployed [96], demonstrating the feasibility of this solution. Change in behaviours can also be inferred by the speech of learners engaged in various activities [119].

Constructivism focuses on understanding how learners construct their knowledge. It emphasises the active and autonomous role of learners while understanding and building knowledge. In a typical classroom, a context with multiple learners, constructivism can be achieved, for instance, by performing collaboration, information exchange and inquiring activities among peers. Activity recognition algorithms, employing audio, video [113] and other qualitative sensor-based data [55,107], can help categorise the specific activities each learner can perform individually or in group such as 'speaking', 'gestures', 'listening', 'reading', 'writing' and many others while constructing information. An instructor, through the individual and group analysis of these categories can design and converge to more engaging, democratic and student-centered constructivist activities that enhance learning and maximise engagement and experience. Dialogs and discussions among learners represent the most powerful constructivist teaching method, thus promoting it is key [93]. By employing natural language processing techniques from qualitative speech or textual data can help reconstruct the flows of arguments made among learners as part of discussions while constructively building knowledge [56]. Additionally, it can help identify clusters of qualitative data and help instructors build formal representations for an argumentative piece of text [100] that can be used for assessing learning.

The key objective of cognitivism is to understand how the human brain actually learns. It involves the exploration of how we memorise and represent knowledge, how we tackle problem solving and acquire skills, as well as how we develop intelligence. It is connected to the notion of cognitive development and it relies on five principles: remembering, understanding, applying, evaluating, and creating. An instructor can assess each of these skills for each learner in the classroom, for example by means of text production. Initially, a learner can be asked to write a paragraph to sketch the content of the previous class (remembering). Subsequently, s/he can focus on writing another paragraph towards interpreting that content through a textual debate, a list of examples and any form of classification of information (understanding). The learner can then focus on problem solving and, for instance, textually answer questions and solve a specific problem (applying). Then, the instructor can ask each student to graphically illustrate certain information, as for instance employing concept maps [92] or producing a pros and cons list (creating). Eventually, the learner can be asked to write a short manual or guidebook to demonstrate important information (creating). Once the textual content for each student is available, the instructor can then employ natural language processing techniques from artificial intelligence to discover similarities against own expert textual information. For instance, semantic similarity techniques can be employed on the text produced by learners and the instructor [19,82,118] and can be used as a form of formative assessment for each of the five aforementioned principles. Similarly, the instructor can employ graph-based methods for assessing conceptual similarity [57,78] of own representations of instructional material and those produced by learners [115]. Another interesting application of cognitivism is the assessment of cognitive load of learners during various educational activities [73,95]. Assessing cognitive load can be done through the analysis of the speech from learners [144], the evolution of their emotional states as assessed via emotion recognition [5] as well as their pupil dilation and movements [16] via image processing. Many other methods exist for assessing cognitive load but they are mainly quantitative [68,69,71,86,112].

Fig. 2. A summary of the solutions developed within artificial intelligence for empowering qualitative research, its data analysis methods in education, teaching and learning in the classroom

In summary, qualitative research methods can greatly benefit from the application of theoretical methods, practical tools and solutions developed within the field of artificial intelligence. Catching up with the plethora of theoretical and practical advances of this fast growing field is very difficult even for those researchers working in the field itself, thus it is not expected that a qualitative research practitioner and scholar can fully cope with it. However, given the fact that artificial intelligence is omnipresent and fully part of our daily life, the suggestion is to encourage educational practitioners, instructors and social scientists, mainly employing traditional qualitative research methods, to embrace it and further explore and employ its methods. To achieve this goal, on one hand, scholars in artificial intelligence should work on providing researchers with less technical knowledge with richer explanations and qualitative descriptions of their own methods. On the other hand, researchers, mainly adopting qualitative research methods, should expand their technical knowledge and devote effort on learning those formal concepts and methods offered by artificial intelligence.

References

1. Arora, S., Ge, R., Halpern, Y., Mimno, D., Moitra, A., Sontag, D., Wu, Y., Zhu, M.: A practical algorithm for topic modeling with provable guarantees. In: International Conference on Machine Learning, pp. 280–288 (2013)
2. Auerbach, C., Silverstein, L.B.: Qualitative Data: An Introduction to Coding and Analysis. NYU Press, New York (2003)
3. Bahrammirzaee, A.: A comparative survey of artificial intelligence applications in finance: artificial neural networks, expert system and hybrid intelligent systems. Neural Comput. Appl. **19**(8), 1165–1195 (2010)
4. Bartlett, M.S., Littlewort, G., Frank, M.G., Lainscsek, C., Fasel, I.R., Movellan, J.R.: Recognizing facial expression: machine learning and application to spontaneous behavior. In: 2005 IEEE Computer Society Conference on Computer Vision and Pattern Recognition (CVPR 2005), San Diego, CA, USA, 20–26 June 2005, pp. 568–573 (2005)
5. Berggren, N., Koster, E.H., Derakshan, N.: The effect of cognitive load in emotional attention and trait anxiety: an eye movement study. J. Cogn. Psychol. **24**(1), 79–91 (2012)
6. Bishop, C.M.: Pattern Recognition and Machine Learning. Information Science and Statistics, 5th edn. Springer, New York (2007)
7. Bogdan, R., Biklen, S.K.: Qualitative Research for Education. Allyn & Bacon, Boston (1997)
8. Brannen, J.: Mixing Methods: Qualitative and Quantitative Research. Routledge, London (2017)
9. Buchanan, B.G.: A (very) brief history of artificial intelligence. AI Mag. **26**(4), 53 (2005)
10. Busso, C., Deng, Z., Yildirim, S., Bulut, M., Lee, C.M., Kazemzadeh, A., Lee, S., Neumann, U., Narayanan, S.: Analysis of emotion recognition using facial expressions, speech and multimodal information. In: Proceedings of the 6th International Conference on Multimodal Interfaces, pp. 205–211. ACM (2004)
11. Caridi, M., Cavalieri, S.: Multi-agent systems in production planning and control: an overview. Prod. Plann. Control **15**(2), 106–118 (2004)
12. Chandler, R., Anstey, E., Ross, H.: Listening to voices and visualizing data in qualitative research: hypermodal dissemination possibilities. Sage Open **5**(2) (2015). https://doi.org/10.1177/2158244015592166
13. Chang, R., Wang, F., You, P.: A survey on the development of multi-touch technology. In: 2010 Asia-Pacific Conference on Wearable Computing Systems, pp. 363–366. IEEE (2010)
14. Charmaz, K., Belgrave, L.L.: Grounded Theory. In: The Blackwell Encyclopedia of Sociology (2007)
15. Chen, N.C., Drouhard, M., Kocielnik, R., Suh, J., Aragon, C.R.: Using machine learning to support qualitative coding in social science: shifting the focus to ambiguity. ACM Trans. Interact. Intell. Syst. (TiiS) **8**(2), 9 (2018)

16. Chen, S., Epps, J.: Using task-induced pupil diameter and blink rate to infer cognitive load. Hum. Comput. Interact. **29**(4), 390–413 (2014)
17. Cimiano, P., Hotho, A., Staab, S.: Learning concept hierarchies from text corpora using formal concept analysis. J. Artif. Intell. Res. **24**, 305–339 (2005)
18. Cohen, L., Manion, L., Morrison, K.: Research Methods in Education. Routledge, London (2002)
19. Corley, C., Mihalcea, R.: Measuring the semantic similarity of texts. In: Proceedings of the ACL Workshop on Empirical Modeling of Semantic Equivalence and Entailment, pp. 13–18. Association for Computational Linguistics (2005)
20. Creswell, J.W.: Educational Research: Planning, Conducting, and Evaluating Quantitative. Prentice Hall, Upper Saddle River (2002)
21. Crowston, K., Allen, E.E., Heckman, R.: Using natural language processing technology for qualitative data analysis. Int. J. Soc. Res. Methodol. **15**(6), 523–543 (2012)
22. Cruz, J.A., Wishart, D.S.: Applications of machine learning in cancer prediction and prognosis. Cancer Inform. **2**, 59–77 (2006). https://doi.org/10.1177/117693510600200030
23. El-Dahshan, E.S.A., Hosny, T., Salem, A.B.M.: Hybrid intelligent techniques for MRI brain images classification. Digit. Signal Process. **20**(2), 433–441 (2010)
24. Ertmer, P.A., Newby, T.J.: Behaviorism, cognitivism, constructivism: comparing critical features from an instructional design perspective. Perform. Improv. Q. **6**(4), 50–72 (1993)
25. Fairchild, K.M., Poltrock, S.E., Furnas, G.W.: SemNet: three-dimensional graphic representations of large knowledge bases. In: Cognitive Science and Its Applications for Human-computer Interaction, pp. 215–248. Psychology Press (2013)
26. Fasel, B., Luettin, J.: Automatic facial expression analysis: a survey. Pattern Recogn. **36**(1), 259–275 (2003)
27. Filko, D., Martinović, G.: Emotion recognition system by a neural network based facial expression analysis. Automatika **54**(2), 263–272 (2013)
28. Forsyth, D.A., Ponce, J.: Computer Vision: A Modern Approach. Prentice Hall Professional Technical Reference, Upper Saddle River (2002)
29. Garrido, A., Onaindia, E., Sapena, O.: Planning and scheduling in an e-learning environment. A constraint-programming-based approach. Eng. Appl. Artif. Intell. **21**(5), 733–743 (2008)
30. Gee, J.P.: An Introduction to Discourse Analysis: Theory and Method. Routledge, London (2004)
31. Ghallab, M., Nau, D., Traverso, P.: Automated Planning: Theory and Practice. Elsevier, Amsterdam (2004)
32. Gil, Y., Greaves, M., Hendler, J., Hirsh, H.: Amplify scientific discovery with artificial intelligence. Science **346**(6206), 171–172 (2014)
33. Glaser, B.G.: The Grounded Theory Perspective III: Theoretical Coding. Sociology Press, Mill Valley (2005)
34. Goldie, J.G.S.: Connectivism: a knowledge learning theory for the digital age? Med. Teach. **38**(10), 1064–1069 (2016)
35. Graves, A., Mohamed, A.R., Hinton, G.: Speech recognition with deep recurrent neural networks. In: 2013 IEEE International Conference on Acoustics, Speech and Signal Processing, pp. 6645–6649. IEEE (2013)
36. Greene, S., Resnik, P.: More than words: syntactic packaging and implicit sentiment. In: Proceedings of Human Language Technologies: The 2009 Annual Conference of the North American Chapter of the Association for Computational Linguistics, pp. 503–511. Association for Computational Linguistics (2009)

37. Grimmer, J., Stewart, B.M.: Text as data: the promise and pitfalls of automatic content analysis methods for political texts. Polit. Anal. **21**(3), 267–297 (2013)
38. Hamet, P., Tremblay, J.: Artificial intelligence in medicine. Metabolism **69**, S36–S40 (2017)
39. Harasim, L.: Learning Theory and Online Technologies. Routledge, London (2017)
40. Hassabis, D., Kumaran, D., Summerfield, C., Botvinick, M.: Neuroscience-inspired artificial intelligence. Neuron **95**(2), 245–258 (2017)
41. Heath, H., Cowley, S.: Developing a grounded theory approach: a comparison of Glaser and Strauss. Int. J. Nurs. Stud. **41**(2), 141–150 (2004)
42. Hinton, G., Deng, L., Yu, D., Dahl, G., Mohamed, A.R., Jaitly, N., Senior, A., Vanhoucke, V., Nguyen, P., Kingsbury, B., Sainath, T.: Deep neural networks for acoustic modeling in speech recognition. IEEE Signal Process. Mag. **29**, 82–97 (2012)
43. Huang, W., Nakamori, Y., Wang, S.: Forecasting stock market movement direction with support vector machine. Comput. OR **32**, 2513–2522 (2005)
44. Illeris, K.: Contemporary Theories of Learning: Learning Theorists... In Their Own Words. Routledge, London (2018)
45. Cooper, P.A.: Paradigm shifts in designed instruction: from behaviorism to cognitivism to constructivism. Educ. Technol. **33**(5), 12–19 (1993)
46. Johnson, B.E.: The speed and accuracy of voice recognition software-assisted transcription versus the listen-and-type method: a research note. Qual. Res. **11**(1), 91–97 (2011)
47. Jordan, M.I., Mitchell, T.M.: Machine learning: trends, perspectives, and prospects. Science **349**(6245), 255–260 (2015)
48. Kalina, C., Powell, K.: Cognitive and social constructivism: developing tools for an effective classroom. Education **130**(2), 241–250 (2009)
49. Kane, M., Trochim, W.M.: Concept Mapping for Planning and Evaluation, vol. 50. Sage Publications, Thousand Oaks (2007)
50. Kempen, G.: Natural Language Generation: New Results in Artificial Intelligence, Psychology and Linguistics, vol. 135. Springer, Netherlands (2012)
51. Koehn, P., Hoang, H., Birch, A., Callison-Burch, C., Federico, M., Bertoldi, N., Cowan, B., Shen, W., Moran, C., Zens, R., Dyer, C., Bojar, O., Constantin, A., Herbst, E.: Moses: open source toolkit for statistical machine translation. In: Proceedings of the 45th Annual Meeting of the ACL on Interactive Poster and Demonstration Sessions, ACL 2007, pp. 177–180. Association for Computational Linguistics, Stroudsburg (2007). http://dl.acm.org/citation.cfm?id=1557769.1557821
52. Kok, S., Domingos, P.: Extracting semantic networks from text via relational clustering. In: Joint European Conference on Machine Learning and Knowledge Discovery in Databases, pp. 624–639. Springer, Heidelberg (2008)
53. Kontopoulos, E., Vrakas, D., Kokkoras, F., Bassiliades, N., Vlahavas, I.: An ontology-based planning system for e-course generation. Expert Syst. Appl. **35**(1–2), 398–406 (2008)
54. Krippendorff, K.: Content Analysis: An Introduction to Its Methodology. Sage Publications, Thousand Oaks (2018)
55. Lara, O.D., Labrador, M.A.: A survey on human activity recognition using wearable sensors. IEEE Commun. Surv. Tutor. **15**(3), 1192–1209 (2012)
56. Lawrence, J., Reed, C.: Argument mining using argumentation scheme structures. In: COMMA, pp. 379–390 (2016)
57. Leake, D.B., Maguitman, A.G., Cañas, A.J.: Assessing conceptual similarity to support concept mapping. In: FLAIRS Conference, pp. 168–172 (2002)

58. LeCun, Y., Bengio, Y., Hinton, G.: Deep learning. Nature **521**(7553), 436 (2015)
59. Lee, C.M., Narayanan, S.S., et al.: Toward detecting emotions in spoken dialogs. IEEE Trans. Speech Audio Process. **13**(2), 293–303 (2005)
60. Legard, R., Keegan, J., Ward, K.: In-depth interviews. In: Qualitative Research Practice: A Guide for Social Science Students and Researchers, vol. 6(1), pp. 138–169 (2003)
61. Lewis, S.C., Zamith, R., Hermida, A.: Content analysis in an era of big data: a hybrid approach to computational and manual methods. J. Broadcast. Electron. Media **57**(1), 34–52 (2013)
62. Liew, J.S.Y., McCracken, N., Zhou, S., Crowston, K.: Optimizing features in active machine learning for complex qualitative content analysis. In: Proceedings of the ACL 2014 Workshop on Language Technologies and Computational Social Science, pp. 44–48 (2014)
63. Lincoln, Y.S., Denzin, N.K.: The Handbook of Qualitative Research. Sage, Thousand Oaks (2000)
64. Liu, B.: Sentiment analysis and opinion mining. Synth. Lect. Hum. Lang. Technol. **5**(1), 1–167 (2012)
65. Liu, B., et al.: Sentiment analysis and subjectivity. In: Handbook of Natural Language Processing, vol. 2, pp. 627–666 (2010)
66. Liu, F., Lin, G., Shen, C.: CRF learning with CNN features for image segmentation. Pattern Recogn. **48**(10), 2983–2992 (2015)
67. Longo, L.: Argumentation for knowledge representation, conflict resolution, defeasible inference and its integration with machine learning. In: Machine Learning for Health Informatics, pp. 183–208. Springer, Cham (2016)
68. Longo, L.: Subjective usability, mental workload assessments and their impact on objective human performance. In: IFIP Conference on Human-Computer Interaction, pp. 202–223. Springer, Cham (2017)
69. Longo, L.: On the reliability, validity and sensitivity of three mental workload assessment techniques for the evaluation of instructional designs: a case study in a third-level course. In: Proceedings of the 10th International Conference on Computer Supported Education, CSEDU 2018, Funchal, Madeira, Portugal, 15–17 March 2018, vol. 2, pp. 166–178 (2018)
70. Longo, L., Dondio, P.: Defeasible reasoning and argument-based systems in medical fields: an informal overview. In: 2014 IEEE 27th International Symposium on Computer-Based Medical Systems (CBMS), pp. 376–381. IEEE (2014)
71. Longo, L., Dondio, P.: On the relationship between perception of usability and subjective mental workload of web interfaces. In: 2015 IEEE/WIC/ACM International Conference on Web Intelligence and Intelligent Agent Technology (WI-IAT), vol. 1, pp. 345–352. IEEE (2015)
72. Longo, L., Dondio, P., Barrett, S.: Enhancing social search: a computational collective intelligence model of behavioural traits, trust and time. In: Transactions on Computational Collective Intelligence II, pp. 46–69. Springer, Heidelberg (2010)
73. Longo, L., Orru, G.: An evaluation of the reliability, validity and sensitivity of three human mental workload measures under different instructional conditions in third-level education. In: McLaren, B.M., Reilly, R., Zvacek, S., Uhomoibhi, J. (eds.) Computer Supported Education, pp. 384–413. Springer, Cham (2019)
74. Lowrance, J.D., Garvey, T.D., Strat, T.M.: A framework for evidential-reasoning systems. In: Classic Works of the Dempster-Shafer Theory of Belief Functions, pp. 419–434. Springer, Heidelberg (2008)
75. Lyon, R.F.: Machine hearing: an emerging field [exploratory DSP]. IEEE Signal Process. Mag. **27**(5), 131–139 (2010)

76. Uschold, M., Gruninger, M.: Ontologies: principles, methods and applications. Knowl. Eng. Rev. **11**, 93–136 (1996)
77. Mandler, G.: Origins of the cognitive (r)evolution. J. Hist. Behav. Sci. **38**(4), 339–353 (2002)
78. Marshall, B., Chen, H., Madhusudan, T.: Matching knowledge elements in concept maps using a similarity flooding algorithm. Decis. Support Syst. **42**(3), 1290–1306 (2006)
79. Mayring, P.: Qualitative content analysis. A companion to qualitative research, vol. 1, pp. 159–176 (2004)
80. McCulloch, W.S., Pitts, W.: A logical calculus of the ideas immanent in nervous activity. Bull. Math. Biophys. **5**(4), 115–133 (1943)
81. McLeod, G.: Learning theory and instructional design. Learn. Matt. **2**(3), 35–43 (2003)
82. Mihalcea, R., Corley, C., Strapparava, C., et al.: Corpus-based and knowledge-based measures of text semantic similarity. In: AAAI, vol. 6, pp. 775–780 (2006)
83. Miller, T.: Explanation in artificial intelligence: insights from the social sciences. Artificial Intelligence (2018)
84. Mitra, S., Acharya, T.: Gesture recognition: a survey. IEEE Trans. Syst. Man Cybern. Part C (Appl. Rev.) **37**(3), 311–324 (2007)
85. Mitra, T., Gilbert, E.: The language that gets people to give: phrases that predict success on kickstarter. In: Proceedings of the 17th ACM Conference on Computer Supported Cooperative Work & Social Computing, pp. 49–61. ACM (2014)
86. Moustafa, K., Longo, L.: Analysing the impact of machine learning to model subjective mental workload: a case study in third-level education. In: International Symposium on Human Mental Workload: Models and Applications, pp. 92–111. Springer, Cham (2018)
87. Muller, M., Guha, S., Baumer, E.P., Mimno, D., Shami, N.S.: Machine learning and grounded theory method: convergence, divergence, and combination. In: Proceedings of the 19th International Conference on Supporting Group Work, pp. 3–8. ACM (2016)
88. Murakami, K., Taguchi, H.: Gesture recognition using recurrent neural networks. In: Proceedings of the SIGCHI Conference on Human Factors in Computing Systems, pp. 237–242. ACM (1991)
89. Musante, K., DeWalt, B.R.: Participant Observation: A Guide for Fieldworkers. Rowman Altamira (2010)
90. Nau, D.S.: Current trends in automated planning. AI Mag. **28**(4), 43 (2007)
91. Ng, H.W., Nguyen, V.D., Vonikakis, V., Winkler, S.: Deep learning for emotion recognition on small datasets using transfer learning. In: Proceedings of the 2015 ACM on International Conference on Multimodal Interaction, pp. 443–449. ACM (2015)
92. Novak, J.D.: Learning, Creating, and Using Knowledge: Concept Maps as Facilitative Tools in Schools and Corporations. Routledge, New York (2010)
93. Orru, G., Gobbo, F., O'Sullivan, D., Longo, L.: An investigation of the impact of a social constructivist teaching approach, based on trigger questions, through measures of mental workload and efficiency. In: Proceedings of the 10th International Conference on Computer Supported Education, CSEDU 2018, Funchal, Madeira, Portugal, 15–17 March 2018, vol. 2, pp. 292–302 (2018)
94. Orru, G., Longo, L.: Direct instruction and its extension with a community of inquiry: a comparison of mental workload, performance and efficiency. In: Proceedings of the 11th International Conference on Computer Supported Education, CSEDU 2019, Heraklion, Crete, Greece, 2–4 May 2019, vol. 1, pp. 436–444 (2019)

95. Orru, G., Longo, L.: The evolution of cognitive load theory and the measurement of its intrinsic, extraneous and germane loads: a review. In: Longo, L., Leva, M.C. (eds.) Human Mental Workload: Models and Applications, pp. 23–48. Springer, Cham (2019)
96. Pan, S.J., Yang, Q.: A survey on transfer learning. IEEE Trans. Knowl. Data Eng. **22**(10), 1345–1359 (2009)
97. Pang, B., Lee, L., et al.: Opinion mining and sentiment analysis. Found. Trends® Inf. Retr. **2**(1–2), 1–135 (2008)
98. Parsons, S.: Qualitative Methods for Reasoning Under Uncertainty. MIT Press, Cambridge (2001)
99. Patton, M.Q.: Qualitative research. In: Encyclopedia of Statistics in Behavioral Science (2005)
100. Peldszus, A., Stede, M.: From argument diagrams to argumentation mining in texts: a survey. Int. J. Cogn. Inform. Nat. Intell. (IJCINI) **7**(1), 1–31 (2013)
101. Petushi, S., Garcia, F.U., Haber, M.M., Katsinis, C., Tozeren, A.: Large-scale computations on histology images reveal grade-differentiating parameters for breast cancer. BMC Med. Imaging **6**(1), 14 (2006)
102. Pierre-Yves, O.: The production and recognition of emotions in speech: features and algorithms. Int. J. Hum. Comput. Stud. **59**(1–2), 157–183 (2003)
103. Pinedo, M.L.: Planning and Scheduling in Manufacturing and Services. Springer, New York (2005)
104. Puppe, F.: Systematic Introduction to Expert Systems: Knowledge Representations and Problem-Solving Methods. Springer, Heidelberg (2012)
105. Rahwan, I., Simari, G.R.: Argumentation in Artificial Intelligence, vol. 47. Springer, US (2009)
106. Ramesh, A., Kambhampati, C., Monson, J.R., Drew, P.: Artificial intelligence in medicine. Ann. R. Coll. Surg. Engl. **86**(5), 334 (2004)
107. Ravi, N., Dandekar, N., Mysore, P., Littman, M.L.: Activity recognition from accelerometer data. In: AAAI, vol. 5, pp. 1541–1546 (2005)
108. Reiter, E., Dale, R.: Building Natural Language Generation Systems. Cambridge University Press, New York (2000)
109. Rizzo, L., Longo, L.: Inferential models of mental workload with defeasible argumentation and non-monotonic fuzzy reasoning: a comparative study. In: Proceedings of the 2nd Workshop on Advances in Argumentation in Artificial Intelligence, Co-located with XVII International Conference of the Italian Association for Artificial Intelligence, AI³@AI*IA 2018, Trento, Italy, 20–23 November 2018, pp. 11–26 (2018)
110. Rizzo, L., Longo, L.: A qualitative investigation of the explainability of defeasible argumentation and non-monotonic fuzzy reasoning. In: Proceedings for the 26th AIAI Irish Conference on Artificial Intelligence and Cognitive Science Trinity College Dublin, Dublin, Ireland, 6–7 December 2018, pp. 138–149 (2018)
111. Rizzo, L., Majnaric, L., Longo, L.: A comparative study of defeasible argumentation and non-monotonic fuzzy reasoning for elderly survival prediction using biomarkers. In: International Conference of the Italian Association for Artificial Intelligence, pp. 197–209. Springer, Cham (2018)
112. Rizzo, L.M., Longo, L.: Representing and inferring mental workload via defeasible reasoning: a comparison with the NASA task load index and the workload profile. In: Proceedings of the 1st Workshop on Advances in Argumentation in Artificial Intelligence AI3@AI*IA. CEURS (2017)
113. Robertson, N., Reid, I.: A general method for human activity recognition in video. Comput. Vis. Image Underst. **104**(2–3), 232–248 (2006)

114. Rogers, N., Longo, L.: A comparison on the classification of short-text documents using latent Dirichlet allocation and formal concept analysis. In: Proceedings of the 25th Irish Conference on Artificial Intelligence and Cognitive Science, Dublin, Ireland, 7–8 December 2017, pp. 50–62 (2017)
115. Rye, J.A., Rubba, P.A.: Scoring concept maps: an expert map-based scheme weighted for relationships. Sch. Sci. Math. **102**(1), 33–44 (2002)
116. Saldaña, J.: The Coding Manual for Qualitative Researchers. Sage, Thousand Oaks (2015)
117. Schunk, D.H.: Learning Theories an Educational Perspective, 6th edn. Pearson, Boston (2012)
118. Seco, N., Veale, T., Hayes, J.: An intrinsic information content metric for semantic similarity in WordNet. In: ECAI, vol. 16, p. 1089 (2004)
119. Shaikh Nilofer, R., Gadhe, R.P., Deshmukh, R., Waghmare, V., Shrishrimal, P.: Automatic emotion recognition from speech signals: a review. Int. J. Sci. Eng. Res. **6**(4), 636–639 (2015)
120. Shen, W.: Distributed manufacturing scheduling using intelligent agents. IEEE Intell. Syst. **17**(1), 88–94 (2002)
121. Shen, W., Wang, L., Hao, Q.: Agent-based distributed manufacturing process planning and scheduling: a state-of-the-art survey. IEEE Trans. Syst. Man Cybern. Part C (Appl. Rev.) **36**(4), 563–577 (2006)
122. Smith, A.E.: Automatic extraction of semantic networks from text using leximancer. In: Companion Volume of the Proceedings of HLT-NAACL 2003-Demonstrations, pp. 23–24 (2003)
123. Sonka, M., Hlaváč, V., Boyle, R.: Image Processing, Analysis and Machine Vision, 3. edn. Thomson (2008)
124. Sowa, J.F.: Knowledge Representation: Logical, Philosophical, and Computational Foundations. PWS (2000)
125. Sowa, J.F.: Principles of Semantic Networks: Explorations in the Representation of Knowledge. Morgan Kaufmann, San Mateo (2014)
126. Spradley, J.P.: Participant Observation. Waveland Press, Long Grove (2016)
127. Staab, S., Studer, R.: Handbook on Ontologies. Springer, Heidelberg (2010)
128. Steffe, L.P., Gale, J.E.: Constructivism in Education. Lawrence Erlbaum, Hillsdale (1995)
129. Stern, P.N., Kerry, J., et al.: Glaserian grounded theory. In: Developing Grounded Theory, pp. 55–85. Routledge, London (2016)
130. Strauss, A., Corbin, J.M.: Grounded Theory in Practice. Sage, Thousand Oaks (1997)
131. Supratak, A., Li, L., Guo, Y.: Feature extraction with stacked autoencoders for epileptic seizure detection. In: 2014 36th Annual International Conference of the IEEE Engineering in Medicine and Biology Society, pp. 4184–4187. IEEE (2014)
132. Taboada, M., Brooke, J., Tofiloski, M., Voll, K., Stede, M.: Lexicon-based methods for sentiment analysis. Comput. Linguist. **37**(2), 267–307 (2011)
133. Tang, D., Qin, B., Liu, T.: Document modeling with gated recurrent neural network for sentiment classification. In: Proceedings of the 2015 Conference on Empirical Methods in Natural Language Processing, pp. 1422–1432 (2015)
134. Tanimoto, S.: Structured Computer Vision: Machine Perception through Hierarchical Computation Structures. Elsevier, New York (2014)
135. Taylor, S.J., Bogdan, R., DeVault, M.: Introduction to Qualitative Research Methods: A Guidebook and Resource. Wiley, Hoboken (2015)
136. Tierney, P.: A qualitative analysis framework using natural language processing and graph theory. Int. Rev. Res. Open Distrib. Learn. **13**(5), 173–189 (2012)

137. Vaughn, S., Schumm, J.S., Sinagub, J.M.: Focus Group Interviews in Education and Psychology. Sage, Thousand Oaks (1996)
138. Ververidis, D., Kotropoulos, C.: Emotional speech recognition: resources, features, and methods. Speech Commun. **48**(9), 1162–1181 (2006)
139. Wadsworth, B.J.: Piaget's Theory of Cognitive and Affective Development: Foundations of Constructivism. Longman Publishing, New York (1996)
140. Wallach, H.M.: Topic modeling: beyond bag-of-words. In: Proceedings of the 23rd International Conference on Machine Learning, pp. 977–984. ACM (2006)
141. Watson, J.B.: Behaviorism. Routledge, London (2017)
142. Williams, A., Katz, L.: The use of focus group methodology in education: some theoretical and practical considerations, 5(3). IEJLL Int. Electron. J. Leadersh. Learn. **5** (2001)
143. Witz, K.G., Goodwin, D.R., Hart, R.S., Thomas, H.S.: An essentialist methodology in education-related research using in-depth interviews. J. Curric. Stud. **33**(2), 195–227 (2001)
144. Yin, B., Chen, F., Ruiz, N., Ambikairajah, E.: Speech-based cognitive load monitoring system. In: 2008 IEEE International Conference on Acoustics, Speech and Signal Processing, pp. 2041–2044. IEEE (2008)
145. Zouaq, A., Nkambou, R.: Building domain ontologies from text for educational purposes. IEEE Trans. Learn. Technol. **1**(1), 49–62 (2008)

What Qualitative Researchers Must Do When Ethical Assurances Disintegrate? Recognise Internal Confidentiality, Establish Process Consent, Reference Groups, Referrals for Participants and a Safety Plan

Martin Tolich(✉) 🔟

University of Otago, Dunedin, New Zealand
martin.tolich@otago.ac.nz

Abstract. Informed consent and confidentiality are the two mainstays of qualitative research ethics, yet they have a propensity to disintegrate in an emergent, iterative research design. This chapter examines how to approach this uncharted territory by having researchers take full responsibility for ethical considerations by using more robust forms of consent like process consent; recognising the dual faces of confidentiality, distinguishing external confidentiality from internal confidentiality. Other responsibilities in post ethics review environment include recognising and addressing big ethical moments. At times, participants and researchers ethical protections disintegrate too. When participants are at risk, furnish referrals (i.e. suicide watch phone numbers). When researchers are at risk work off a safety plan. Additionally, given this unpredictability, researchers should create a standing reference group to assist answering the fourth question above: what to do when the project raises ethical questions not foreseen in formal ethics review or by the researcher.

Keywords: Internal confidentiality · Process consent · Anonymity · Reference groups

1 Introduction

Informed consent and confidentiality are the two mainstays of qualitative research ethics. Paradoxically they are strong with propensity to disintegrate, requiring qualitative researchers to take responsibility for their practice by having a thorough understanding of what qualitative research ethics are and what they are not. Qualitative researchers cannot rely on anonymity and when they do, (often) using pseudonyms, they put participants at risk. Nor can they rely on formal ethics committee review to bolster consent and confidentiality because formal ethics review is incomplete. Ethics review committees only pose speculative questions for qualitative researchers. They can ask researchers to (1) describe their research, (2) outline the ethical issues that will arise in this research and (3) how the researcher will address those ethical issues [25]. Formal ethics review cannot ask the fourth and most important question: what the

© Springer Nature Switzerland AG 2020
A. P. Costa et al. (Eds.): WCQR 2019, AISC 1068, pp. 22–32, 2020.
https://doi.org/10.1007/978-3-030-31787-4_2

researcher will do when their project's emergent research question transforms, making consent and confidentiality assurances take on an altered, less robust, character [25]. Not only do formal ethics committees not ask this question, a researcher cannot predict with any certainty how their iterative research will change in the field affecting how consent and confidentiality ethical assurances are practiced.

This chapter examines how to approach this uncharted territory by having researchers take full responsibility for ethical considerations by using more robust forms of consent like process consent; recognising the dual faces of confidentiality, distinguishing external confidentiality from internal confidentiality [21] (or deductive disclosure [9]). Other responsibilities in this post ethics review environment include recognising and addressing big ethical moments [6]. At times, participants and researchers ethical protections disintegrate too. When participants are at risk, furnish referrals (i.e. suicide watch phone numbers). When researchers are at risk work off a safety plan. Additionally, given this unpredictability, researchers should create a standing reference group to assist answering the fourth question above: what to do when the project raises ethical questions not foreseen in formal ethics review or by the researcher.

Conceptualising deficiencies in qualitative research ethics involves reframing notions of the primacy of the method. In quantitative research, the linear research instrument (the questionnaire) holds the primacy of the method. In mixed methods, primacy is the dictatorship of the research question [20]. In qualitative research, the researcher embodies this primacy simultaneously collecting and analysing data. This chapter accepts this definition but enlarges primacy to include the ethical responsibilities a qualitative researcher must take on to address the fourth question above. Qualitative researchers are not only accountable for the data collection and analysis they are also solely responsible for the practice of ethics fundamental to qualitative research. These fundamentals are consent and confidentiality. While both concepts are robust, they are also fragile. Qualitative researchers' promises that all recorded conversations are confidential is false. These conversations are always subject to subpoena. To appreciate the strength of qualitative research ethics researchers must simultaneously embrace the limits of confidentiality when interview participants are relational e.g. when interviewing family members will other family members be able to identify what others said.

Consent too has limits; what a researcher tells informants in a focus group or an unstructured interview about the nature of the research can change during the data collection as the researcher inductively asserts their primacy over the method, prompting questions outside the scope of the previous consent. In these cases, consent is malleable. What is not malleable is the qualitative researcher's obligation to protect those that volunteer to take part in their research. If, and when confidentiality and informed consent disintegrate there are steps the researcher must take.

2 Background

This chapter contrasts the ethical responsibilities practised by a survey researcher with those practiced by qualitative researchers. Quantitative researchers' tie two ethical assurances to their research instrument. Prior to filling in a questionnaire, a respondent reads "filling in this questionnaire implies your informed consent." The participant information sheet also informs them that the information they supply is anonymous. Essentially, when the respondent submits their completed (or uncompleted) questionnaire this person's identity and what information they shared becomes unretrievable. This definition of anonymity is central to comprehending the frailties of qualitative research ethics. The definition of anonymity offered to a respondent should not be seen as an academic definition, but one found in everyday discourse and in a dictionary.

> **anonymous** *a. of unknown name; of unknown or undeclared source of authorship; impersonal; adv.* **Anonymity** *nameless*

The ethics contained within the primacy of the method in survey research are watertight. A test of anonymity would be if a respondent had second thoughts about taking part in the research and wanted to withdraw after they submitted their questionnaire. They could not. Neither the researcher nor the respondent could identify their particular questionnaire to permit extraction. Plus, there is no signed consent form to record the respondent's participation. The relationship between the respondent filling in the questionnaire and the researcher is ephemeral. It does not matter if the researcher knows the identity of the persons who took part in the survey, as once the questionnaire is submitted the researcher does not know how any individual person responded to the survey questions.

Anonymity assurances have caveats; it assumes the data collection instrument acquires no unique identifiers such as the respondent's name, social security number, or driver's license number [19]. If the survey sample size is small in numbers or based on a region or an occupation demographic, questions can threaten to expose the identity of respondents and disclosure of their data. For example, a questionnaire asking military personnel to provide their rank, gender, or theatres of war served may identify the very few females in the military's upper echelons. If, however, these caveats are controlled the dictionary definition above is assured by the ephemeral consent process. Qualitative researchers do not have a similar ephemeral consent process. Their consent process is long lasting creating unique ongoing ethical considerations for the researcher.

Qualitative researchers ask their informant (as opposed to a respondent) to reveal their identity by signing a consent form actively demonstrating their willingness to accept the ethical provisions offered by the researcher. Anonymity is not a valid provision as at least one other person, the researcher, knows the identity of the person and what the person said. This knowledge can never be unknown or anonymised and offering this ethical surety is ethically flawed and methodologically clumsy. A historic example illustrates the nuance of this embedded relationship. The following excerpt taken from an informant decades ago, as part of the author's previous research [22], continues to resonant audibly. This supermarket clerk tells her story:

> My job involves checking out customers. Talking with them. I know most of the ladies, and a few of my men come through. This is on the morning shift, and I know most of their personal habits. I know how they like their water bagged. We have an old people's home near us.

The fact that I can still hear the voice of this woman as I silently read the quote means that this data is not of an unknown source and never can be as I will always know who said this quote and for the past 26 years this women's quote has remained *confidential*. This was the assurance given to this supermarket clerk at the time, and this assurance still stands.

3 Confidentiality

On first reading the definition of confidentiality is simple: it refers both to the identity of the person and the information disclosed. The researcher knows the name of the person who said the quote and promises not to tell other persons the identity of the person when reporting this information. This supermarket clerk cannot be anonymised yet common definitions of anonymity by qualitative researchers exacerbate misunderstanding with imprecision.

The British Sociological Association Statement of Ethical Practice conflates confidentiality and anonymity by using *and* when they could have used *or*. Guideline 18 states:

"Research participants should understand how far they will be afforded anonymity *and* confidentiality..."

Scott [18] also conflates anonymity. "Anonymity is a continuum (from fully anonymous to very nearly identifiable)." What Scott means is that anonymity can very nearly be unknown but that is not a sound ethical assurance with the potential to harm informants. Saunders et al. [15] claim 'anonymity' has commonly been used either interchangeably with, or conflated, with 'confidentiality'. They conflate the known and unknown stating "anonymity is one form of confidentiality – that of keeping participants' identities secret." The essence of this confusion is separating the identity of the person and information they shared. The problem is that the person's identity and any statement the person makes cannot be separated.

Qualitative data can be de-identified by redacting names and context, but it cannot be anonymised. The researcher will always know the source of de-identified data. At no time should a qualitative researcher promise participant's anonymity. The term anonymity must be used in its dictionary sense when discussing consent with participants to collect qualitative data as that is their comprehension. In other words, there are limits to confidentiality. These are also exposed by the threat posed by a subpoena [13]. Anonymised survey data is not subject to subpoena. It is not retrievable.

Clarity of ethical concepts is essential; failure to do so potentially puts qualitative research participants in harm's way. Researchers cannot offer participants ethical assurances of both confidentiality and anonymity interchangeably; as if the double assurance were better than one. It is not; the concepts of anonymity and confidentiality are mutually exclusive.

Confidentiality is an essential qualitative research ethics assurance. Anonymity is not. Confidentiality should be thought of like an iceberg; only the tip is known but what lurks unseen, below the surface is also a source of potential harm. The easily identified aspect of confidentiality, the tip above the surface is external confidentiality [21]. It is well known to researchers and found in any ethical code. External confidentiality is traditional confidentiality where the researcher acknowledges they know what the person said but promises not to identify them in the final report. The less apparent aspect of confidentiality is internal confidentiality [21] or deductive disclosure [9]. The threat of internal confidentiality is the ability for research participants involved in a single study to identify each other in the final publication of the research. Internal confidentiality is the part of the iceberg that lies below the surface, going unacknowledged in ethical codes. If a researcher interviews family members, fellow workers, or a member of their small town the threat to confidentiality is sourced not by strangers (i.e. external confidentiality) but fellow residents/occupants/workers. Each of these can identify themselves and by default others. Internal confidentiality is predictable and when overlooked it has the potential to generate what Guillemin and Gillam [6] call big ethical moments.

4   Big Ethical Moments

Historical exemplars of big ethical moments generated by breaches of internal confidentiality are common in the qualitative sociology and anthropology and routinely sourced to a naive belief that pseudonyms provide robust ethical assurances. They do not. Street Corner Society [27], William Whyte's seminal text, is a case in point. In Whyte's original text, he gave pseudonyms to the region (Cornerville) and its inhabitants (e.g., "Doc"), thus protecting them with external confidentiality. The appendix of his 1981 edition captures the everyday world of doing ethnographic study, but also provides an insight into the harm caused by breaches of internal confidentiality. Participants told Whyte about how insiders recognized themselves and other insiders in the text:

> Pecci (Doc) did everything he could to discourage local reading of the book for the possible embarrassment it might cause a number of individuals, including himself [27].

Despite promises of external confidentiality, when Whyte's participants read the book, they saw themselves and those close to them.

Solutions to problems posed by internal confidentiality often suggest anonymizing participants by using pseudonyms. Yet at no time can qualitative researchers conjure the known to be unknown as Wiles suggests:

> The primary way that researchers seek to protect research participants from the accidental breaking of confidentiality is through the process of anonymization, which occurs through the use of pseudonyms [29].

Pseudonyms are a short-sighted solution causing exponential harm. Relational persons take great delight in breaking the code. In an anthropological study in a rural United States town a researcher caused anger and dissension among those whom they studied when residents broke the code. Munchmore [12] reports:

> When the [anthropology] book was published, many townspeople were highly disturbed to see some of the most intimate details of their lives recorded in print. Even though the author had attempted to protect his informants by using pseudonyms, their true identities were easily recognizable to anyone familiar with the area. Fifteen years later, another anthropologist who

visited the town was surprised to discover that the local library's copy of the book had *the real names of all the individuals pencilled in next to their pseudonyms.* Even after all those years, some of the community members were still visibly upset about the ways in which they had been portrayed.

In Ellis' [3, 4] *Emotional and Ethical Quagmires in returning to the field* she presents an account of dealing with her own pain when she realizes the distress her study of Fisher Folk, a study of a Chesapeake fishing community, has caused her informants. In returning to the fishing village Ellis discovered the illiterate research participants had had the book read to them by another researcher. The residents were outraged.

The pseudonyms used to secure confidentiality had failed to work and key informants felt they could identify themselves and others in the text. Ellis reports the residents felt the book had made them look stupid. Ellis' strategy of inventing pseudonyms was basic, starting with the same letters as the double names of the Fishneckers and having other similarities in sound. This made it easy to keep names straight, but at the cost of making it convenient for Fishneckers to figure out the characters in my story.

Vidich and Bensman's [26] 1968 book *Small Town in Mass Society: Class, Power and Religion in a Rural Community* used pseudonyms for some of 3000 inhabitants of a town they called Springdale. The town's response to this invasion of privacy was an uprising. Writing an editorial in *Human Organisation* Whyte [28] cited a "Springdale" newspaper account of the episode:

> The people of the village (Springdale) waited quite a while to get even with Art Vidich, who wrote a Peyton Place type book about their town recently. The featured float of the annual Fourth of July parade followed an authentic copy of the jacket of the book, *Small Town in Mass Society,* done large scale by Mrs Beverly Robinson. Following the book cover came residents of (Springdale) riding masked in cars labelled with fictitious names given them in the book. But the payoff was the final scene, a manure spreader filled with very rich barnyard fertiliser, over which was bending an effigy of "The Author".

Scheper-Hughes' [16, 17] experience in an Irish village was not dissimilar to Vidich or Ellis. She was eventually run out of town. When she returned to the site of her 1979 study of the mental health in an isolated Irish village, she found villagers had deciphered her attempts to provide pseudonyms as ethical assurances. She described her use of pseudonyms as ineffective:

> I would be inclined to avoid the 'cute' and 'conventional' use of pseudonyms. Nor would I attempt to scramble certain identifying features of the individuals portrayed on the naive assumption that these masks and disguises could not be rather easily decoded by villagers themselves [16].

This cannon of stories about how researchers like Whyte, Ellis, Vidich, Scheper-Hughes made their informants vulnerable to harm by failing to grasp the ever present threat posed by internal confidentiality. For researchers this represents an ethical *own goal.* This big ethical moment presents itself not at the origins of the research but within a project's dissemination of results. What is tragic is how each of these cases of harm was avoidable if recognising how qualitative research ethics can disintegrate.

When they do, the researcher remains responsible for protecting participants but others, including participants, can be made to share this responsibility.

5 Reference Groups

Creating a reference group is one solution to shoring up ethical assurances. This could be a group made up of supervisors, colleagues and in the case of graduate students, fellow graduate students [14]. The role of the reference group is to provide dispassionate advice for the researcher, to think outside the box. Edwards and Weller [2] created a reference group prior to conducting longitudinal interviews with rural youth. They made use of their reference group when a big ethical moment developed; one of their informants passed away at age 17 after having previously taken part in interviews at age 11 and 14. They tasked the reference group with giving advice on what interview recordings or transcripts they could share with the deceased boy's family, especially the boy's grieving mother.

Consider joining a reference now, charged with giving advice to a researcher whose big ethical moment with confidentiality dilemmas is causing loss of sleep. Drawing on advice given above in this chapter about the limits of anonymity, what advice would the reader give? The researcher (cited in [10]) said:

> I'm doing multi-site case study research, in a small number of institutions in a small country where the number of such institutions is relatively small. In spite of my best efforts to anonymize my sites, projects and respondents (using aliases, codes, and general role descriptors) any informed reader would have little difficulty identifying the sites, even the individual respondents. Deductive disclosure is a real concern. I'm assured by others that these people, given their professional roles are not naïve and have verified their transcript in full knowledge of my intention to cite or quote them... As a consequence, I have decided that each individual respondent will verify (and amend if necessary) their own transcript... I'm reluctant to rock the boat by exploring into much detail, unless asked, what they actually understand by anonymity. I've spelt it out in writing, and they seem to realise what they are signing up to. Still keeps me awake at night though!

What keeps this researcher up at night is seeking ethical solutions with the unworkable concept of anonymity. The researcher's persistence to anonymise the data set fails to grasp how qualitative research ethics have disintegrated under the threat of internal confidentiality. The researcher's reference group did not refocus attention away from anonymity to consent.

> I'm assured by others that these people, given their professional roles are not naïve and have verified their transcript in full knowledge of my intention to cite or quote them.

Confidentiality and anonymity were the wrong options. Nothing this researcher could do could make the known unknown. It was too late to put the genie back in the bottle. This researcher and their reference group needed to find alternative ethical solutions. What the researcher should have been asking informants was an ongoing form of consent, known as process consent, not for additional confidentiality assurances.

Clandinin and Connelly's [1] ethics guidelines widely cited in the qualitative research literature claim ethical considerations can and must be negotiated throughout the research process. Process consent [5] is the most common definition of this negotiation. It is an active form of consent and taking the participant's right to withdraw beyond a passive construction. Rather than leaving it up to the participant to withdraw at any time the researcher repeatedly invites the participant to volunteer to be part of each phase of the project. Without process consent the right of a participant to withdraw from the research project initially written in the consent form appears to be written in disappearing ink. The narrative researcher Josselson [7] astutely labels the informed consent process "a bit oxymoronic, given that participants can, at the outset, have only the vaguest idea of what they might be consenting to". With some candour Josselson [8] says process consent "strikes terror into researchers because it means just what it says." The researcher could have been losing sleep because offering process consent meant potentially losing the data. This is a risk the researcher must bear.

Explicitly offering informants' process consent in the loss of sleep scenario would have been best practice. It would have expanded participant autonomy. The participant, not the researcher, then decides whether they want to remain in the research or not. In other words, the confidentiality ethical assurances had disintegrated to such an extent that the researcher had no ethical assurances they could give the participant. This situation is not unusual in qualitative research and it is remedial. This set of circumstances routinely happens in focus group research where research participants must take responsibility for their own safety; this may entail them withdrawing from the study even though it strikes terror into researchers.

Focus groups offer participants few ethical assurances [11]. Focus group researchers cannot offer participants internal confidentiality because it is outside of their control: researchers can place few restrictions on focus group members. Researchers hold no ethical sanction over a participant should they reveal outside the focus group what was disclosed by another focus group member. Thus, promises of confidentiality must be limited to external confidentiality: that is, that the researcher will not identify any participant or what they said in any publication. If focus group participants are known to each other, for example if they are drawn from within the same organization, internal confidentiality is especially problematic, setting up particular ethical issues. Expect anything said in the focus group to be gossiped outside the focus group.

Focus groups pose more substantial ethical problems than one-on-one interviews. A participant in a one-on-one interview has opportunities to withdraw a remark during the interview or sometimes, if the participant reads an interview transcript, they can delete the remark during process consent. In focus groups verbal statements cannot be taken back. The bell, once rung, cannot be unrung. Thus, to use the word confidentiality without clarification may be taken as offering a layperson more than the concept can deliver. A warning in the participant information sheet could read:

> Please note there are limits on confidentiality as there are no formal sanctions on other group participants from disclosing your involvement, identity or what you say to others in the focus group. There are risks in taking part in focus group research and taking part assumes that you are willing to assume those risks [23].

The researcher who lost sleep above chose the wrong ethics option as confidentiality assurances had disintegrated. When the researcher was asking participants for their approval to publish the material, they were not using the correct ethical assurance.

This was not anonymity or confidentiality, it was consent. What the researcher should have been doing was asking each participant if they still wanted to take part in the research. In other words, rather than relying on the previous informed consent process they should have used process consent. This would have caused her fewer sleepless nights. But it would put the research at risk. This exposes a conflict of interest. The rights of participants versus the rights of the researcher.

While focus group researchers may offer participants ethical assurances such as confidentiality and informed consent these assurances are unenforceable. The principle of caveat emptor (let the buyer beware) [23] may be a more useful tool for those involved in focus group research: that is, let the researcher, the participants and the ethics committee beware that the only ethical assurance that can be given to focus group participants is that there are few ethical assurances.

6 Researcher/Participant Safety

The disintegration of qualitative research ethics takes other forms for both the participant and for the researcher. Guillimen and Gillam's [6] provide the now classic example of Sonia, a research participant who reveals during an interview about rural health services in Australia that her husband is sexually abusing her daughter. What should a researcher do next? Neither the researcher nor the ethics committee predicted anything like this during the formal ethics review. Guillemen and Gillam suggest reflexivity is needed to address big ethical moments like this. A practical solution is to enter every research interview with a list of social services informants can follow up. These would include women's refuge, suicide watch, rape crisis. Providing these referrals is a researcher's responsibility, they are not responsible for providing counsel for the participant.

Big ethical moments can also envelop the researcher making them unsafe. Stories they hear from informants can be emotionally draining, pulling them into a precarious space. A postgraduate student studying adolescent poverty shared this description of uncertainty [24].

> Recently one of my participants ran away from a violent situation in her home and found herself homeless. She reached out to me in a text stating that "if I wasn't desperate, or that if it wasn't my last resort, I wouldn't contact you". It was then, I realised that I had become part of these unstable housing stories. That's when I panicked. I did not know how to respond. I read this message as serious. I tried calling her immediately, but her phone was turned off or out of battery.

The student relied on her reference group to make sense of her responsibilities in this situation.

Collecting data in unsafe spaces can unsettle researchers and researchers should have a safety plan. Seiber and Tolich [19] highlighted steps any researcher could take when they found themselves in a risky situation. These steps were drawn from "A Code of Practice for the Safety of Social Researchers" (n.d.). A most basic safety strategy is telling some person where and when the research will take place as well as contacting them at the end of the interview. In other words, assume ethical assurances can disintegrate for the researcher as well. Another strategy is a rapid exit protocol any researcher could use when they feel sufficiently threatened in an interview. Consider this and practice it:

> Pull a cell phone from one's pocket as if it had vibrated, stand up answering the phone speaking with some urgency to an imaginary family member repeating the news that a relative is seriously ill and their presence was required. The researcher need only say, "that is tragic, I will be right there" and the researcher leaves the site saying to the participant they will be in touch.

In these situations, the researcher must remain proactive by creating a safety plan, imaging ethics can disintegrate for them too.

7 Discussion

Expect the unexpected should be a basic assumption for qualitative researchers. They must know the limitations of qualitative research ethics, what to do if and when they disintegrate. Researchers need to be proactive offering participants consent at both the beginning and at the end of a project. Anonymity is never an option. A second proactive stance is knowing there are limits to confidentiality. These limits should not come as a surprise; they are known before the research begins when researching relational persons. Relational participants need to be made aware of the threat posed by internal confidentiality. When the number of research sites diminishes so too do confidentiality assurances. Yet even when ethical assurances disintegrate totally, like happens routinely in focus group research, the researcher's responsibility is to switch ethical assurances toward a participant's autonomy, allowing them to make the decision to stay or withdraw from the research. This action, while ethical, places the researcher's data at risk.

The primacy of the method is broader for qualitative research than data collection and analysis and occurs after ethics review. This primacy means that the researcher is solely responsible for the protection of participants who volunteer to take part in this study. Yet in process consent the burden of responsibility for risk can be shared with the participant; they must be given the opportunity to withdraw at the last minute.

References

1. Clandinin, D.J., Connelly, F.M.: Narrative inquiry: experience and story in qualitative research (2000)
2. Edwards, R., Weller, S.: Ethical dilemmas around anonymity and confidentiality in longitudinal research data sharing: the death of Dan. In: Tolich, M. (ed.) Qualitative Ethics in Practice, pp. 97–108. Routledge, New York (2016)
3. Ellis, C.: Fisher Folk: Two Communities on Chesapeake Bay. University of Kentucky Press, Lexington (1986)
4. Ellis, C.: Emotional and ethical quagmires in returning to the field. J. Contemp. Ethnogr. **24**, 711–713 (1995)
5. Ellis, C.: Telling secrets, revealing lives: relational ethics in research with intimate others. Qual. Inq. **13**(1), 3–29 (2007)

6. Guillemin, M., Gillam, L.: Ethics, reflexivity, and "ethically important moments" in research. Qual. Inq. **10**(2), 261–280 (2004)
7. Josselson, R.: Ethics and Process in the Narrative Study of Lives, vol. 4. Sage, Thousand Oaks (1996)
8. Josselson, R.: The ethical attitude in narrative research: principles and practicalities. In: Clandinin, J. (ed.) Handbook of Narrative Inquiry: Mapping a Methodology, pp. 537–566. Sage, Thousand Oaks, CA (2007)
9. Kaiser, K.: Protecting respondent confidentiality in qualitative research. Qual. Health Res. **19** (11), 1632–1641 (2009)
10. Macfarlane, B.: Researching with Integrity: The Ethics of Academic Enquiry. Routledge, New York (2010)
11. Morgan, D.: The Focus Groups Guidebook. Sage, Thousand Oaks (1998)
12. Muchmore, J.A.: Methods and ethics in a life history study of teacher thinking. Qual. Report **7**(4), 1–17 (2002)
13. Palys, T., Lowman, J.: A belfast project autopsy: who can you trust? In: Tolich, M. (ed.) Qualitative Ethics in Practice. Routledge, New York (2016)
14. Pollard, A.: Field of screams: difficulty and ethnographic fieldwork. Anthropol. Matters **11** (2), 1–23 (2009)
15. Saunders, B., Kitzinger, J., Kitzinger, C.: Anonymising interview data: challenges and compromise in practice. Qual. Res. **15**(5), 616–632 (2015)
16. Scheper-Hughes, N.: Ire in Ireland. Ethnography **1**(1), 117–140 (2000)
17. Scheper-Hughes, N.: Saints, Scholars, and Schizophrenics: Mental Illness in Rural Ireland. Univ of California Press, Berkeley (1979)
18. Scott, C.R.: Anonymity in applied communication research: tensions between IRBs, researchers, and human subjects. J. Appl. Commun. Res. **33**(3), 242–257 (2005)
19. Sieber, J., Tolich, M.: Planning Ethically Responsible Research, 2nd edn. Sage, Los Angeles (2013)
20. Tashakkori, A., Teddlie, C.: Issues and dilemmas in teaching research methods courses in social and behavioural sciences: US perspective. Int. J. Soc. Res. Methodol. **6**(1), 61–77 (2003)
21. Tolich, M.: Internal confidentiality: when confidentiality assurances fail relational participants. Qual. Sociol. **27**(1), 101–106 (2004)
22. Tolich, M.: Alienating and liberating emotions at work supermarket clerks' performance of customer service. J. Contemp. Ethnogr. **22**(3), 361–381 (1993)
23. Tolich, M.: The principle of caveat emptor: confidentiality and informed consent as endemic ethical dilemmas in focus group research. J. Bioethical Inq. **6**(1), 99–108 (2009)
24. Tolich, M., et al.: Researcher emotional safety as ethics in practice. In: Iphofen, R. (ed.) Handbook of Research Ethics and Scientific Integrity. Springer, Cham, in Press
25. Tolich, M., Fitzgerald, M.: If Ethics Committees Were Designed For Ethnography. J. Empir. Res. Hum. Res. Ethics **1**(2), 71–78 (2006)
26. Vidich, A.J., Bensman, J.: Small Town in Mass Society: Class, Power, and Religion in a Rural Community. University of Illinois Press, Champaign (1968)
27. Whyte, W.F.: Street corner society: The social structure of an Italian slum. University of Chicago Press, Chicago (1943/1981)
28. Whyte, W.F.: Freedom and responsibility in research: the Springdale case. Hum. Organ. **17** (2), 1 (1958)
29. Wiles, R.: What are qualitative research ethics?, vol. 2. A&C Black, London (2012)

The Researcher as an Instrument

Safary Wa-Mbaleka[(⊠)]

Asian Qualitative Research Association, Adventist University of Africa,
Advent Hill Road, Ongata Rongai, Nairobi, Kenya
Wa-MbalekaS@aua.ac.ke

Abstract. In qualitative research, there are many different sources of data. Qualitative research data are collected using many different methods. Interestingly, one of these data collection methods is the researcher himself or herself. This is the reason why most experts consider the researcher as an instrument. The question always asked is "What does it really mean?" This chapter explains what it is and what is expected from the researcher in his or her role as an instrument throughout a qualitative research study. The ethical considerations pertaining to this important role are also discussed. This chapter is meant to bring this important role to everyone's awareness so that rigor in qualitative research can be fostered.

Keywords: Rigor in qualitative research · Data collection · Data analysis · Data interpretation · Research instrument · Trustworthiness · Ethical considerations

1 Introduction

Qualitative research (QLR) continues to spread around the world. What once used to be seen as complex, confusing, or undesirable is now considered as a fast-growing, innovative research type around the globe. More and more institutions and research organizations are integrating QLR, more QLR reports are now being produced through theses, dissertations, and scholarly publications. Newer QLR textbooks are becoming more and more practical in guiding novice researchers in carrying on their studies.

A special excitement about QLR is evidently growing in many different fields. Part of it is that QLR provides flexibility in exploring problems in depth [1–4]. Data can be collected from many different sources such as interviews, observations, documents, fieldnotes, videos, audios, photos, artifacts, and many more [5–10]. Many QLR experts consider that everything can be data [11, 12], as long as we are able to explain why that information is needed for our research study.

In fact, the researcher himself or herself is also considered an instrument for data collection [4]. This notion has been held in the field of QLR for several decades. While this notion may be considered beneficial to the researcher, it may also come with challenges. What does the researcher as an instrument really mean? What role does the researcher play in a QLR study? How do we deal with the researcher's bias? Do we even have to deal with the researcher's bias in the first place? If the researcher is a primary instrument in QLR data collection, analysis, and interpretation, how trustworthy are the

© Springer Nature Switzerland AG 2020
A. P. Costa et al. (Eds.): WCQR 2019, AISC 1068, pp. 33–41, 2020.
https://doi.org/10.1007/978-3-030-31787-4_3

outcomes of a QRL study? As a research instrument, what should a researcher do at the different stages of a QLR journey? These and many other similar issues are the reason for this chapter. While all the solutions may not be provided in this chapter, it is meant to generate discussion to dig deeper in this issue of the researcher as an instrument.

2   Defining the Researcher as an Instrument

Most people who have a solid foundation on QLR know that the researcher is considered an instrument. While it is a notion that everyone seems to agree with, manyqualitative researchers may not be able to explain effectively what it means. Yet, it is well-accepted QLR assumption that "the researcher is the primary instrument for data collection and analysis" [4]. In other words, the qualitative researcher uses "his or her eyes and ears and filters" [3] to collect, analyze, and interpret the data.

Although quantitative researchers may at times be quite uncomfortable about it, "in qualitative research scholarship, the researcher is 'the human instrument'...through which we gather data and gain understanding" [13]. The role the researcher plays in QLR is significant throughout the whole research study. It must therefore be taken seriously. This reality has implications from four different angles.

First, research instruments are designed by the researcher because a qualitative researcher cannot use instruments designed by previous researchers [5]. The qualitative researcher must design different tools and instruments used in collecting data from the different sources mentioned earlier in this chapter. For instance, the qualitative researcher must design interview guides, observation protocols or observation guides or observation checklists, and all other means of recording data needed for the execution of a study. While designing these instruments, the researcher has a high likelihood of introducing his or her bias. Without proper care in the process of designing the research instruments, the researcher runs the risk of designing tools that will only confirm his or her own bias. Such instruments would negatively affect the trustworthiness of the study.

Second, our emotions affect the data and our data affect our emotions [14, 15]. In many cases, people undertake a QLR study because they are interested in the issue, the problem, or the topic. It may be a problem that they themselves have experienced or an experience their loved one has had. The fact that qualitative researchers can relate quite well to the issue they are studying puts them in a fairly emotional state. This emotional state has some bearing on the data that is collected and how it is collected. On the other hand, as the qualitative researcher digs deeper in their exploration, new insights about an issue have some impact of the researchers' emotional state. It is almost impossible to conduct a QLR study well and not have any impact on the researcher. The study has an impact on the researcher and vice versa. Again, this reality calls for the need to examine carefully what it really means to be practically a researcher as an instrument.

Third, in connection to the previous point, emotions bring out participants' vulnerability [13]. More than any other means of data collection, this seems to happen more in interviews where the qualitative researcher and participant are involved in a constructive dialogue. The qualitative researcher needs to plan how to deal with participants' vulnerability and the researcher's vulnerability. Sometimes, in the process of conducting the study, the researcher may also experience some emotional challenge that can make him or her vulnerable on a personal level or on the ability to carry on the study effectively. The vulnerability that comes from the researcher as an instrument must therefore be considered carefully in planning and carrying out the QLR study.

Last, QLR is not free from bias. The researcher brings his or her bias to the QLR study. According to [16], "the subjectivity of the researcher is something to be embraced, not controlled for or eliminated" (p. 178). While quantitative research promotes the elimination of subjectivity, although not practically possible in the real world, QLR promotes embracing subjectivity. Indeed, trying to hide the bias, as is common in quantitative research, simply weakens the research study's trustworthiness. Spelling out the researcher's bias clearly helps increase the study's credibility. This is done through the process of researcher's reflexivity; that is, the self-critical analysis of the researcher's bias, assumptions, background, expertise, as pertaining to the proposed QLR study.

3 Importance of Reflexivity

In an attempt to help the researcher to become a good instrument, reflexivity is proposed for all qualitative researchers. While there are many different definitions of reflexivity, it can be simply defined as a critical analysis of self as a researcher [6]. Reflexivity is a conscious process that "forces us to come to terms with...our choice of research problem and with those with whom we engage in the research process, but [also] with ourselves and with the multiple identities that represent the fluid self in the research setting" [6].

This process is important because "reflexivity forces the researcher to re-examine his or her positioning in relation to methodology, theory, participants and the self" [15]. This process must happen throughout the research study, from its design all the way to data interpretation. The qualitative researcher does this to "increase accountability for the knowledge that is produced" (p. 87). It is an important process for the QLR study's trustworthiness.

Reinharz [17] proposes three types of self that are interconnected in a QLR study. First, there are research-based selves. These are biases that come from the person's research training and experience. Basically, the QLR study is affected by how the researcher was trained in research and on his or her past experience conducting research. Second, the researcher is likely to bring some selves from his or her historical, social, academic, and personal backgrounds. These are not to be ignored when writing and taking into account one's reflexivity. Last, there are also situationally-created selves; that is, selves created during the research study. As the researcher carries out the study, he or she can develop new identities. It is the researcher's work in his or her reflexivity to evaluate critically how these selves are affecting the study.

As complex as reflexivity may be, each qualitative researcher needs to know his or her role to play in writing and following it throughout the study. Lincoln, Lynham and Guba [6] proposes five ways of understanding our role. First, as qualitative researchers, we must interrogate each of our selves. We should not take each for granted. Second, we must question how our selves affect the study. Next, we should always question how our selves affect our interaction with the participants. Additionally, we must make an effort to discover the meaning of the phenomenon under exploration while discovering our self in the process. Last, it is the qualitative researcher's role to discover progressively how their new identity is shaped throughout the study.

This section has provided a general overview of reflexivity and the role it plays in QLR. For more exploration on reflexivity, readers are directed to Corlett and Mavin [18], Haynes [15] and Palaganas, Sanchez, Polintas and Caricativo [19]. Haynes proposes the following questions to be considered as part of reflexivity (p. 78):

- What is the motivation for undertaking this research?
- What underlying assumptions am I bringing to it?
- How am I connected to the research, theoretically, experientially, emotionally?
- And what effect will this have on my approach?

While missing in many QLR publications, reflexivity is an important process in all QLR studies. It helps establish a strong claim for the researcher as a credible instrument. It is useful in the critical analysis of the researcher, his or her effect on the study and vice versa. While some preliminary work of reflexivity can be done in the design process of a QLR study, it needs to continue throughout the actual study. Indeed, qualitative researchers must be involved in "continuous reflexivity and self-scrutiny" [20]. Reflexivity is more than just writing the reflexivity statement.

4   The Researcher in Data Collection

When collecting QLR data, it is important to consider four main points. First, it is already established that our biases affect the data if we do not have a certain mechanism of controlling them. It is therefore our responsibility to keep our biases in view, from the reflexivity statement, while data collection is ongoing. Doing so can help prevent us from being driven mainly by our bias. While bias cannot practically be avoided fully in QLR, its effect can be lowered by focusing the data collection on the stated research problem, research purpose, and research questions.

Second, our biases may prevent us from catching some important data. It is important to understand how strong the researcher's biases, background knowledge and experience can affect data collection [1, 5, 14] to the point that the researchers may miss some important data. This weakness can be due to the fact that the researcher fails to capture the reality as presented in data collection or the data collection tools are mainly designed to follow the researcher's bias. In this case, the validation of the data collection instruments before using them can be of help [7, 20]. Experts on the topic who are requested to check the instruments before they are used in the data collection can help catch some of the obvious biases and weaknesses of those instruments.

Third, the positive side of the researcher's background is that it can also help understand some intricacies of the research problem. Having some background knowledge about the topic under exploration is not all bad. For someone who may have had an experience similar to the one related by the research participants, it may be easier and more effective to make quick connections. In this case, the researcher with limited acquaintance with the topic under exploration may be at disadvantage.

Last, since the qualitative researchers have longer contact with the research participants than is common in quantitative research, the qualitative researchers must build rapport with the participants [21]. According to [22], linguistic competence coupled with pragmatics (or proper usage of language in the target context) leads to communicative competence. From this perspective, four assumptions are proposed for the qualitative researchers to build the needed rapport in data collection.

- *Assumption 1: The researcher must know the language of the participants and use it appropriately to communicate with them.* For sure, the use of a research assistant to conduct the interviews is not ideal, although it works well when the researcher is not knowledgeable in the language of the participants.
- *Assumption 2: The researcher must respect the culture of the participants.* The researcher needs to contextualize the communication that he or she has with the participants. Having some good foundation knowledge about the culture of the participants can help the researcher prepare and carry on the study well.
- *Assumption 3: The researcher must respect the research participants.* All participants, whether rich or poor, educated or illiterate, male or female, no matter their geographic, socio-economic, linguistic, religious, racial, tribal, gender, or academic backgrounds, must be treated with respect and dignity.
- *Assumption 4: Trust depends on the previous assumptions; failing in one of them can create a dent in the trust between the researcher and his or her participants.* All the four assumptions should therefore be taken seriously. The moment a research participant loses trust in the researcher, the credibility, quality, and quantity of the data collected may be compromised.

5 The Researcher in Data Analysis

The qualitative researcher's role as instrument continues even in data analysis. The researcher plays at least five roles in data analysis. There may be more but these five, proposed by Merriam and Tisdell [4] can be a good start. The paragraphs that follow discuss each of the five roles.

It is the job of the qualitative researcher to seek deeper understanding through both verbal and nonverbal communication. While words carry a significant meaning about a topic, nonverbal communication is equally, if not more, important [23]. The researcher should analyze both the overt and covert meanings of what participants communicate. This depends heavily on the researcher's linguistic competence, cultural understanding, and correct interpretation of what the researcher hears and sees.

Additionally, the qualitative researcher should develop an immediate information processing skill. While data analysis is expected to take quite some time—because data comes in different shapes and sizes—it is still important for the researcher to process information quickly. The researcher should be able to connect data efficiently to the research problem, purpose statement, research questions, and theoretical framework. This skill is needed in both data collection and analysis since data analysis starts during data collection [24].

Next, it is the role of the qualitative researcher to seek clarification and to summarize data. Clarification helps the researcher to avoid misinterpretation of the data. Summarization helps the researcher create a big and complete display of the different connections and ramifications of the topic or problem under exploration. The researcher must present a clear audit trail about how he or she conducted the study, analyzed the data, and interpreted the findings. This practice is also another important step in increasing the study's trustworthiness.

Furthermore, the qualitative researcher must check and recheck the analyses for accuracy. Data analysis needs to be thorough and complete. It should not be done in a hurry because QLR data present a multidimensional perspective on the research problem. The researcher must make sure his or her analysis is acceptable in QLR. He or she must see to it that all the data is analyzed as accurately as possible. Whenever there seems to be some contradictory findings, not only must the researcher report on those contradictions, he or she must also find out why those contradictions occurred. Better yet, the researcher should also use triangulation in the process of addressing the contradictions.

Last, the qualitative researcher must explore unusual or unanticipated responses. When a researcher is focused and careful in both data collection and analysis, it is not uncommon to come across some unusual or unanticipated responses. In quantitative research, these can be simply thrown away. In QLR, it is important to explore them further and to report them. In fact, this practice is known as strength of QLR because QLR gives the possibility to "probe into responses or observations as needed and obtain more detailed descriptions and explanations of experiences, behaviors, and beliefs" [25]. This is a QLR advantage that quantitative research does not provide.

6   The Researcher in Data Interpretation

According to Rivera and Tracy [26], our emotions and our biases "fundamentally construct and refract meaning" (p. 215). Basically, it is likely for qualitative researchers' interpretation to be fairly constructed or erroneously made. Again, it is the job of the researcher to ensure that data interpretation is accurate and aligned with the purpose of the study. Four ways to make data interpretation focused on the research problem rather than the researcher's bias are recommended below.

First, the focus of data interpretation must be on the theoretical framework [4]. Most QLR reports have a theoretical framework. This framework is supposed to be the lens through which the researcher sees the phenomenon under exploration. During the data collection, analysis, and interpretation, the researcher must check and recheck how the data contribute to the meaning of the phenomenon from the perspective of the prepared theoretical framework. The researcher must also connect the discussion of the findings to the theoretical framework [7, 8].

Second, the data interpretation must also be aligned with the research problem, purpose statement, and the research questions [7, 8]. These components of any research study provide the clear direction of the study. Deviating from them in data analysis and data interpretation would only create a gap or dissonance in the QLR report. It is actually easy to go off track in QLR because the data is plentiful and diverse. Keeping the interpretation of the data focused on the research problem, research goals, and research questions can help removing off-topic data questions can help interpreting off-topic data. The alignment between the different parts of a research report must be kept throughout the manuscript.

Third, in the data interpretation, the researcher's and participants' subjectivity must be made visible [27]. Through reflexivity, the researcher is able to clearly state his or her biases. Through thick description and direct quotes from the participants, the researcher can reveal the participants' subjectivity. QLR is about the co-construction of reality as experienced and understood by both the researcher and the participants. Demonstrating the subjectivity of both parties can actually help increase the study's credibility.

Last, it is the researcher's job to create reality's objectivity by building on individual subjectivities. It is true that, in QLR, objectivity is considered a fiction [5, 28]. However, by conducting any research study, it is with the goal of coming up with some sort of commonly agreeable conclusion. Yet, it is well established that QLR is subjective [29, 30]. So, it can be agreed that a good QLR is the sum of subjective interpretations of reality that lead to an agreed-upon common understanding of the studied phenomenon.

7   The Researcher in Reporting

Reporting the findings of a QLR study must be truthful. The researcher must align it with the research problem, goal, questions, and theoretical framework. Unfortunately, "our commonsensical notions about the world, that is, what we feel we already know about it, constrain our abilities to change how we see, and think about, the object of our study" [31]. For this reason, the qualitative researchers must strongly stick to the new insights from the study instead of the confirmation of their own bias. The researcher who reports only on findings that confirm what they had in mind prior to the study is simply biased.

To be able to report the findings truthfully, qualitative researchers must regularly "question the assumptions about what [they] take to be normal and seek ways to upset conventionalized ways of thinking" [31]. Research is not done only to confirm what other researchers have found or what the researcher believes to be true. It is about getting insights on a topic as expressed by the research participants. A good qualitative researcher must be able to report the findings truthfully, even if they challenge his or her assumptions, background knowledge, belief systems, and even values. After all, through reflexivity, the researcher will have had his or her opportunity to present his or her beliefs or bias.

8   Conclusion

Saying that a researcher plays the role of an instrument is probably more complex than it has ever been discussed before. It is true that a qualitative researcher is an instrument. If this instrument is not well monitored, it can have a devastating impact on QLR studies. At each stage of a QLR study, the researcher needs self-critique and self-scrutiny to ensure the study is not driven solely by his or her bias.

When we look carefully at this concept of the researcher as an instrument, we need to keep the researcher's reflexivity in mind throughout the study. In fact, we may even "need to unlearn what we have learned with regard to 'good' research as being exclusively rational, linear, and objective" [13]. While QLR may be messy, we need to explain what we did, how we did it, and why we did it. We must somehow embrace our subjectivity; that is, we must "identify [our biases] and monitor them in relation to the theoretical framework and in light of [our] own interests, to make clear how they may be shaping the collection and interpretation of data" [4]. We must be aware of our biases, how they may affect our study, how to prevent them from significantly affecting our study, and how the study shapes us. After all, there is no value-free QLR study.

References

1. Creswell, J.W., Poth, C.N.: Qualitative Inquiry and Research Design: Choosing Among Five Approaches. SAGE, Thousand Oaks (2016)
2. Denzin, N.K., Lincoln, Y.S. (eds.): The SAGE Handbook of Qualitative Research. SAGE, Los Angeles (2017)
3. Lichtman, M.: Qualitative Research in Education: A User's Guide. SAGE, Thousand Oaks (2013)
4. Merriam, S.B., Tisdell, E.J.: Qualitative Research: A Guide to Design and Implementation. Wiley, New York (2016)
5. Creswell, J.W.: Qualitative Inquiry and Research Design: Choosing Among Five Approaches. SAGE, Thousand Oaks (2013)
6. Lincoln, Y.S., Lynham, S.A., Guba, E.G.: Paradigmatic controversies, contradictions, and emergin confluences, revisited. In: Denzin, N.K., Lincoln, Y.S. (eds.) The SAGE Handbook of Qualitative Research, pp. 213–263. SAGE, Los Angeles (2017)
7. Wa-Mbaleka, S.: Writing Your Thesis and Dissertation Qualitatively: Fear No More. Oikos Biblios Publishing House, Silang (2018)
8. Wa-Mbaleka, S.: Student Advising in Qualitative Research: Fear No More. Oikos Biblios Publishing House, Silang (2019)
9. Yin, R.K.: Case Study Research: Design and Methods. SAGE, Newbury Park (2014)
10. Yin, R.K.: Qualitative Research from Start to Finish. Guilford Publications, New York (2015)
11. Brinkmann, S.: Qualitative Interviewing. Oxford University Press, Oxford (2013)
12. Glaser, B.G., Strauss, A.: The Discovery of Grounded Theory: Strategies for Qualitative Research. Aldine, Chicago (1967)
13. Rivera, K.D.: 'Use Your Feelings': emotion as a tool for qualitative research. In: Cassell, C., Cunliffe, A.L., Grandy, G. (eds.) The SAGE Handbook of Qualitative Business and Management Research Methods, pp. 450–467. SAGE, Thousand Oaks (2018)
14. Haynes, K.: Tensions in (re)presenting the self in reflexive autoethnographical research. Qual. Res. Organ. Manag. 6, 134–149 (2011)
15. Haynes, K.: Reflexivity in qualitative research. In: Symon, G., Cassell, C. (eds.) Qualitative Organizational Research: Core Methods and Current Challenges, pp. 72–89. SAGE, Los Angeles (2012)

16. Grandy, G.: An introduction to constructionism for qualitative researchers in business and management. In: Cassell, C., Cunliffe, A.L., Grandy, G. (eds.) The SAGE Handbook of Qualitative Business and Management Research Methods, pp. 173–184. SAGE, Thousand Oaks (2018)

17. Reinharz, S.: Who am I? The need for a variety of selves in the field. In: Hertz, R. (ed.) Reflexivity and Voice, pp. 3–20. SAGE, Thousand Oaks (1997)

18. Corlett, S., Mavin, S.: Reflexivity and researcher positionality. In: Cassell, C., Cunliffe, A.L., Grandy, G. (eds.) The SAGE Handbook of Qualitative Business and Management Research Methods, pp. 377–399. SAGE, Thousand Oaks (2018)

19. Palaganas, E.C., Sanchez, M.C., Polintas, M.V.P., Caricativo, R.D.: Reflexivity in qualitative research: a journey of learning. Qual. Rep. **22**, 426–438 (2017)

20. Pryett, P.M.: Validation of qualitative research in the "real world". Qual. Health Res. **13**, 1170–1179 (2003)

21. Daymon, C., Holloway, I.: Qualitative Research Methods in Public Relations and Marketing Communications. Taylor & Francis, New York (2011)

22. Genc, B., Bada, E.: Culture in language learning and teaching. Read. Matrix **5**, 73–84 (2005)

23. Willing, C.: Interpretation and analysis. In: Flick, U. (ed.) The SAGE Handbook of Qualitative Data Analysis, pp. 136–149. SAGE, Thousand Oaks (2013)

24. Miles, M.B., Huberman, A.M., Saldaña, J.: Qualitative Data Analysis: A Methods Sourcebook. Sage, Los Angeles (2014)

25. Guest, G., Namey, E.E., Mitchell, M.L.: Collecting Qualitative Data: A Field Manual for Applied Research. SAGE, Los Angeles (2013)

26. Rivera, K.D., Tracy, S.J.: Embodying emotional dirty work: a messy text of patroling the border. Qual. Res. Organ. Manag. Int. J. **9**, 201–222 (2014)

27. Lather, P., Pierre, E.A.: Post-qualitative research. Int. J. Qual. Stud. Educ. **26**, 629–633 (2013)

28. Merriam, S.B.: Qualitative research: A guide to design and implementation. John Wiley & Sons, San Francisco, CA (2009)

29. Bogdan, R.C., Biklen, S.K.: Qualitative research for education: An introduction to theories and methods. Ally and Bacon, New York (2007)

30. Taylor, S.J., Bogdan, R.C., DeVault, M.: Introduction to Qualitative Research Method: A Guidebook and Resource. Wiley, New York (2016)

31. Freeman, M.: Thinking for Qualitative Data Analysis. Taylor & Francis, New York (2017)

The Use of Selected CAQDA Software Examples in a Research Project Based on the Grounded Theory Methodology

Jakub Niedbalski[(⊠)] and Izabela Ślęzak

The Faculty of Economics and Sociology, University of Lodz, Łódź, Poland
jakub.niedbalski@gmail.com, iza.slezak@gmail.com

Abstract. The purpose of the paper is to show how computer-aided qualitative data analysis (CAQDA) tools can be used in research practice. Based on particular research projects, the authors attempt to show how to carry out research in accordance with the procedures of grounded theory methodology using the functions offered by programs such as Audacity, WeftQDA, and CmapTools. The paper presents the toolkit of a qualitative researcher who uses computer software to aid the research process on a daily basis, and it is structured as follows. Firstly, the emphasis is put on the form of the analytical process in the context of a chosen method and using certain types of computer software. At the same time, the authors highlight the improvements that have been made, but they also stress the consequences and potential difficulties related to applying CAQDA in qualitative research. This speech (article) is both a review and educational, allowing the reader to become familiar with the possibilities provided by CAQDA tools and their application in research projects based on qualitative methods.

Keywords: Computer-aided qualitative data analysis · CAQDA ·
Qualitative research methods · Methodology of grounded theory · Audacity ·
WeftQDA · CmapTools

1 Introduction

The paper is intended to show how the use of three computer programs, available free of charge, makes it possible to carry out the most important actions related to the formal preparation, development, and analysis of data in accordance with the grounded theory (GT) methodology. It will also show both the current findings of the researchers and the final output from their analysis in the form of a diagram that integrates and presents the generated categories and the relationships between them. Furthermore, the article aims to show the advantages but also some restrictions related to the use of those programs.

The programs were deliberately selected to complement each other in terms of their functions when it comes to carrying out research based on the grounded theory methodology [3, 6, 7, 13]. It does not mean, however, that the information we describe may be useful only to a narrow group of researchers because the presented programs may be successfully employed in research based on research methods other than GT. In the case of the grounded theory methodology (GT), we mainly use data that are firstly

© Springer Nature Switzerland AG 2020
A. P. Costa et al. (Eds.): WCQR 2019, AISC 1068, pp. 42–57, 2020.
https://doi.org/10.1007/978-3-030-31787-4_4

unstructured; secondly, they exist as text; and thirdly, they are acquired through research techniques such as interviews or observations. Therefore, since text plays a key role in a GT-based analysis (it is, of course, impossible to ignore other current varieties of the described methodology and the various types of related data, e.g., images, among others), it is this element that our attention should focus on. This is supported by another, strictly technical argument, as the presented programs mostly allow for the analysis of text files. This excludes the possibility of applying other types of data. However, as already mentioned, it is the textual data that provides the primary source of information analyzed within GT. And this may somehow justify the emphasis we put on the programs that support this kind of data.

Nevertheless, it is possible to carry out a simple analysis of audio and visual data in those programs, although in a simplified scope. The former is supported by Audacity (where we can, e.g., add labels to soundtrack fragments), and the latter may be included in a mind map under development in CmapTools (where we can use images as a visualization of the mind map elements). However, since there are significant limitations related to the application of this type of data, the possibilities of these programs in this area will be only outlined in the paper. The main stress will be placed on the preparation and analysis of text data. However, it must be emphasized that text data need to be converted into the .txt format, meaning we are deprived of the possibility of applying formatting to the font, or using different typefaces, etc. (although some nuances of speech can be successfully rendered through a system of diacritical signs). Furthermore, there is no possibility to transfer data between the programs, which means certain actions must be carried out separately – at least partially – and in each program individually.

Nevertheless, the presented programs, despite some limitations, may significantly improve the work comfort of a researcher, and therefore contribute to more efficient and perhaps better preparation of data (although this mostly depends on the researchers themselves and their analytical skills and knowledge), allowing them to obtain satisfactory conclusions on that basis.

2 Preparing Data for Analysis with Audacity

Before starting to analyze data, a researcher must first process it, i.e., prepare previously gathered material. As such material will usually come from interviews recorded on various carriers of audio files, it becomes necessary to transform them in the most faithful way possible into text, i.e., transcription. Modern technology can greatly facilitate this process. It provides the researcher with extremely helpful tools, significantly speeding up the data preparation process. Today, devices that record audio are in widespread use. This is of great significance from the perspective of a researcher's work, because it makes it possible not only to make a good quality recording, or to archive the recordings easily, but mostly it allows them to be transcribed much more easily. Such possibilities are provided by many commonly used devices, such as smartphones. However, the professional work of a researcher requires tools specifically designed to record audio, i.e., voice recorders. Their advantage over other devices is that they make it possible to set numerous parameters that improve the recording quality and, furthermore, they are built in such a way as to "collect" specific sounds interesting for the researcher in the best way possible. It is worth mentioning that by using this type of device, we can install and use special programs intended not only to transfer the audio or change the file format but also to facilitate transcription. They are usually relatively simple tools which do not allow any complex operations on audio files; therefore, if a researcher would like to use slightly more sophisticated tools equipped with particular options that are helpful while transcribing, they need to seek suitable computer programs. Their selection is gradually growing; however, for the most part, they require the user to purchase a license, often at great expense. Not everyone can afford to buy such software or to pay to have the audio transcribed. Fortunately, there are alternative solutions available in the form of free, yet powerful software intended for data transcription. Audacity is one such tool, equipped with several interesting and helpful features that greatly facilitate and accelerate the transcription process.

Particularly noteworthy is the transcription toolbar, which makes it possible to play the sound at a slower or faster speed, depending on the user's needs, by adjusting the speed with a slider. When you move the slider to the right, towards the plus sign, you increase the speed; when you move it to the left, towards the minus sign, you slow the speed down. Normal speed is 1.00; below this value, the playback speed will be lower (e.g., $0.01\times$ means one-hundredth of the normal speed) and above, the playback speed is greater (e.g., $3.00\times$ means three times the playback speed). In order to set the exact speed of playing the recording, you need to click twice on the transcription toolbar and enter a specific speed value in the window that opens [10] (Fig. 1).

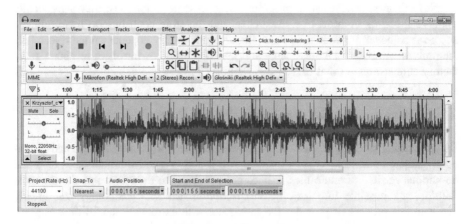

Fig. 1. Audacity window.

A disadvantage of this option is the fact that it is impossible to change the playback speed while the recording is being played. In order to change the speed, you need to choose the desired speed value and then click *Play* on the transcription toolbar; a single click means that the recording will be played from the beginning (or from the point marked previously).

What greatly facilitates transcription is that it is possible to set a fragment to be repeated (loop playback). Listening to the same piece of interview repeatedly without needing to play, stop, and rewind the recording will not only make the work easier and improve the transcription quality, but it will also reduce the time of this task. In order to use this function, you need to choose *Play in a loop!* from the *Transport* menu or use the keyboard shortcut Shift+Space (you can also click Play while holding the Shift key).

The program also has an extensive selection of effects which can be used to modify the audio recording, and some features may serve someone who works with recordings of interviews. It is worth paying attention, in particular, to those options that are intended to improve the quality of the recording. In the case of a researcher who uses audio materials, it is even more important when the recordings are not of the best quality, or they have some defects which make it difficult to decipher the sense of the interviewee's statement. One tool that can solve the problem of "deciphering" an audio recording is the *Noise reduction* option. This tool serves to remove – or at least reduce – the background noise that may appear in a recording. This means that we can quite effectively neutralize some clearly audible sounds such as whistling, the whimpering of a dog, or a ringing bell; however, it is much more difficult to remove some irregular noises that area encountered, for example, in traffic. Another helpful tool intended to improve the audio recording quality is *Click removal*, which helps to correct a recording by reducing the audible pops (similar to those you can hear when playing a vinyl disc) and clicks or din [10].

It should also be stressed that the preparation process, i.e., data transcription, may be a part of the analysis itself, depending on the research methodology adopted. Furthermore, the act of transcribing not only changes the form of research data but it also involves modifying them and, to some extent, interpreting them. It is therefore worth noting one additional feature available in the program, namely the ability to create labels. This feature might be particularly useful for someone who uses Audacity as a tool for transcribing interviews. Labels can be used to make annotations to mark those fragments which are our questions and those which are the respondent's answers, but it can also help us to manage a simple coding system. Thus, it is easier to move throughout the material and to search for the fragments of your interest faster, but you can also make the first attempts of coding directly on an audio file (Fig. 2).

When it comes to technicalities, it is worth mentioning that the labels can be added in several ways, and importantly, this can be done both after having listened to a fragment or instantly while the recording is playing. Moreover, a user may modify the labels (e.g., enter names with Polish characters) with a label editor. In order to open the editor, you need to go to the *Edit* menu and choose the item *Labels -> Edit Labels*. As a result, a dialog window will open where you can, among other things, provide the labels with names, change and correct those names, set when they begin and end, as well as use the label import and export option.

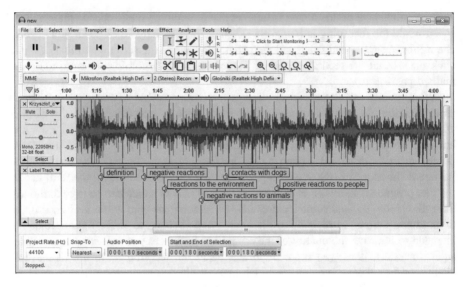

Fig. 2. Labels ascribed to fragments of a soundtrack in Audacity.

Audacity is a universal tool with functions that are helpful for qualitative researchers. Within the course of its development, various features have been added to the program that aid the transcription of audio recordings. It is thus a tool which, with some effort from a new user, will contribute to improving work efficiency.

3 Data Analysis with WeftQDA

With properly developed data, namely some transcribed files with audio recordings of the interviews, a researcher may start analyzing them. It needs to be borne in mind that while transcribing the data, a researcher may come up with some interesting ideas, and certain fragments of particular words by the interviewee may seem especially significant from the perspective of the problem they develop. However, there is nothing to prevent such information from being recorded by the researcher. The labels available in Audacity mentioned above can be helpful, but a researcher can simply create memos in the form of notes, e.g., in MS Word or OpenOffice Writer, indexing and archiving them accordingly. Later, they can be included in the analysis in any computer program that aids qualitative data analysis, such as WeftQDA. This program may serve to group together and later process all notes made by the researcher. WeftQDA is much more powerful than programs such as Word, and generally supports the data development process. Among the available functions, the following possibilities should be highlighted: creating databases with text materials, searching through texts for specific words, ascribing codes to specific text fragments, creating a tree of codes and their clear structure, searching for phrases and words in a text, searching within codes, creating memos with short information or the researcher's analytical thoughts, and creating simple cross tables.

The program has a very simple interface, and all of the important features are available at your fingertips. At the top of the main window there is the main menu where you can access most of the functions. The largest part of the program is the working window which displays the source materials, namely the texts from the interviews, memos from observations, etc. There is a smaller project window (*Documents and Categories*) next to the main window that displays imported documents and created categories. A list of all documents imported into the project is displayed in the upper part of this window. In turn, the lower part is intended to demonstrate the "category tree," namely the categories organized in a hierarchical structure. In the category tree, you can also find the recorded results of searches and inquiries previously generated by the researcher (Fig. 3).

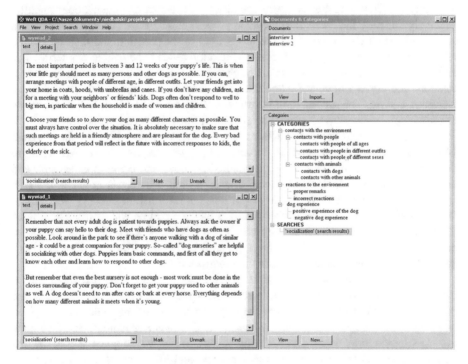

Fig. 3. WeftQDA windows.

Using particular features is similarly intuitive. However, as is the case with every new program, WeftQDA requires a certain amount of experience and that the user acquire several of the most important principles related to operating the program. Before you can start working with the program, you need to consider the following issues. Firstly, you need to create and save a new project, which is actually a database where all further text materials will be collected. We do this by using the *New Project* option located in the *File* menu. Secondly, after creating (or opening) a project, the researcher must add documents to the database. These may be transcriptions from interviews, memos from field observations, as well as various kinds of texts from press releases, leaflets, posters, etc. It must be mentioned here that the program is limited to text data saved in the .txt format, i.e., without formatting. Furthermore, the program does not make it possible to edit any data; thus, there is a need to check them carefully before they are imported. Thirdly, it is worth arranging the windows properly, because this will make working more comfortable and, at the same time, all of the information will be more visible for us. To do this, it is a good idea to set the size of the windows and their location on the screen in such a way that the *Documents and Categories* window is situated next to the main window of the program which displays the coded contents (as well as information such as memos or data search results). It is worth adding here that a researcher may use the *Window* settings so they can not only view several documents at once but also adjust the way they are displayed to suit their own needs. This is how you can compare fragments of selected documents conveniently, which significantly facilitates the analytical process in the case of using GT. Finally, a user should learn the meanings of the terms used in the program and then be able to reflect them in the language of GT. Special attention needs to be paid to two terms. The first is *marking*. This option allows the researcher to register the connection between a category and a text fragment by "marking" the document section. This procedure is, therefore, coding. The second term is *category* ("categories"), which hides collections of individual "marked" (i.e., coded) fragments of texts from various documents.

Having completed the above actions and become familiar with how the program works, the researcher may proceed to carry out proper analytical work. According to GT, a researcher should first focus on generating codes and creating categories, and later on, the order of their connections. The basic unit of analysis in GT is the data fragments, which may take the form of "portions" of information selected according to the issues of interest to the researcher or topics which are more widely discussed by an interviewee. In structural terms, they can be individual phrases, lines, sentences, or paragraphs to which a researcher may assign particular codes. These, in turn, provide a sort of "label" of various conceptualization degrees [9].

In WeftQDA, the coding process involves assigning particular text fragments to categories. First, the researcher needs to mark a fragment of a text then select the category he wants to assign a given data portion to. To do this, you mark a given *category* in the category tree in the *Documents and Categories* window and then click the *Mark* button in the lower part of the document. In this way, a given area of the text will be marked as being related to this category, and the name of the category will appear in the window of the document currently being processed, next to the *Mark* button. As a result, you can then encode the selected fragment, and the font will change from black to blue.

At the beginning of any research, Glaser and Strauss [7] recommend creating concepts of a low level of abstraction – the generality of the generated concepts should not exceed the local framework. Based on those concepts, more general categories appear within the course of the research. Such categories are elements of the project that group the themes, places, people, and other issues which interest the researcher.

In WeftQDA, to create a new category, first, select a specific category in the category tree and then click the *New* button. A small dialog box will then appear where you should enter the name of the new category and confirm it by clicking "Ok." The new category will be subordinate to the category that was last selected in the category tree. If the researcher does not select a given category, and the new category is in the wrong place in the tree, it can be easily changed. Simply drag it with the mouse and drop onto the appropriate category. In the same way, by dragging particular categories, you can organize the category tree structure according to the process in the researcher's analytical work [10, 11].

According to GT, the process of writing theoretical memos should progress along with coding and generating analytical categories. The memos are a record of the researcher's concepts related to codes, and they are intended to make the applied categories more specific and give the coding process some direction [12, 14]. As suggested by Gibbs [5], they also provide a connection between two stages of analysis – coding and writing a report. The writing of memos is a process that accompanies an analyst who applies GT from the beginning of the research process.

In WeftQDA, a researcher can create memos which can refer both to whole documents and particular categories. This is how the researcher can record various ideas and concepts on the connection between particular codes or more general hypotheses which result from the ongoing data analysis. In order to create a memo, you need to click on a category or document name in the *Documents and Categories* window. In the first case, the entire document is displayed, and in the second, only those fragments of the text appear that have been coded with a selected category. In both situations, a "details" tab also appears in the new window, with a field for a memo.

We will use an example of research related to the socialization of a puppy to explain the presented notions. The first action was open coding, and in the meantime, there were some changes introduced to the coding tree structure which gradually arranged the categories in a particular manner according to the emerging concept of the researcher (Fig. 4).

The tree structure made it possible to reflect a degree of hierarchy (superior-subordinate relation) among the categories, and thanks to adding appropriate information in the form of memos, it was possible to determine the links between them.

The main recommendation from the authors of grounded theory [7] is to perform ongoing comparisons, namely, to compare fragments described with a similar code or to compare the way of coding one case with another [4]. Increasingly more general categories that reveal underlying uniformities are generated on this basis [8].

In practice, while using the WeftQDA application, the comparative procedure is performed with the data search option. This process involves reviewing fragments of a text and other data which were coded with a particular code. To do this, use the option to view all sections of individual documents coded with category data. Thus, you can compare and check what is common and different for the content which was assigned to specific categories.

To see text segments in particular documents which were coded with a given category, find their names in the category tree (in the *Documents and Categories* window) and double click or select their name and click the *View* button.

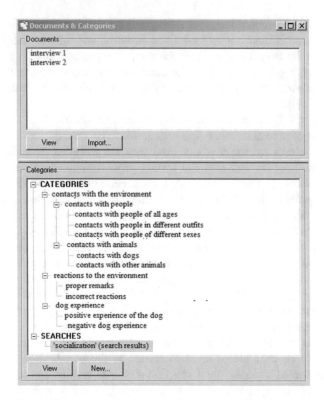

Fig. 4. The structure of the category tree created in WeftQDA.

The search tools are also useful for comparing data. They are significant in the data comparing process, both in terms of threads which are of interest to the researcher and the particular categories which have already been generated. Searching also makes it possible to check codes that may co-exist or exclude each other. This is done using the OR, AND, AND NOT logical operators available in the *Query* option, which refers to the ability to search the encoded content of different documents with two or more categories or phrases using the listed operators.

In the discussed example of the socialization of a puppy, based on the generated categories of "contacts with people" and "adverse reaction of a dog," we can check whether and in which situations puppies respond negatively towards particular categories of people (Fig. 5).

Analysis of contents coded with the above-mentioned categories of interview fragments suggests that dogs exhibit aggressive behavior towards big men, especially those who are aggressive towards the dog, and weaker people, including the elderly, the ill, and children.

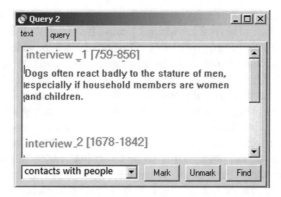

Fig. 5. Search results for selected categories in WeftQDA.

WeftQDA has a *Code Reviews* search tool which makes it possible to create category lists. This feature is an answer to the question of how often and in what scope the same text (or text segment) is assigned to particular categories.

To illustrate the *Code Review* function, an example may be presented where the researcher generated the following categories: "contacts with people" and "contacts with animals," and two other categories, "aggressive behaviors" and "positive behaviors". This can help a researcher check the category of actors (human or animals) towards which the puppy will show particular reactions (positive or negative) (Fig. 6).

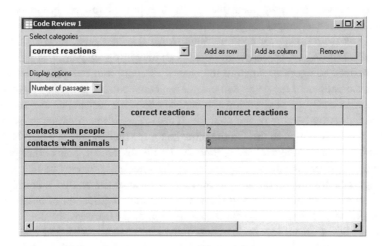

Fig. 6. Results of listing selected categories in WeftQDA.

Based on the list, it can be concluded that a puppy is more likely to react negatively towards other animals than people.

Finally, the search tools can also help with "semi-automatic" coding. The *Search* or *Query* options can be used for this purpose. The first one makes it possible to search text data for the occurrence of a given phrase or word. The second makes it possible to query the coexistence of several keywords by using the "CONTAINS WORD" feature. In both cases, those fragments will be found where the words which are searched for appear. The researcher should look through all text segments marked this way and decide on an individual basis whether they can be coded with a specific code.

WeftQDA has numerous features which make it suitable for use in projects based on the procedures of GT, including the options of coding, creating a tree structure, and searching or writing memos. Furthermore, its unquestionable advantages are that it is free and it is intuitive to use. These features make WeftQDA a program for virtually every user. However, the program has some limitations which might impact its usefulness for some users. First of all, it must be borne in mind that WeftQDA only supports the .txt format; thus, it is impossible to import documents without losing their formatting. Moreover, the program supports only the analysis of text data, so it is not possible to analyze other types of materials, such as images, audio, or video. Furthermore, WeftQDA does not make it possible to visualize the data interpretation results, i.e., to create graphs or diagrams, which makes the analysis less rich, in a way, and restricts its control. Nevertheless, it must be stressed that as a code-and-retrieve tool [1], it completely fulfills its function. It certainly is very useful for researchers who want to effectively organize their data and arrange new information arising from their analytical work.

4   Visualizing the Concept with CmapTools

A researcher who analyzes data based on the procedures of grounded theory methodology often attempts to present the total or partial output of their analysis visually, apart from coding or creating memos. Most of the time, such visualizations are conceptual maps, and the most important is an integrating diagram that incorporates all the generated categories related to the central category into one piece. All kinds of visual representations of the generated categories, connections, and links between them that are continuously created during the analysis help the researcher to have a clear picture of the current status of the analysis. This makes it much easier to assess and verify the results obtained so far. On the other hand, the final diagram, which is, in fact, the crowning of the whole analysis carried out by the researcher, helps other people to better understand the nature and character of the interdependencies among the categories, significantly facilitating insight into the essence of the analyst's concept [5].

As the essence of the analysis in GT is the multiple modification of categories and their connections, in other words, the transformation of the concept, the diagram that is intended to reflect all project components must be constantly changed and updated. Therefore, the tools which can serve to develop such maps must be easy to use. Not only this, but that they should also make it possible to reflect the researcher's concept accurately, present the diagram in the most transparent way for everyone, and make it easy and quick to modify. These are the advantages of CmapTools and, importantly, these features are completely free of charge.

CmapTools makes it possible to create very complex and advanced integrating models – thus, it is even recommended for people adopting GT, as it makes it possible to present and distribute the output of concepts using the export and transfer data function (Fig. 7).

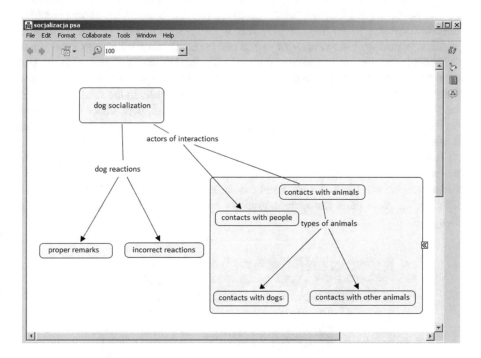

Fig. 7. CmapTools window with an open conceptual map.

We start working with the program by creating a new Cmap model. To do this, select *New Cmap* from the *File* main menu (or use the Ctrl+N keyboard shortcut). This will open a new window with the default name *Untitled 1,* where the number indicates each subsequent project created in CmapTools. Now we can start creating a model, which is done by adding more elements to the map project. To do this, double-click anywhere in the working area. This will display a default shape with question marks inside; these should be replaced with the proper name of the generated category. To do this, simply double-click on the shape and then enter the proper name, replacing the question marks.

As the idea of creating the model is to illustrate the researcher's analytical concepts, it is natural that each element can be linked to several other categories at the same time, and that it can contain more than one phrase that combines them. To do this, drag the arrow between the selected elements and enter the name of this connection (category relation) which will generate a link between the project elements that represent particular categories [10].

Extended maps, such as integration diagrams, may include a series of elements, which makes them illegible. However, CmapTools can also help in this case because it features some helpful tools to "nest" the categories and connect the elements within different diagrams. In the first case, in order to create such a group of model elements (which represent specific categories within our concept), we need to select the model components that we are interesting in by clicking the left mouse button with the Ctrl key pressed (or you can choose all model components by pressing Ctrl+A). Then right-click to get the context menu and choose *Nested Node/Create*. In the second case, it is enough to open two model projects saved in the program folders, choose a particular element in one of them, and then by clicking the mouse button, drag the line which should connect this element with another element located in the second model.

In our research example related to the socialization of a dog, a conceptual map was divided into several smaller ones. This was done using several separate maps in CmapTools, where particular elements can be connected, thereby providing a possibility to achieve a more transparent structure based on separated superior categories (Fig. 8).

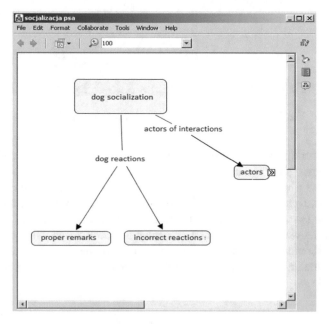

Fig. 8. An example of "nesting" the elements of a conceptual map developed in CmapTools.

Editing options can be used to make the relationships and connections of various elements of the map more transparent and more accurate when it comes to their characteristics. These options make it possible to format the line, shapes, font, etc. Furthermore, it is possible to insert graphics files into the structured models, which complement and illustrate the presented concepts, at the same time visually enriching a given project of a conceptual map [14]. Such visual elements are introduced in a simple way, through a mechanism that involves dragging a selected illustration and dropping it onto any element of the map.

Finally, it is worth mentioning the possibility to export data and whole diagrams in the form of jpg photos, among others, which can later be inserted into a report or other publication as an illustration. This option means that the maps and diagrams created in the program can not only serve the work of an analyst, but they are also a perfect source of figures that enrich the content of papers and publications.

CmapTools has several important advantages as a tool that is helpful in a qualitative researcher's work. First of all, the program is designed to develop conceptual maps and interpret data with diagrams, arrange and integrate codes, and graphically reconstruct the analyzed processes. It will undoubtedly satisfy the expectations of even relatively demanding users in this respect. CmapTools will certainly be a perfect program both as a basic tool for an analyst's work as well as a complementation to WeftQDA, which does not have options to visualize data or integrate the development of diagrams.

5   Conclusions

The paper is first of all intended to introduce CAQDAS and show the way of using common computer programs that are available free of charge, which enable the process of preparation and implementation of a research project based on the grounded theory methodology. Therefore, it presents those functions of the software that support this process to the greatest extent.

In the case of Audacity, we have the option of processing audio materials, above all related to transcription of audio files. Presented functions to improve the process of rewriting the listened recordings, but also to improve their quality. The use of Audacity in the context of the implementation of GT-based research can be important in the process of initial coding of empirical material (open coding) and the possibility of data comparison already at the stage of their listening or rewriting (continuous comparison method).

However, the proper analysis can be carried out using the WeftQDA program, which offers a set of tools for working with text documents. Although this program has not been designed for its application to a particular research method, due to the support of activities such as coding, sorting, categorizing and searching data, it can be successfully used by researchers using GT. What deserves special attention in this program is the ability to create a tree of codes and categories, which is a key issue from the position of research based on GT. In addition, thanks to the data search option, you can perform a continuous comparison procedure in the WeftQDA program and also use the search system to select and "ground" test hypotheses.

In turn, CmapTools is a comprehensive tool for creating conceptual maps, as well as presenting data in graphic form. For this reason, it is a program for people who need to visualize their analytical concepts or simply improve the data interpretation process.

The CmapTools program, in accordance with the needs of people using GT, allows creating complex and advanced models integrating, organizing and integrating codes, as well as graphically reconstructing the analyzed processes.

However, it should be borne in mind that despite the number of helpful options that the presented programs offer, as well as their complementarity, they are not perfect and faultless solutions [11]. Nevertheless, the potential benefits that the researcher can derive from using them make them recommendable solutions. And even if it takes some time to become familiar with them, it is worth making this effort; it may pay off in the future with much more efficient and effective work by the researcher [2].

References

1. Bieliński, J., Iwańska, K., Rosińska-Kordasiewicz, A.: Analiza danych jakościowych przy użyciu programów komputerowych. ASK. Społeczeństwo. Badania. Metody **16**, 89–114 (2007)
2. Bringer, J.D., Johnston, L.H., Brackenridge, C.H.: Using computer-assisted qualitative data analysis software to develop a grounded theory project. Field Methods **18**(3), 245–266 (2006)
3. Charmaz, K.: Grounded theory. Objectivist and constructivist methods. In: Norman, D., Lincoln, Y. (eds.) Handbook of Qualitative Research, pp. 213–222. Sage Publications, Thousand Oaks (1994)
4. Charmaz, K., Mitchell, R.G.: Grounded theory in ethnography. In: Atkinson, P. (ed.) Handbook of Ethnography, pp. 160–176. Sage, London (2001)
5. Gibbs, G.: Analizowanie danych jakościowych. PWN, Warszawa (2011)
6. Glaser, B.: Doing Grounded Theory: Issues and Discussions. Sociology Press, Mill Valley (1978)
7. Glaser, B., Strauss, A.: The Discovery of Grounded Theory. Strategies for Qualitative Research. Aldine Publishing Company, New York (1967)
8. Gorzko, M.: Procedury i emergencja. O metodologii klasycznych odmian teorii ugruntowanej. Wydawnictwo Naukowe Uniwersytetu Szczecińskiego, Szczecin (2008)
9. Niedbalski, J.: Komputerowe wspomaganie analizy danych jakościowych. Zastosowanie oprogramowania Nvivo i Atlas.ti w projektach badawczych opartych na metodologii teorii ugruntowanej. Wydawnictwo Uniwersytetu Łódzkiego, Łódź (2014)
10. Niedbalski, J.: Odkrywanie CAQDAS Wybrane bezpłatne programy komputerowe wspomagające analizę danych jakościowych. Wydawnictwo Uniwersytetu Łódzkiego, Łódź (2013)

11. Niedbalski, J., Ślęzak, I.: Analiza danych jakościowych przy użyciu programu NVivo a zastosowanie procedur metodologii teorii ugruntowanej. Przegląd Socjologii Jakościowej **8** (1), 126–165 (2012)
12. Richards, L.: Using NVivo in Qualitative Research. Sage, London (1999)
13. Strauss, A., Corbin, J.: Basics of Qualitative Research. Grounded Theory Procedures and Techniques. Sage Publications, Newbury Park (1990)
14. Weaver, A., Atkinson, P.: Microcomputing and Qualitative Data Analysis. Avebury, Aldershot (1994)

Parliamentary Amendment Mobile Application: A Qualitative Approach About E-Government

Dayse Karenine de Oliveira Carneiro[1,2]([✉])(iD),
Mauro Célio Araújo dos Reis[2]([✉])(iD),
Maria Eugênia Diniz Figueirêdo Cireno[2]([✉])(iD),
Bruno Henrique Oliveira Lima[2]([✉])(iD),
Ana Paula Rodrigues dos Santos[2]([✉])(iD),
Jório Mendes de Lima Ayres[2]([✉])(iD), and Dárcio Guedes Junior[2]([✉])(iD)

[1] University of Brasília (UnB), Asa Norte, Brasília, DF, Brazil
[2] Ministry of Health, Esplanada dos Ministérios, Bloco G, Brasília, DF, Brazil
{dayse.carneiro,mauro.reis,maria.cireno,bruno.lima,
ana.psantos,jorio,darcio.guedes}@saude.gov.br

Abstract. The adoption of innovative technologies by the Public Administration has been increasing in recent years. In this sense, e-government emerges as a strategic way of using modern information and communication technologies to provide citizens and organizations with rapid and convenient access to information on governmental activities. From this perspective, the Brazilian Ministry of Health (MH) has developed the Parliamentary Amendment Application to facilitate and streamline, safely and functionally, the management of amendments to health resources made by representatives, senators and citizens. This article, therefore, seeks to analyze the implementation of the Parliamentary Amendment Application in the management of public health resources and services. As a result, it proposes a framework to explain the implementation of the application, to further empirically and conceptually the knowledge related to this phenomenon. The results reveal it to be an opportunity for understanding e-government in the context of the Brazilian public health sector, based on the identification of four dimensions and 22 variables. In addition, the work presents new variables concerning the implementation of the application: its use as a management tool; its characterization as a technological innovation; and the possibility of exercising social control over public health services.

Keywords: E-government · Parliamentary amendment · Ministry of Health · Public health sector · Innovation

1 Introduction

Beginning in the 1990s, Brazil started implementing changes on both the national and local levels in terms of the interactions between the State and society [29]. This scenario led to the government's perception of innovation as a strategic and indispensable factor

A. P. Costa et al. (Eds.): WCQR 2019, AISC 1068, pp. 58–71, 2020.
https://doi.org/10.1007/978-3-030-31787-4_5

necessary for the development of an interface in the supplying of public services, as well as in dealing with social challenges, and increasing the potential of the State's capacity to respond to local and individual needs [2].

An important issue we need to emphasize is the increased interest in innovation over the last years [8]. Another issue concerns pressure for a change in State policy in terms of making information open to the public. These changes empower citizens to become more involved in the public environment as actors with political and social rights [11].

These matters in question emerge from factors that permeate the society-centric view of Public Administration, in which the nature of public action goes beyond state intervention exclusively and begins with the idea that action can be initiated by society as well [24]. As a result, the public service sector started to seek, develop, apply and support new technologies and better ways of solving problems, mainly oriented towards collecting better information from, and providing better service to, the population [2], and this became known as electronic or e-government [21].

2 Aim of This Study and Research Questions

E-government has grown in Public Administration through the adoption of Information and Communications Technologies (ICT) in order to optimize public services through digital transformation. The public health sector is an example of this and has adopted technologies that help it manage countless quantities of data related to procedures and policies [3, 7, 16].

Given this perspective, the Brazilian Ministry of Health (MH), the institution responsible for the National Health System (SUS) and public policies aimed at promoting and assisting public health, has developed the Parliamentary Amendment Mobile Application, which is a practical e-government application, available on Android and IOS platforms. This application aims to make management clear, agile, safe and functional through the implementation of parliamentary amendments related to public health actions and resources made by representatives, senators and citizens. Therefore, entering information into a mobile application, while providing congressmen and citizens with access to all of the data related to their amendments in terms of monitoring their execution represented a challenge to the MH's project team, particularly in terms of delivering a large volume of information on public health resources.

Researchers have carried out studies that analyze the implementation of e-government. However, these studies are scarce, especially in developing countries [38], which reflects a need to analyze the implementation of Parliamentary Amendment Mobile Applications as an e-government initiative to better manage public health resources and achieve their potential in terms of improving social welfare.

Based on the above context, we have set out to answer the following research question: what are the attributes, functions, outcomes and interactions related to the implementation of a Parliamentary Amendment Mobile Application in the management of public health resource, actions and services? Thus, we aim to analyze the implementation of the Parliamentary Amendment Mobile Application by the Brazilian Ministry of Health.

3   Importance of Research

The increasing need to develop new ways of delivering public services has presented a challenge to public management due to its traditionally top-down methods and instead it has introduced new methods and control tools for providing services. Public agents, therefore, have started to encourage the introduction of technology to offer more data and broaden access to it, to facilitate these new government services [32]. Specifically, in health services many of their successful experiences have stemmed from the introduction of new technological processes. In addition, there is evidence that technology in health services directly affects the quality and efficiency of public health management [15, 34].

It is worth noting that factors such as a scarcity of resources, laws and regulations, the economic crisis, political turnover, political and social barriers, have negatively influenced the development and supply of public services. On the other hand, the exchange of knowledge and experiences among public servants has had a positive effect on the development process and service delivery [20]. The need to understand the challenges of implementing this application in the public health sector has become relevant in view of the indispensability of the efficient management of resources for investing and maintaining the public health network, which in their absence, can cause serious harm to the public's health.

4   Theoretical Background

This section presents a brief theoretical assumption about innovation in public services and e-government. Afterward, we will present a framework for analyzing e-government in the Brazilian public sector.

4.1   Innovation in the Providing of Public Services

Since the beginning of the 21st century, there has been a change in the role of public service users [33], who have become increasingly connected with other users, informing themselves about the characteristics of service delivery, its benefits and various characteristics. Thus, they have become active in the creation of value in services rendered by the government and also concern themselves with issues such as transparency, efficiency and effectiveness [18, 31].

It is relevant to emphasize the importance of obtaining sources of information that are external to the government [4, 5, 19] and adopting technologies as crucial elements for increasing the performance and legitimacy of the services delivered to society [9, 10]. However, in addition to technological development, adoption or implementation, innovation in public services involves connections and crosscutting interactions, which include governmental relationships with other governments, businesses, public servants, citizens, social groups and nonprofit organizations [13, 26].

Public organizations are looking for ways to establish a culture of innovation in their environment, since an innovative workplace promotes greater satisfaction and significant organizational commitment, positively affecting organizational results [37], which reflects directly on the quality of services provided to citizens [9]. Therefore, public sector innovation represents both the processes of change intended in the quest for efficient management and the democratization of state action [14], and it aims to increase the volume of services offered to the population and increase popular participation in the formulation, institution and control of public policies.

The adoption of Information and Communications Technologies (ICTs) favors innovations in public services, for example, the use of information systems, which represents significant support for decision-making in the development of government programs [1, 12, 17]. Especially in health services, many of their successful experiences have resulted from innovation in technological processes. There is evidence that technological innovation in health services has a direct impact on the quality and efficiency of the management of public health services [15, 34]. In this case, innovation depends on the connections established and that they work as expected, since any technology involves a complex number of critical elements that include people, equipment, instruments, the organization of work and organizational models [36].

It is fundamental that the State play the principal role of authority in which the various agents seek to exert their influence. It is then up to the State, to address the challenges in managing these relationships, resources and interactions established for the development of technologies and innovations that aim efficient and effective delivery of public services.

4.2   E-Government

The process of innovation in government has changed due to the increasing incorporation of technologies, due to the demand for higher productivity of public service providers and due to the complexity of deepening specialized knowledge [30]. These changes require government agents to articulate their activities and interventions with the actions of other agents who are interested in the activities carried out by the public sector [20].

E-government, therefore [11], is associated with State reform movements, as well as transformations in Public Administration thinking, and is a milestone for the creation of democratic institutions compatible with the needs and characteristics of the information age. From this perspective, government institutions have begun a transformation process in order to use technology as an integral part of their modernization strategies in meeting the public's growing demand for quality information [29].

However, the development of e-government is a challenging task for the public sector, and it involves cultural issues and a change in processes. It is also necessary to manage joint efforts with different approaches in the development and implementation of e-government [23]. Nevertheless, e-government is a way for governments to use ICT to provide citizens and businesses with convenient access to government information and services in order to improve the quality of services and provide more opportunities for participating in democratic processes [13].

E-government has become a phenomenon in which public institutions around the world have adopted new ways of harnessing ICTs to serve better their constituents [25]. This trend has led to the suggestion that public organizations need to adopt processes of continuous nonlinear innovation in order to meet the emerging demands for better services, greater transparency, and control and openness of government activities in order to strengthen relationships and build new partnerships within society [26]. Accordingly, e-government is the providing of government services through technological means that allow access to government information at any time and with equal access. This new perception has required development of the potential of the public sector to reformulate and build a new model of relationships between citizens and government [13, 26].

The literature, therefore, presents the dimensions and their respective characteristics concerning e-government, as shown in Table 1. Each dimension represents a set of crucial features for e-government implementation [13]. The attribute dimension presents the essential elements observed in developing countries in e-government initiatives [13]. The functions dimension highlights the implications of the government regarding its ability to provide government services through non-traditional electronic means [13]. The interactions dimension portrays the network of relationships between the six main actors: government, citizens, civil servants, non-profit organizations, private companies and social groups [13, 26]. And finally, the results dimension configures the main e-government deliverables that aim to improve their citizens' experience in the use of public services [38].

Table 1. Dimensions and characteristics of E-government

Attributes [13]	Functions [13]	Main interactions [13, 26]	Outcomes [38]
Comprehensive	To allow users online access to personal benefits	Government-to-Government (G2G)	Increased trust in government
Integrated	To promote online access to bidding, purchasing, sales and payments	Government-to-Citizens (G2C)	Improvement in ethical behavior
Ubiquitous	To promote citizen participation	Government-to-Public Servants (G2S)	Capacity building
Transparent	To promote interactions between services	Government-to-Social Groups (G2SG)	Open government
Easy to use	To allow access to government information	Government-to-Business (G2B)	Improvement of administrative efficiency
Accessible	To facilitate compliance.	Government-to-Non-profit Organizations (G2N)	Improvement of public services
Secure			Increase in professionalism
			Increase in social value

Based on the literature mentioned above, the next section will present an e-government framework as a tool to analyze the implementation of the Parliamentary Amendment Application developed by the Brazilian Ministry of Health.

5 Data Collection and Analysis Methods

To accomplish our objectives, we are using a qualitative multi-method approach [6] to combine the research processes of the Parliamentary Amendment Application development and implementation through a triangulation of techniques such as a literature review [35], participant observation [22] and a focus group [27].

For the literature review, we collected, read and interpreted scientific articles and books on public sector innovation and e-government. We selected 57 publications with these themes present in their titles, keywords and abstracts. After reading them, we directly used 37 of them in the construction of a theoretical background, as well as a framework for the analysis of e-government initiatives in the public health sector.

We conducted desk research on meeting reports, process manuals, access, and application performance indicators made available to us by members of the implementation project to further our understanding, and analysis [35]. From January to March 2019, we conducted participant observations of the working processes with the MH team to gain insight into the context, behavior, and relationships between those involved in the application development and the implementation project [22]. We took note of all of this information to obtain material for the elaboration of the focus group and the analysis framework.

From the information collected in the literature review, documentary research and participant observations, we prepared the focus group script. We applied this script to a group of six members in April 2019 to gather detailed information about the development of the amendment application. It was an essential technique for identifying participant perceptions, feelings, attitudes and ideas. Focus group is a crucial instrument that enables researchers to grasp how group members interpret reality, the extent of their knowledge and their experiences related to the application's implementation [28].

In order to conceptually and empirically further the knowledge related to this phenomenon, we analyzed the data using content and document analysis [27] to propose a framework for the analysis of e-government initiatives in the public health sector, as shown in Fig. 1, with the following dimensions and characteristics as defined in the literature [13, 26, 38]. In the orange dimension, we deal with the attributes of electronic government. The blue arrows show the functions. The white color shows the main interactions between the actors involved, and finally, the gray circle shows the results related to the implementation of e-government initiatives.

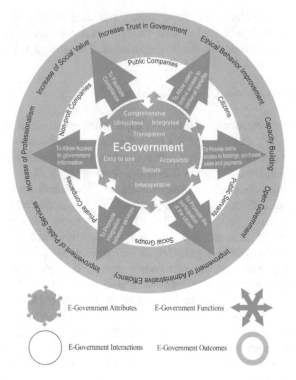

Fig. 1. Framework of E-Government (Source: elaborated by the authors)

To apply the framework to the data collected from the focus group, we defined two vectors: vector one corresponds to the statements that reflect the presence of the analysis categories and variables previously identified in the literature, while vector two groups the analysis categories and new variables observed in this phenomenon.

The results section presents Tables 2, 3, 4 and 5, and summarizes the systematic results of the focus group data analysis. In brackets, we indicate the categories, followed by the variables identified in the literature by the authors mentioned above. In addition, the codes in parentheses, in agreement with the transcribed statements, represent the quotes from the staff members, namely public employees E1 to E6, who worked on the application development.

6   Results

The results of using this framework, developed based on the literature [13, 26, 38], to analyze the implementation of the Parliamentary Amendments Application reveal the opportunity to understand e-government within the context of the Brazilian public health sector from the identification of the categories and variables present in this phenomenon. In addition, this study reports new variable related to the characteristics of this government innovation.

The analysis of the focus group data for the e-government attributes revealed the presence of the variables easy to use, accessible, comprehensive and integrated [13]. The new variables found are the mobile application used as a management tool and technological innovation, as described in Table 2.

Table 2. Results of the application of the framework: Attributes of the Parliamentary Amendments Application, according to the project team

Vector 1	Vector 2
(…) Nowadays, you must have everything at hand. Before we developed the web system, we made improvements to enable access to information with greater credibility and functionality, in real-time. [Attributes - Easy to use; Accessible] - (E1)	The application makes it possible to monitor proposed amendments to the budget, and it allows you to monitor the process at its current stage. [Attributes - Management tool] - (E1)
Another fact is that besides the political force that the amendments have, according to the budget used in this process, it is the constitutional obligation that half of these resources must be used for health services. We manage around 20,000 to 30,000 projects a year. Therefore, we need to create tools to facilitate the monitoring of amendments indicated in the budget. Thus, the congressmen get to follow all twelve steps of the proposal (…) One thing is adding an amendment itself to the budget, and another is that an amendment turns into a health service that offers a straight connection to citizens. [Attributes - Comprehensive - Integrated] - (E2)	In my perception, the application is not innovative just because it exists, but rather because of the direct and indirect actions that it makes possible. (…) The speed with which updates must be made makes the financial resources arrive faster at the end of the provision of the health service and the health account lifetime. Here we are talking about budgets, and when we talk about budgets, the numbers are large. The time a proposal takes to be processed, with the advent of the app can be different and quicker, and this is innovation. [Attributes - Technological innovation] - (E2)
	It is the first tool given by the government that makes it possible to monitor the values of congressional amendments online and provides an integrated system for all of the generated data. However, today, we have other government applications, but they are not online. Our application gathers all the information from the representatives and their amendments in an integrated manner, and we pass along information on a bidding and awarding basis. In the past, all other information was delayed or was lacking in information. [Attributes - Technological innovation] - (E6)

E-government should enable all users to operate websites and government websites promptly in order to find the information they seek and perform the operations they require [13, 26], taking into account the individual needs of users, and enabling them to utilize all the resources of a clearly presented system. It is important to note that all e-government systems must be integrated, so users can avoid the need to provide the same data repeatedly, and this saves time.

For the government, this represents a cost reduction because it does not need to develop multiple systems that manage the same data separately [13]. In addition, e-government is a management tool and a technological innovation, according to the interviewees. That is, the application is an e-government solution designed through software that processes data on Parliamentary Amendments, dealing with the resources for public health actions and services, facilitating and reducing the time of implementation and the management by representatives and senators.

Table 3 shows the e-government functions identified in the participants' statements. The content analysis performed on the transcribed data from the focus group reveals the presence of the functions to allow access to government information, to allow users online access for their own personal benefit, and to promote citizen participation [13]. There were no new features identified for this dimension.

Table 3. Results of the application of the framework: Functions of the Parliamentary Amendments Application, according to the project team.

Vector 1	Vector 2
The application facilitated meeting advisor requests. There were periods here in the Ministry of the Health when you arrived at the reception of the Congressional Advisory Department on the fifth floor and (…) it was absurd to see congressional advisors waiting in line just to ask if the resource was or would be committed [Functions – To allow access to government information] - (E1)	
One thing that is possible is to follow the registration situation of the beneficiary of a health resource, monitoring to see if it is active or if there is any impediment. [Functions – facilitate compliance] - (E1)	
The tendency is that this will improve the management of the resources indicated by the congressmen. So, it is possible to monitor the process, anywhere there is access to the internet. [Functions – To allow users online access to personal benefits] - (E3)	
Originally, the application was designed to meet the needs of congressmen and their advisors. Later, the project evolved to encompass the needs of citizens as well, but we need to advance more on the amendment and proposal levels to make them more related to the health services, because the idea is for citizens to see where and how the health resources are being applied [Functions – To promote the participation of the citizen] - (E1) (E2)	

Providing access to government information must be one of the most common e-government initiatives. Therefore, the Parliamentary Amendment Application fulfills the function of providing users with online monitoring of amendments to the budget from their formalization to their execution.

Very soon, they will grant access to citizens so that they can follow the process of transferring health resources through the application, as this information is already available on the Ministry of Health's website.

The main interactions identified in the application development process for the Parliamentary Amendments occurred from Government-to-Government (G2G), Government-to-Public Servants (G2S) and Government-to-Business (G2B) [13]. We identified no new variables for this dimension of the e-government framework, as can be seen in Table 4.

Table 4. Results of the application of the framework: Main Interactions of the Parliamentary Amendments Application, according to the project team

Vector 1	Vector 2
When the demand for the app came up, we did initial research to determine the feasibility of developing it. After that, a private institution decided what features it would have, and we made a prototype base on these functionalities. Once the prototype and the project were approved, we started the development with a team of two developers and a requirements analyst. [Interactions – Government-to-Business] - (E5)	
The amendments app I see as not only a useful tool for congressmen and their advisors, but also as an integrative institutional way to promote health services. [Interactions – Government-to-Government] - (E2)	

The creation of an online database of cooperation and communication between government agencies results in the formation of a database that has a direct impact on the efficiency and effectiveness of government activities. This connection provides agents with direct access to standard information of interest to all stakeholders, whether they are government agents or members of the private sector [13].

It is essential that the relationships between Ministry of Health agents, congressmen and beneficiaries of health resources, as well as between government agencies and private companies, be friendly, convenient and transparent in ways that result in governmental benefits in the form of improved management and delivery of services as well as a reduction in operating costs.

Table 5 shows the outcomes of the Parliamentary Amendments Application. According to the project team discourse, it was possible to address the perceived results with the implementation of the Parliamentary Amendment Application in the Ministry of Health such as increased social value, improved administrative efficiency, increased trust in government and open government.

Increased trust in government, or social trust, is indicative of the occurrence of increased social value [38]. This confidence is a result of government engagement, well-being, the sharing of databases, skills and resources, ensuring that system users have access to how the government manages the economy, public resources, and the provision of services. The implementation of e-government by the Ministry of Health promotes increased trust in government through transparency, participation and control of decisions, as well as administrative efficiency and social control, which is another new variable pointed out by the focus group participants.

Table 5. Results of the Application of the Framework: Outcomes of the Parliamentary Amendments Application, according to the Project Team

Vector 1	Vector 2
However, I think the application promotes social control because congressmen are legitimate representatives of society and play the role of social controllers of budget execution [Outcomes – Increase in Social Value] - (E2)	However, I think the application promotes social control because congressional are legitimate representatives of society and play the role of social controllers of the budget execution by the executive branch of the Brazilian government. [Outcomes – Social Control] - (E2)
The application is the first government tool that provides the values of online monitoring of the Congressional Amendments and the integration of all of the systems that generate this data, and it is innovative. [Outcomes – Improvement of administrative efficiency] - (E6)	
The application gathers all the systems I use, where I have information from representatives, where I have full amendments, and I pass the information on bidding and awards on… [Outcomes – Improvement of public services; Increase trust in government] - (E5)	
The Ministry of Health needed to process the information on Congressional Amendments, such as their formalization and implementation, for example, in a much faster and more effective way so that congressmen could have this information in the palms of their hands. The same amendment that came out yesterday, is proposed today, is committed, and tomorrow the money is in the account. So how fast this happens today is far superior and different from how it happened two years ago [Outcomes – Capacity building; Open Government] - (E4)	

7 Final Remarks and Future Agenda

This study analyzes the implementation of the Parliamentary Amendment Mobile Application by the Brazilian Ministry of Health (MH). To perform the analysis of this innovation within the context of the MH, we used a qualitative approach through triangulation between a review of the literature, participant observations, documentary analysis and focus group. Based on the content analysis and bibliographic study, we have proposed a framework for the analysis of e-government experiences.

To identify the dimensions and variables of innovation present in the studied phenomenon, we carried out content analysis with a focus group of members who developed this innovation, which revealed the opportunity to identify, through the discourses of these stakeholders, fours dimensions and 22 variables.

In addition, we have identified three new variables related to the attributes and the outcomes of innovation developed respectively: the mobile application used as a management tool; the application as a technological innovation; and the possibility of exercising social control by the people who use the application.

It is necessary to emphasize that the analysis of the experience of e-government, in the light of the presented results, contains some limitations. The data presented refers exclusively to the reality of the MH and, therefore, may not reflect the particularities of the management of other public organizations.

However, due to the growing interest in e-government, we suggest quantitative studies that test the relationships between the variables presented, considering the perceptions of a more significant number of stakeholders related to the innovations generated in the public sector. We also suggest studies involving the creation of indicators to measure the performance of these innovations, as well as studies dealing with the tendencies of mobile government or m-government.

Furthermore, we conclude that this study reveals that this is an opportunity to understand the e-government phenomenon in the context of the public health sector, especially for the efficient management of applications related to information technology innovations in developing countries.

References

1. Barras, R.: Towards a theory of innovation in services. Res. Policy **15**, 161–173 (1986). https://doi.org/10.1016/0048-7333(86)90012-0
2. Bloch, C.: Measuring Public Innovation in the Nordic Countries: Copenhagen Manual. MEPIN, Copenhagen (2011)
3. Baldwin, J., Lin, Z.: Impediments to advanced technology adoption for Canadian manufacturers. Res. Policy **31**(1), 1–18 (2002). https://doi.org/10.1016/S0048-7333(01)00110-X
4. Bwalya, K.J.: Handbook of Research on E-Government in Emerging Economies: Adoption, E-Participation, and Legal Frameworks: Adoption, E-Participation, and Legal Frameworks. IGI Global, Hershey (2012)
5. Bugge, M., Mortensen, P.S., Bloch, C.: Measuring Public Innovation in Nordic Countries. Report on the Nordic Pilot studies-Analyses of methodology and results (2011)

6. Creswell, J.W., Clark, V.L.P.: Designing and Conducting Mixed Methods Research, 2nd edn. SAGE Publications, Thousand Oaks (2011)
7. Das, A., Singh, H., Joseph, D.: A longitudinal study of e-government maturity. Inf. Manag. **54**(4), 415–426 (2017). https://doi.org/10.1016/j.im.2016.09.006
8. Damanpour, F., Walker, R.M., Avellaneda, C.N.: Combinative effects of innovation types and organizational performance: a longitudinal study of service organizations. J. Manage. Stud. **46**(4), 650–675 (2009). https://doi.org/10.1111/j.1467-6486.2008.00814.x
9. Demircioglu, M.A., Audretsch, D.B.: Public sector innovation: the effect of universities. J. Technol. Transf. **44**, 596–614 (2019). https://doi.org/10.1007/s10961-017-9636-2
10. Denzin, N.K.: The Research Act. Prentice Hall, Englewod Cliffs (1989)
11. Dias, T.F., Garcia, A.B.R., Camilo, N.L.F.S.: Um olhar sobre o governo aberto no nível subnacional: o índice institucional do governo municipal aberto nas principais cidades do Brasil. In: GIGAPP Estudios Working Papers, vol. 6, no. 115, pp. 83–100 (2019)
12. Djellal, F., Gallouj, F., Miles, I.: Two decades of research on innovation in services: which place for public services? Struct. Change Econ. Dyn. **27**, 98–117 (2013). https://doi.org/10.1016/j.strueco.2013.06.005
13. Fang, Z.: E-government in digital era: concept, practice, and development. Int. J. Comput. Internet Manage. **10**(2), 1–22 (2002)
14. Farah, M.F.S.: Gestão pública municipal e inovação no Brasil. In: Andrews, C.W., Bariani, E. (org.) Administração pública no Brasil: breve história política. Unifesp, São Paulo (2010)
15. Farias, J.S., Guimarães, T.A., Vargas, E.R., Albuquerque, P.H.M.: Adoção de prontuário eletrônico paciente em hospitais universitários de Brasil e Espanha. A percepção de profissionais de saúde. Revista de Administração Pública **45**(5), 1303–1326 (2011). https://doi.org/10.1590/s0034-76122011000500004
16. Field, T., Muller, E., Lau, E., Gadriot-Renard, H., Vergez, C.: The case for e-government: excerpts from the OECD report "The E-Government Imperative". OECD J. Budgeting **3**(1), 61–96 (2003). https://doi.org/10.1787/16812336
17. Gallouj, F.: Innovating in reverse: services and the reverse product cycle. Eur. J. Innov. **1**(3), 123–138 (1998). https://doi.org/10.1108/14601069810230207
18. Gallouj, F., Rubalcaba, L., Toivonen, M., Windrum, P.: Understanding social innovation in services industries. Ind. Innov. **25**(6), 551–569 (2018). https://doi.org/10.1080/13662716.2017.1419124
19. Guerzoni, M., Taylor Aldridge, T., Audretsch, D.B., Desai, S.: A new industry creation and originality: insight from the funding sources of university patents. Res. Policy **43**(10), 1697–1706 (2014). https://doi.org/10.1016/j.respol.2014.07.009
20. Gomes, A.C., Machado, A.G.C.: Fatores que influenciam a inovação nos serviços Públicos: o caso da secretaria municipal de Saúde de Campina Grande. Cadernos Gestão Pública e Cidadania **23**(74), 47–68 (2018). https://doi.org/10.12660/cgpc.v23n74.68005
21. Hernandéz-Bonivento, J., Gandur, M.P., Najles, J.: Gobierno Municipal Abierto: de la proximidad administrativa a la acción colaborativa. Departamento de la Gestión Pública Efectiva, Secretaría de Asuntos Políticos, Organización de los Estados Americanos OEA. Washington DC, EEUU (2014)
22. Jorgensen, D.L.: Emerging trends in the social and behavioral sciences: an interdisciplinary, searchable, and linkable resource. Participant Obs. **1**(15) (2015). https://doi.org/10.1002/9781118900772.etrds0247
23. Jussila, J., Sillanpää, V., Lehtonen, T., Helander, N., Frank, L.: An activity theory perspective on creating a new digital government service in Finland. In: Hawaii International Conference on System Sciences (2019)
24. Keinert, T.M.M.: Administração Pública no Brasil: Crises e Mudanças de Paradigmas. Annablume/FAPESP, São Paulo (2000)

25. Marchionini, G., Samet, H., Brandt, L.: Digital government. Association for computing machinery. Commun. ACM **46**(1), 24 (2003)
26. Ndou, V.: E-government for developing countries: opportunities and challenges. Electron. J. Inf. Syst. Dev. ctries. **18**(1), 1–24 (2004). https://doi.org/10.1002/j.1681-4835.2004. tb00117.x
27. Miles, M.B., Huberman, A.M.: Qualitative Data Analysis: An Expanded Sourcebook. SAGE Publications, Inc., Thousand Oaks (1994)
28. Morgan, D.L.: Focus Groups as Qualitative Research, vol. 16. SAGE Publications, Thousand Oaks (1996)
29. OECD: Digital Government Review of Brazil: Towards the Digital Transformation of the Public Sector. OECD Digital Government Studies (2018). https://doi.org/10.1787/24131962
30. Peduzzi, M.: Mudanças tecnológicas e seu impacto no processo de trabalho em saúde. Trabalho, Educação e Saúde **1**(1), 75–91 (2002). https://doi.org/10.1590/s1981-77462003000100007
31. Pestoff, V.: Citizens and co-production of welfare services. Public Manage. Rev. **8**(4), 503–519 (2006). https://doi.org/10.1080/14719030601022882
32. Pestoff, V.: Co-production and third sector social services in Europe: some concepts and evidence. Int. J. Volunt. Nonprofit Organ. **23**(4), 1102–1118 (2012)
33. Prahalad, C.K., Ramaswamy, V.: The Future of Competition: Harvard Business School Press, Boston, Massachusetts (2004)
34. Queiroz, A.C.S., Albuquerque, L.G., Malik, A.M.: Gestão estratégica de pessoas e inovação: estudos de caso no contexto hospitalar. Revista de Administração **48**(4), 658–670 (2013). https://doi.org/10.5700/rausp1112
35. Scott, J.: A matter of record: documentary sources in social research (2014)
36. Tidd, J., Bessant, J., Pavitt, K.: Gestão da Inovação. Bookmann, São Paulo (SP) (2008)
37. Torugsa, N.A., Arundel, A.: The nature and incidence of workgroup innovation in the Australian public sector: evidence from the Australian 2011 State of the Service Survey. Aust. J. Public Adm. **75**(2), 202–221 (2016). https://doi.org/10.1111/1467-8500.12095
38. Twizeyimana, J.D., Andersson, A.: The public value of E-Government – a literature review. Gov. Inf. Q. **36**, 167–178 (2019). https://doi.org/10.1016/j.giq.2019.01.001

From CAQDAS to Text Mining. The Domain Ontology as a Model of Knowledge Representation About Qualitative Research Practices

Grzegorz Bryda(✉) [ID]

Institute of Sociology, CAQDAS TM Lab,
Jagiellonian University, ul. Gołebia 24, 31-007 Krakow, Poland
grzegorz.bryda@uj.edu.pl

Abstract. The nature of qualitative research practices is multiparadigmaticity which creates coexistence of different research and analytical approaches. This paper is a methodological reflection on how the process of qualitative data analysis is developing, moving from traditional CAQDAS coding procedures through Content Analysis dictionary-based approach towards the textual data exploration for knowledge discovery in corpora using Natural Language Processing and Text Mining procedures. This change is described on the example of the process of analyzing and discovering the ways through which qualitative research practices are conceptualized and represented in the vivid language of scholarly articles. Taking into account the problem of a "curse of abundance" in the present-day field of qualitative research I try to organize and articulate these practices in a legible system of knowledge representation employing the information concept of domain ontology. In the process of building the ontology of the contemporary field of qualitative research practices, I link know-how drawn from sociology, social science computing, NLP and text mining, digital humanities and corpus linguistics.

Keywords: Qualitative research practices · Corpus linguistics · CAQDAS · Content analysis · Text Mining · Natural Language Processing · Dictionary-based classification approach · Knowledge representation · Domain ontology

1 Knowledge Discovery in Textual Data

From the methodological point of view, a qualitative researcher may use primary (evoked) data which emerge in time of fieldwork study or secondary (existing) data which can be obtained from various data sources. Moreover, those data can be structured and unstructured, visual or textual such as in qualitative sociology. In social sciences we analyze different data types: statistical and demographic, transactional, sales, registers, official reports, technical documentation, chronicles, parish books and other archival information, bibliographical and library, websites, personal documents, i.e. blogs, letters, diaries, autobiographies, transcripts of interviews, observation notes, geographical (GIS information), images, movies etc. We use data from population

© Springer Nature Switzerland AG 2020
A. P. Costa et al. (Eds.): WCQR 2019, AISC 1068, pp. 72–88, 2020.
https://doi.org/10.1007/978-3-030-31787-4_6

censuses conducted for administrative or public purposes, for example, to examine the structure of households, distribution of income and expenditure, patterns of immigration and migration, changes in the family structure, mobility social or c echoes of rural, urban and metropolitan areas or data collected by opinion polling centers, scientific institutes or non-governmental organizations are used to analyze changes in public opinion, political attitudes, quality of life, migration, social activity etc. Currently, with the development of Digital Humanities (DH), Computational Social Science (CSS), Data Science and Data Mining approaches, procedures and algorithms help us to understand, explain and predict human behavior by studying large volumes of unstructured data stored in databases, warehouses, repositories or corpora. Those extremely large data sets may be analyzed computationally to reveal patterns, trends, and associations, especially relating to human behavior and interactions. Along with the development of digitalization and new technologies, we have an increasing number of unstructured, textual or visual, qualitative data on the richness of social life. We are living in an era of Big Qualitative Data [5]. However, the complexity and multidimensionality of those data sets require a specific methodological and analytical approach, the appropriate techniques, algorithms and tools capable of processing a large number of textual or visual qualitative, unstructured data. One of these is the Knowledge Discovery in Databases (KDD), which includes Natural Language Processing (NLP) and Text Mining approach to enrich traditional qualitative data analysis in social sciences [38]. The consequence of this process is the shift from traditional CAQDAS coding procedures through Content Analysis dictionary-based approach towards the data exploration for the knowledge discovery in textual corpora with using the NLP and Text Mining procedures. This paper is the methodological reflection how development of data analysis in the field Digital Humanities (DH) and Computational Social Science (CSS) influence on the area of computer-aided qualitative data analysis (CAQDAS). I try to show the character of this shift in the context of the project I realize now: The domain ontology as a model of knowledge representation about the contemporary field of qualitative research.[1]

2 CAQDAS, Text Mining and the Knowledge Discovery

In the field of qualitative data analysis, the idea of knowledge discovery does not have a fixed tradition. In some sense, this idea is partly represented in the Glaserian version of Grounded Theory as well as in Qualitative/Quantitative Content Analysis approach and partly by searching functionalities of CAQDAS tools (Atlas TI, NVivo or QDA-Miner). The Grounded Theory (GT) methodology call this process serendipity (the

---

[1] The domain ontology as a model of knowledge representation about the contemporary field of qualitative research is the project funded by Polish National Science Center (Competition: OPUS 12; Panel Description: HS6_13: Theoretical sociology, methodological orientations and variants of empirical research). The project duration is 2017–2020.

discovery context).[2] According to GT, which is mainly a data-driven approach we can search for and discover phenomena or relationships that we were not looking for at the beginning of our research. So, serendipity can occur in time of fieldwork, in time of data analysis process or time of theory building. Serendipity is an immanent feature of the well-established theory whose clear procedures allow for the processual discovery of the structure of social phenomena. In traditional qualitative, analytical practice serendipity emerges by comparing coded passages of texts or by using CAQDAS software for explorative data analysis. The context of discovery can be a consequence of reasoning: deductive (logical inference, from general to specific), inductive (inference from rationale, from detail to general), abductive (reasoning about probable causes on the basis of knowledge of effect, explaining what known, the reverse of deduction) or even in special conditions heuristic reasoning (inference without strict rules, based on associations or analogies with something previously known). In the traditional sense, serendipity refers to the sequence of actions, during which researcher "discovers" by accident dependency, regularity or the property of the phenomenon or social process in time of raw data investigation. Knowledge discovery is carried out naturally by following the logic and procedures of the given methodological paradigm. Hence, serendipity in qualitative sociology appears as a natural context for discovering knowledge in the field of research and data analysis that can be supported or not by CAQDAS. In this sense, serendipity is the result of the sociological imagination, knowledge, personal experience, fieldwork and analytical skills of the researcher. Therefore, is it possible to refer the idea of serendipity to the Text Mining analysis and the knowledge discovery of qualitative data? Currently, NLP and Text Mining approach is associated with the process of searching based on a simple or advanced query, looking for the relationships and dependencies, text parsing, tokenization, the process of text stemming or lemmatizing, word/phrase/text extracting, disambiguation, annotation, tagging (coding), categorizing, classifying, grouping, model building and visualization [17, 33]. Of course, not all these tasks are similar to the traditional CAQDAS analytical steps and procedures. In the case of Text Mining, Knowledge discovery is purely analytical and involves the interaction with analyzed data. The context of the discovery is experimentally constructed in the process of data extraction by using different analytical techniques or algorithms. In this sense, discovering knowledge (mining, drilling) in qualitative data is an interactive and iterative process of extracting, coding/tagging and searching for new, unexpected configurations, regularities and patterns. In practice, all qualitative data such as transcriptions of individual interviews or group interviews, diaries, biographies, journals, books and any written documents may be analyzed with NLP and Text Mining procedures.

From methodological point of view, one should distinguish between data exploration and discovery of knowledge in qualitative datasets. KDD process consist of a

---

[2] Serendipity is used as a sociological method in grounded theory, building on ideas by sociologist Robert K. Merton, who referred to the "serendipity pattern" as the fairly common experience of observing an unanticipated, anomalous and strategic datum which becomes the occasion for developing a new theory or for extending an existing theory.

few stages: data cleaning (checking data quality), data integration (combining different data into single dataset, corpora), data selection (choosing data for further analysis), data transformation/consolidation (transforming data for further analysis), data mining/exploration (searching for patterns and data modeling)[3], patterns/models' evaluation and knowledge representation (visualizing patterns or models). The data mining (exploration) is only one step in the process of data analysis for a better understanding of relationships hidden in data. The essence of the exploration is the selection of appropriate analytical techniques or algorithms that help to find relationships or patterns in the semantic context. If data exploration is supported or automated by NLP procedures and Text Mining techniques then appear new possibilities in the scope of the interaction with text data and building the knowledge representation models. For example, using NLP text formalization (presentation of statements created in natural language in a formal language) with the algorithms of induction rules reflect the idea of knowledge discovery. However, they do not allow for a generalization but help to better understand the semantic context and linguistic structure of analyzed data. Because qualitative data are representations of empirical objects of the socio-cultural world: statements, facts, events, etc. By reference to what they represent, they are meaningful, and thus they are carriers of specific information. Therefore, it is extremely important to choose the proper method of exploration for the analysis of the different type of qualitative data such as press articles, blogs, individual interviews, group interviews. Knowledge discovery approach has an active role in the process of data interpretation. Relations between data, information and knowledge can be described similarly as the relationship between object, symbol and idea in the semiotic triangle Ogden and Richards [37]. In turn, data, information and knowledge are three vertices of the triangle referred to as an epistemic triangle, reflecting the character of representing and discovering knowledge in qualitative data. If we place these triangles in reverse to each other, turning the basis of the epistemic triangle so that it is closer to the top of the semiotic triangle, we will obtain a diagram showing the relations between cognitive structures (knowledge) and language structures (linguistics, NLP) that occur in Text Mining analyzes. Consequently, each epistemic structure discovered in the process of analyzing textual data derives from the structure of the semiotic natural language. The distinction between data, information and knowledge is therefore not only important for determining the relationship between them but above all for understanding the role of CAQDAS combined with Text Mining in the process of discovering knowledge in qualitative data.

The rapid development of information technologies and the availability of large volumes of textual and web data also causes that many CAQDAS programs are currently heading towards online coding and exploratory analysis with Text Mining

---

[3] At present exploration KDD covers different types of analytical procedures i.e. descriptive analysis, matrix and cross-tabs analysis, discovery of association or inductive rules, classification techniques, grouping techniques (factor analysis, cluster analysis, k-means, two-stage grouping), prediction and statistical analysis (i.e. logistic regression, discrimination), discovering sequence patterns, searching for deviations, anomalies, traditional searching and content extraction etc. They are connected not only with the understanding of the language of data and the content of documents but also with the skill of their multidimensional analysis, synthesis of knowledge, meaning or interpretation.

methodology [44]. Regardless of which functionalities are present in CAQDAS programs, a qualitative researcher should be critical at every stage of the research process as to the effects of new technologies or functionalities, and keep in mind that the CAQDAS programs and Text Mining algorithms are only a tool in the data analysis process. Considering the process of discovering knowledge in qualitative data, one can refer it to the relation between the world of empirical facts and theoretical constructions. In this relation, computer software, NLP, data mining processes, methods of scientific reasoning, and algorithms and analytical techniques play a mediating role between these worlds. The use of knowledge about the syntax of language, semantics and logic of connections between elements of expression is an attractive area in exploration and analysis of text data in CAQDAS programs. In these programs, there are many algorithms or analytical techniques for extracting information from text data. Each method of text analysis, however, has its strengths and weaknesses. The majority of CAQDAS tools currently are based on one approach to text analysis, which is usually based on a specific methodological paradigm (grounded theory, mixed methods, qualitative or quantitative content analysis, discourse analysis, etc.) [18]. At the same time, the development of CAQDAS is the implementation of new functionalities, algorithms or analytical techniques, such as the procedure of automation coding based on data coding patterns (Qualrus, Atlas TI, NVivo, QDAMiner, WordStat). It is not a surprise to many CAQDAS users to build contingency tables or matrixes in the construction of typology or data classification. For some qualitative researchers' implementation of statistical techniques or Text Mining algorithms in CAQDAS programs is contradictory with the essence of traditional qualitative data analysis. The usage of quantitative techniques in the qualitative data analysis, e.g. correspondence analysis, cluster analysis, multi-dimensional scaling enriches the process of discovery of knowledge in qualitative data, revealing new areas of qualitative sociology development. In qualitative sociology, due to the multiplicity of available text data and, in principle, unlimited possibilities of their collection, there is a field to use both Text Mining techniques and the CAQDAS approach in the process of discovering knowledge. Both solutions are based on the techniques of unstructured text data exploration and analysis. Text mining is defined as an extraction process information in document collections by identifying and searching for patterns, regularities, relationship structures in text data [16]. Likewise, it is a computer-aided analysis of qualitative data. The paradigm of the grounded theory in CAQDAS programs requires the exploration or discovery of knowledge in qualitative data with an open mind, and the identification of categories, concepts and constructions that explain specific social processes should not be imposed in any way from outside [21–23]. A qualitative researcher, as well as a Text Mining analyst, allow to "emerge" codes, categories or analytical constructions from the analyzed data in the process of continuously comparing coded object: statement, paragraphs or documents. The possibility of using text data mining algorithms in the computer-aided analysis of qualitative data and automation of initial data coding prevents the researcher from imposing any structure, other than that which is contained in the text data itself. This does not mean, of course, that automatic tagging or analytical procedures supporting text data pre-processing are better than traditional coding procedures used by researchers associated with the paradigm of well-established theory. Text Mining has a big impact on CAQDAS users and can be a good supplementation of traditional methods of qualitative data analysis and their analytical workshop, as well as the idea of knowledge discovery. Text Mining and its techniques and algorithms are the extensions of content analysis techniques in which analytical categories "emerge" through the usage of frequency lists or in the process of creating the categorization key. In Text Mining, as in the content analysis, a frequent effect of the analysis are analytic dictionaries, sets of keywords used to classify the analyzed texts or documents. Content analysis and Text Mining use computer algorithms for counting keywords, building stop lists, etc., but the drilling of text data goes further towards discovering network relations among words or phrases.

3   CAQDAS, Text Mining and the Problem of a "Curse of Abundance" in the Present Field of Qualitative Research

The nature of qualitative research practices is its multiparadigmaticity which legitimizes the coexistence of different methodologies in the field of qualitative research and data analysis practices. Thus, this field is, quite naturally, accompanied by exceptional creativity in milieus which, since the 1980s, have been resourcefully overcoming the limitations of their approaches. This was made especially manifest in the course of a triple crisis affecting the spheres of representation, legitimization and the impact of qualitative research practices [13–15]. A consequence was the elaboration of ever more novel means of conceptualization and realization of qualitative research which, however, mean that the drawing-up of a consistent and coherent image of the field of qualitative research has become an increasingly reckless venture. Even the cutting-edge shapers of innovative trends in this sub-discipline admit to the difficulties inherent in exhaustively and comprehensively reviewing the latest variations of qualitative research practices [12]. Already in the early 1990s, opinions were expressed that the situation in this field bore the signs of a "curse of abundance" [15]. As Norman Denzin and Yvonne Lincoln stated at the time "Researchers have never before had so many paradigms, strategies of inquiry, and methods of analysis to draw upon and utilize" [15:11]. Succeeding years of further evolution in the ways by which qualitative research can be conceptualized and practiced have brought the exponential expansion of pertinent knowledge (above all, in the form of scholarly articles published in journals dedicated to qualitative research). Still, this accrued material has yet to be comprehensively systematized. However, this has also meant that the drawing-up of a consistent and coherent image of the field of qualitative research has become an increasingly reckless venture [39]. Nevertheless, a need for orientation in the field of one's research practice, augmented by a desire to be "on top of the game," leads to undertake an attempt to reduce the complexity. The examination of numerous individual acts of qualitative research practice will permit the distinguishing of certain, reproduced patterns of behavior. So, my research and analytical project tackles the problem of a "curse of abundance" in the present-day field of qualitative research. I try to answers to the following questions: which elements (theoretical paradigms, methodological trends, research methods and techniques, topics and areas of investigation, etc.) are constitutive for research practices; what is the structure of relations among these elements; which elements form currently dominant research patterns; what is the nature and the dynamics of these patterns; how these patterns are expressed in the language spoken by qualitative researchers; and is it even possible to accurately trace the diversity of qualitative research practices while concurrently reducing the complexity of this field's semantic space? The goal of this project is to discover (analytically recognize) the dominant ways by which qualitative research designs and practices are conceptualized and represented in methodological language, to organize and articulate this knowledge in a legible system of network representation [6].

3.1   Research Methodology

The project methodology is the combination of know-how from social sciences (especially CAQDAS and qualitative sociology), digital humanities, social informatics, corpus linguistics, Natural Language Processing and Text Mining [8, 19, 20, 27, 32]. This research follows the principles of the mixed methods paradigm [10, 11, 41] according to which – in the process of knowledge discovery pertaining to the contemporary field of qualitative research – I combine qualitative methodology and data analysis approach [2, 16, 27, 35] with the procedures of content analysis [1, 3, 28, 40, 43] and Text Mining [4, 26, 31]. The project research design is based upon a process of the discovery of knowledge in textual corpora and the knowledge systematization by searching for relations between paradigms, methodological trends, methods and techniques of qualitative research and data analysis as they manifest in the scientific, written language. In other words, I analytically explore connections and relations formed among paradigms, methodological trends, methods and techniques for conducting qualitative research which manifest in the articles' vivid language, published by different authors in five qualitative, mainstream methodological journals. Employing qualitative and quantitative content analysis approach as well as the Natural Language Processing procedures and Text Mining methodology of data analysis, the project examine English written articles, published between 1990–2018 in Qualitative Inquiry (QI), Qualitative Research (QR), International Journal of Qualitative Methodology (IJQM), The Qualitative Report (TQR), and Forum: Qualitative Social Research (FQSR). Those articles are a representation of how qualitative research is conceptualized and practiced at present. These specific periodicals have been selected to the corpora due to their positioning in the milieus of qualitative researchers around the world; this is evidenced by their high Impact Factor index values as well as their transdisciplinary nature. The articles published in these journals convey representation of various conceptualizations and implementations of qualitative research – including those which – due to their novelty or exceptionality do not (yet) belong to the canon shaped in anthologies and textbooks. However, because of their relatively briefer publishing cycle, journals react much faster to changes taking place in the field of qualitative research, and, therefore, they facilitate the distinction of tendencies as yet unsanctioned by tradition and well-known methodological books. To a much greater extent, their contents also reflect the polyphony which underlies the diversity found in qualitative research practices which entail the thoughts and actions of both the authority figures as well as the novitiates, not to mention representatives of several academic disciplines. As a repository of individual testimonies to research experiences, the journal articles comprise an opulent (and increasingly wealthier) source of information on the subject of the current condition of the field of qualitative research. The intention is to create a model representing up-to-date knowledge about qualitative practices of research – that is, the ways by which those practices manifest in the contemporary field of qualitative research are conceptualized and operationalized. The model of methodological knowledge representation refers to the information technology concept of domain ontology known from the field of knowledge engineering. This model will be rendered in the form of a legible scheme of the classification of means by which qualitative research is conceptualized and realized; it will encompass a comprehensive catalogue of up-to-date variations in qualitative research practices and their specifications. Moreover, the model entails the delineation of a concept scheme epitomizing the structure of this current qualitative research field and facilitate insight into its elements and their semantic configurations.

3.2 What Is Domain Ontology?

In knowledge engineering the term ontology is defined as a hierarchical system of categories and relations with a universal or limited scope. However, it differs from other ways of representing knowledge in that it not only provides a scheme or description of a given field, but using formal tools of logic (axioms, definitions, rules) allows to strictly define the hierarchy of its elements and criteria for their classification.[4] Ontology in knowledge engineering pertains to such presentations of select academic fields of scholarship which encompass a field's domain dictionary, as well as the set of semantic and logical rules that describe the interrelationships among its elements. Every field creates ontologies to limit its complexity and organize information into data and knowledge. A domain ontology is a representation of some part of reality. It allows a researcher to create a conceptual scheme reflecting the structure of relationships and the possible configurations between basic elements of a given field's knowledge (i.e., its concepts, methods, techniques, etc.). Ontologies are formal specifications of domain concepts. They represent entities of a specific knowledge domain and the relationships that can hold between the entities. Ontologies are formal descriptions of the so-called "body of knowledge" that composes a domain. Regardless of being implicitly or explicitly applied during the modeling, ontologies set the relation between formal signs used in computer simulations and "meaning" as a notion of human minds. The development of ontology means always analyzing and organizing our knowledge about the elements of a given field and the way or ways of expressing it, which sometimes is an end in itself, because it may mean an advantage in the form of creating a certain interdisciplinary basis for communication between experts or just reveal not always coherent ways to deal with certain issues. Also, thanks to its formal structure, ontology allows for systematic inference about what belongs to its domain – this means that we can not only create a model of a given problem, project, process, enterprise, but also provide it with data from information systems. Ontology is,

---

[4] Nowadays, the word ontology is used in a lot of diverse research fields including natural language processing, information retrieval, electronic commerce, Web Semantic, software component specification and information systems integration. In this context, some ontology models and languages have been developed in the last decade.

therefore, more than a close representation of our knowledge because it can provide a flexible platform for building or integrating information and elements of information systems based on them. Underlying this term of the ontology may be different structures of knowledge, and, hence, certain ontological divisions. The most fundamental is the universal (base) ontologies, those including core concepts of scientific knowledge, versus the limited (domain) ontologies, those describing a slice or piece of reality within a select range of knowledge. Due to the degree of formalization, also distinguished are informal ontologies developing based on predefined vocabularies, dictionaries, thesauri, and taxonomies versus formal ontologies grounded in data structure or logic. As a means by which the semantic space of knowledge representation is represented, domain ontology supports scholarly communication between researchers and facilitates the formation of a coherent means of expressing its problematics. A consequence of the creation of domain ontology is a knowledge base development that allows the monitoring of trends in a given field's theory and methodology; this, in turn, becomes the foundation for knowledge-based empirical research. The principles underpinning the advancement of domain ontologies constitute the core of our endeavor [45]. My project develops domain ontology of the field of qualitative research, anchored in the written language of communication applied in qualitative research practices. The key stages of the analytical procedure of developing domain ontology according to the idea knowledge discovery are listed below.

3.3   Analytical Approach in Building Domain Ontology

The analytical strategy for the process of building domain ontology of the field of qualitative research combines know-how from the sociology (mainly qualitative sociology), Data and Text Mining, social science computing, corpus and computational linguistics. The analytical frame is grounded in traditional CAQDAS coding procedures and textual data analysis with a combination of a dictionary-based approach, mixed content analysis, Natural Language Processing procedures and Text Mining approach [17]. The starting point of analysis was building article's corpora with the intention of data management, data analysis and afterwards the reconstructing a semantic-logical network which will illustrate relationships between basic elements of the contemporary field of qualitative research and can be the analytical model of the knowledge representation of qualitative research designs and practices. All articles in the pilot study have been downloaded from websites as documents PDF transformed to TXT and XML formats and imported to the QDAMiner program. Then the articles were checked by us in terms of their use as research material. We excluded reviews, editorial entries, sets of abstracts, notes from the body of the texts post-conference and publication reports. After selection of articles they have been described by the following variables: article identifier, author, year of publication, magazine, type of article, a fact the occurrence of keywords, the fact of occurrence of an abstract. The final corpora in the pilot study contained 2043 documents: traditional articles, articles in the form of essays, articles in the form of poems and other written forms. According to Data Mining methodology, articles in corpora were randomly divided for two subsets: learning and testing. Teaching set contained 333 articles that have been used for developing classification categories in the dictionary, and a testing dataset with the rest of articles for checking the accuracy of the classification. By drawing articles to the training dataset, we have kept a balanced distribution of the number of articles according to the magazine and the year of publication.

The first step in building domain ontology is a preparation of terminological dictionary via coding process [36]. All articles have been analyzed in terms of their logical structure and content before coding. Next, words and phrases were established as the coding unit. Moreover, six initial thematic areas for coding have been distinguished: theoretical paradigms, methodological trends, methods and research techniques, problems and challenges, topics and areas of research and authorities. We started from extracting and coding of methodological words and phrases occurring in titles, abstracts and keywords in articles. The coding process was merged with the procedure of building a classification dictionary with the methodology of Content Analysis and Text Mining [24, 25, 42]. The process of construction of the dictionary has been started with the review lists of keywords and key phrases obtained as a result of GT coding. Ultimately, due to the large variety of methodological words and phrases, in the pilot study, we decided to focus on the first three areas: theoretical paradigms, methodological trends, research methods and techniques. All of coded words and phrases in pilot study were assigned to one of these three areas and categorized according to its subtype because of

DATA METHODS/TECHNIQUES
   INTERVIEWING/INTERVIEWS
      ACTIVE_INTERVIEWS
         ACTIVE RESEARCH INTERVIEW*
         ACTIVE INTERVIEW*
      COGNITIVE_INTERVIEWS
         COGNITIVE INTERVIEW*
         SEMISTRUCTURED COGNITIVE INTERVIEW*
      CONSTRUCTIONIST_INTERVIEWS
      CONVERSATIONAL_INTERVIEWS
         CONVERSATIONAL INTERVIEW*
         OPEN-ENDED CONVERSATIONAL INTERVIEW*
      DYADIC_INTERVIEWS
         DYADIC INTERVIEW*
         DYAD*
         INTERVIEW TO THE DOUBLE
      FACE-TO-FACE_INTERVIEWS
         TRADITIONAL FACE-TO-FACE INTERVIEW*
         FACE-TO-FACE INTERVIEW*
         INDIVIDUAL INTERVIEW*
     SIDE SHADOWING_**INTERVIEWS**
         SIDESHADOWING INTERVIEW*
         SIDESHADOW INTERVIEW*

Fig. 1. Excerpt of the classification dictionary

semantic co-occurrence and linguistic similarity. For example Participatory Methodology, Participatory Qualitative Research, Participatory Action Research, Participatory Research are lexical indicators of Participatory Methods; Walking, Mobile Walking, Mobile Exploration are language expressions fixed to Mobile Methods; Biographical Interviewing, Biographical-Narrative Method, Biographical Interview are a part of Biographical Interviews and Digital Narrative Method, Digital Storytelling, Oral Storytelling, Storytelling are a part Storytelling, but on higher level they are in the category of Narrative Methods. The preliminary version of classification dictionary contains 9561 types of predefined lexical-semantic indicators such as words, phrases, and sometimes short phrases extracted in time coding and taken from qualitative, methodological books. All coding areas/categories/subcategories/words or phrases in the codebook and classification dictionary are isomorphic and are hierarchically ordered (see Fig. 1).

Applying a Content Analysis and Text Mining approach based on predefined patterns we can search, code and analyze unlimited number of articles in their natural (lexical and semantic) context what would be difficult to do traditionally. This is known in informatics as a dictionary-based approach or dictionary-based classification – automatic-learning techniques and algorithms that "teach themselves" on a predefined pattern. This approach can be successfully used in analyzing transcriptions, theme extraction, seeking out of regularities, and the discovery of patterns in sets of unstructured, qualitative data. Such a solution augments the process of qualitative content analysis which applies traditional, manual procedures of text data coding [29] by adding the possibility of supervision by the automatic coding of new content via predefined dictionary classification patterns (words, phrases or text passages). This is an iteration (repeated) process: at each phase, new codes or vocabulary categories can be added while old codes are verified. The domain ontology project is still continuing, and the classification dictionary structure is developing. Before we will build domain ontology of the field of qualitative research the final version of classification dictionary will be evaluated by qualitative research and data analysis experts.

The second step in building domain ontology is the reconstruction of the semantic-logical relations among categories and subcategories. In contrast to the dictionary-based approach, we do not assume any code hierarchy or classification. We rather discover relationships by searching for co-occurrences and similarities. In order to reconstruct such relations, we can use a lot of statistical exploratory techniques e.g., correspondence analysis, multidimensional scaling, hierarchical cluster analysis, two-step cluster as well as data mining techniques e.g., decision trees, induction rules, Kohonen self-learning networks or neural networks. In pilot study I decided to use Social Network Analysis (SNA) to present relationships among 47 dominant categories of methods and techniques (words and phrases). SNA plot was done in Gephi (see Fig. 2).

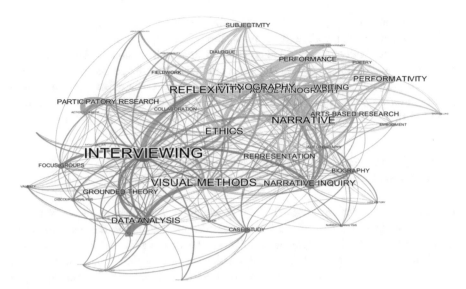

Fig. 2. The pilot study domain ontology network representation of contemporary qualitative research field.

Analyzing this network, we can distinguish four methodological synsets (semantic clusters) centered around Interview methods, Participatory methods, Arts based research methods and Narrative methods. This is only a small example of domain ontology of the contemporary field of qualitative research practices based on the network representation. If we add publication date and journal-title we can use this network of methodological relationships and synsets to discover cutting-edge trends in the theory and methodology of qualitative research, and also predict new methodological paths.

Developing the domain ontology of qualitative research is not only a means by which we can translate – via a classification dictionary or by network relationships – the "living" language of the world of qualitative research practices into algorithms, analytical techniques, and procedures in order to look at the "curse of abundance" from a distance. It is also a way of reducing the complexity of the field of qualitative research practices, and providing a compass for orientation amidst the forest of changing methodological trends, methods and research techniques, and analytical strategies. It also facilitates the identifying of research paths (represented by the problem–method–theory scheme), the monitoring of new trends in methodology, and the crafting of a new typology of research approaches. Domain ontology, as a model of knowledge representation (a system of classification and categorization of concepts and the relationships among them), helps create a metalanguage in the dispersed field of qualitative research; this ontology could become a key to communication between researchers and the further development of qualitative methodology and research practices. From the perspective of the evolution of content analysis methods and techniques in qualitative research, the approach we propose is the result of deliberations on the possibilities for computer-assisted qualitative data analysis, the latest information technology, and text mining algorithms and techniques in the processing of large amounts of textual data

[44]. An important step in this process is the transition from traditional qualitative analysis based on text coding procedures, through comprehensive content analysis which combines both qualitative and quantitative approaches (mixed content analysis), towards the application of methods for exploration and knowledge discovery in textual databases [7, 9, 30]. This process involves a shift from cause-effect analysis to correlation analysis based upon searches for relationships and dependencies in big data sets [5, 34].

4   CAQDAS and Text Mining Are There Alternative or Complementary Approaches?

I am not familiar with any scholarly inquiry which takes advantage of state-of-the-art digital technology methods of the social sciences, and integrating the methods, techniques, and analytical strategies of qualitative and quantitative sociology, computer linguistics, data science, and social informatics to analyze the "living" language of qualitative research practices hidden in methodological articles. This study is also a methodological reflection on the CAQDAS development process that is gradually moving towards the area of Knowledge Discovery in Textual Databases and Text Mining. The use of computer-aided analysis of qualitative data in the field of qualitative sociology has already become a fact. The environment of qualitative researchers using CAQDAS software in their research projects is growing up. But my own experience shows that the use of various CAQDAS tools leads to an increase in methodological awareness, which translates into greater accuracy and precision in the process of qualitative data analysis. However, qualitative data analysis using methodology, algorithms and Text Mining techniques is a kind of novelty in the field of qualitative sociology. In this paper, creating the ontology model of the contemporary field of qualitative research, I tried to show the process of change in CAQDAS analysis passing through Content Analysis towards natural language processing and Text Mining analysis. Text Mining is a set of techniques in which are equipped with algorithms for automatic or semiautomatic extraction of information from textual data (articles, IDI and FGI transcription etc.). Text Mining is based on the use of computer software to find hidden for human, due to the limited perceptual and temporal possibilities, the regularities contained in text data. If the CAQDAS analytical algorithms are used to work with smaller sets of qualitative data, then the Text Mining technique allows conducting analyzes in which the size of the data set is unlimited. Text Mining as a set of techniques and algorithms supporting information extraction processes and inductive search for patterns and interdependencies in data sets (knowledge discovery and representation) is the quintessence of the exploratory approach in the analysis of text data, commonly known as Text Analytics. However, if traditional text analyzes using CAQDAS programs, like Text Mining, allow for a comprehensive exploration of qualitative data in the process of discovering knowledge, the question arises about their mutual relations based on methodology and analysis of qualitative data. Are these approaches competitive or complementary to each other? Why can Text Mining be used in the analysis of qualitative data? What do the Text Mining procedures offer? What problems does Text Mining solve? In the methodological sense, the data mining concerns both the data generated and the existing data. In the analytical sense, it requires knowledge and skills to integrate qualitative and quantitative data, as well as their comprehensive analysis. Thanks to the development of computer science, advanced statistical algorithms, artificial intelligence or machine learning methods, Text Mining also enriches the diagrams of traditional exploratory analysis of qualitative data, which is the effect of recording statements, events or activities of social actors. This approach allows not only a comprehensive understanding of social phenomena and processes in qualitative sociology but also the creation of analytical classification or prediction models based on discovery patterns and regularities. Text Mining analyzes use different explorative techniques and analytical algorithms to find hidden for human regularities contained in the structure of text data, due to human limited perceptual and time capabilities. Computer-aided analysis of qualitative data using Text Mining techniques or natural language processing is a kind of novelty in the field of quality sociology. The application of Text Mining in the area of CAQDAS raises both the reliability of the results of qualitative data analysis and the quality of qualitative research in sociology and social sciences. CAQDAS gains a more versatile character and enormous possibilities for analyzing text data in the linguistic, syntactic, semantic and pragmatic dimensions. On the methodological and analytical level, Text Mining and CAQDAS are not as different as they might seem. The differences mainly come down to the dimension of the automation of the data mining process. Analytic procedures in the field of discovering knowledge are similar.

In practice, the difference between Text Mining and CAQDAS is noticeable in the range of the number of text data processed. If the CAQDAS software is used in work with smaller sets of qualitative data, then Text Mining allows conducting analyzes in which the size of the data set is unlimited. Text Mining algorithms allow you to view and analyze information, the number of which is almost unimaginable for a qualitative researcher and to perform calculations and analyzes in an extremely short time. The possibilities of analyzing and understanding large volumes of text data are smaller due to their multidimensionality. Techniques and Text Mining analytical algorithms complement and enrich not only the traditional functionalities of CAQDAS software but also our analytical skills. In contrast to the traditionally applied a priori techniques in sociology, CAQDAS and Text Mining refer to the discovery of relations between variables in a situation where there are no predetermined expectations or assumptions about the nature of these relations. The hypotheses are generated *a posteriori* from the data than the a priori. In a typical exploratory data analysis process, many variables are taken into account and compared, in many different combinations and configurations, in search of significant relationships between them. These dependencies represent models of text data mining, built based on advanced methods and analytical techniques. The analytical model is created as the effect of data and variable configuration, regardless of the size of data sets and the number of variables. Text Mining, when constructing such models, is based on inductive-deductive reasoning as well as inductive - abductive reasoning in the area of a finite set of text documents. Models represent the structure of empirical associations, which are then tested and interpreted. A collection of qualitative data in CAQDAS is also finite, and inference is most often based on inductive-deductive or inductive - abductive reasoning. Its effect is usually cognitive maps showing relations between documents, codes, categories or concepts.

If you accept the definition of CAQDAS for Ann Lewins referring to the so-called Qualitative analysis of qualitative data [32], rather it would be preferable to use Text Mining techniques and algorithms in the area of computer-aided qualitative data analysis. The comparison of the logic of the Text Mining and CAQDAS analysis shows that on analytical and methodological grounds they are rather epistemologically compatible than competitive ones. As in many qualitative methodological approaches to the process of qualitative data analysis, Text Mining "encourages" the researcher to be open to constructing models describing the structure of data relations. In other words, it can be said that the techniques and algorithms of text data mining fit perfectly into different stages of computer-aided analysis of qualitative data in the process of discovering knowledge.

CAQDAS, Text Mining and modern IT technologies allow for methodological solutions that automate and enrich the analysis of qualitative data. However, in contrast to the popular belief, Text Mining is not an automated, reflexive action. As in the case of CAQDAS, it is an iterative process that requires a conscious approach from the researcher to the data analysis. In practice, Text Mining is usually semi-automated (supervised) methods that require knowledge and knowledge of analytical techniques. Working with CAQDAS programs teaches the researcher methodological rigor, adherence to procedures, accuracy and precision in the process of qualitative data analysis, and Text Mining opens up new areas of knowledge, interdisciplinary and requires additional analytical skills, which positively affects the quality of analyzes and field research. Support for computer-aided analysis of qualitative data with advanced techniques and Text Mining analytical algorithms causes various methodological paradigms to intersect in CAQDAS programs: well-established theory - content analysis or mixed methods. In terms of methodology, Text Mining is in some sense a reflection of the logic of the grounded theory. Analytically, it is similar to content analysis. However, although both approaches use computer algorithms to analyze text data, Text Mining goes further. Characterized by the unique ability to process the natural language and use the knowledge contained in the object and thematic dictionaries in the process of qualitative analysis. Thanks to this, the application of procedures of Natural Language Processing and Text Mining techniques in the field of qualitative sociology and computer-aided analysis of qualitative data leads to the deepening of knowledge about the mechanisms of social activities and processes. It also favors the integration of data from many different sources, existing data and data from field qualitative research. Data integration leads to the systematic development and integration of sociological knowledge and also improves the quality of analyzes and qualitative research. For a social researcher, especially a qualitative researcher, it is extremely important to approach from the data side, discover knowledge from data, build multi-dimensional analytical models, mechanisms and social activities, and consequently test dependencies and hypotheses between variables in these models through the application of traditional methods and techniques of sociological research. The conviction that knowledge is contained in data, in the way it is collected and analyzed, is present in quality sociology forever. The logic of exploring large sets of text data using Text Mining and natural language processing brings to the area of computer-aided analysis of qualitative data new, unprecedented possibilities of discovering relationships in various social systems, and thus broadening and deepening sociological knowledge.

References

1. Bernard, R.H., Ryan, G.W.: Content analysis. In: Bernard, R., Ryan, G. (eds.) Analyzing Qualitative Data: Systematic Approaches, pp. 287–310. SAGE, Los Angeles (2010)
2. Bong, S.A.: Debunking myths in qualitative data analysis. Forum Qual. Sozialforschung 3 (2) (2002). http://www.qualitative-research.net/index.php/fqs/article/view/849
3. Berelson, B.: Content Analysis in Communication Research. Free Press, Glencoe (1952)
4. Berry, M.W.: Survey of Text Mining: Clustering, Classification, and Retrieval. Springer, New York (2004)
5. Brosz, M., Bryda, G., Siuda, P.: Big Data i CAQDAS a procedury badawcze w polu socjologii jakościowej. Przegląd Socjologii Jakościowej 13(2), 6–23 (2017)
6. Bryda, G., Martini, N.: W stronę ontologii pola badań jakościowych. Przegląd Socjologii Jakościowej 12(4), 24–40 (2016)
7. Bryda, G.: Caqdas, Data Mining i odkrywanie wiedzy w danych jakościowych. In: Niedbalski J, Metody i techniki odkrywania wiedzy. Narzędzia CAQDAS w procesie analizy danych jakościowych, pp. 13–40. Wydawnictwo Uniwersytetu Łódzkiego, Łódź (2014)
8. Bryda, G.: Caqdas a badania jakościowe w praktyce. Przegląd Socjologii Jakościowej 10(2), 12–38 (2014). http://przegladsocjologiijakosciowej.org
9. Bryda, G., Tomanek, K.: Od Caqdas do Text Miningu. Nowe techniki w analizie danych jakościowych. In: Niedbalski J, Metody i techniki odkrywania wiedzy. Narzędzia CAQDAS w procesie analizy danych jakościowych, pp. 191–218. Wydawnictwo Uniwersytetu Łódzkiego, Łódź (2014)
10. Collins, K.M.T., Onwuegbuzie, A.J., Jiao, Q.G.: A mixed methods investigation of mixed methods sampling designs in social and health science research. J. Mixed Methods Res. 1(3), 267–294 (2007)
11. Creswell, J.W.: A Concise Introduction to Mixed Methods Research, Kindle edn. SAGE, London (2015)
12. Creswell, J.W.: Qualitative Inquiry and Research Design: Choosing Among Five Traditions. SAGE, Thousand Oaks (1998)
13. Denzin, N., Lincoln, Y.: The SAGE Handbook of Qualitative Research, 4th edn. SAGE Publications, Thousand Oaks (2011)
14. Denzin, N., Lincoln, Y.: The SAGE Handbook of Qualitative Research, 3rd edn. SAGE Publications, Thousand Oaks (2005)
15. Denzin, N., Lincoln, Y., et al. (eds.): Handbook of Qualitative Research. SAGE Publications, Thousand Oaks (1994)
16. Dey, I.: Qualitative Data Analysis: A User-Friendly Guide for Social Scientists. Routledge, London, New York (1993)
17. Feldman, R., Sanger, J.: The Text Mining Handbook: Advanced Approaches in Analyzing Unstructured Data. Cambridge University Press, Cambridge (2007)
18. Fielding, N.G.: The diverse worlds and research practices of qualitative software. Forum Qual. Sozialforschung 13(2) (2012). http://www.qualitative-research.net/index.php/fqs/article/view/1845/3369
19. Fielding, N.G., Lee, R.M.: Computer Analysis and Qualitative Research. SAGE, London (1998)
20. Fielding, N.G., Lee, R.M.: Using Computers in Qualitative Research. SAGE, London (1993)
21. Glaser, B.G., Strauss, A.L.: Odkrywanie teorii ugruntowanej: strategie badania jakościowego, translated by Gorzko M., Zakład Wydawniczy Nomos, Kraków (2009)
22. Glaser, B.G.: Theoretical Sensitivity: Advances in the Methodology of Grounded Theory. Sociology Press, Mill Valley (1978)

23. Glaser, B.G.: Basics of Grounded Theory Analysis: Emergence vs. Forcing. Sociology Press, Mill Valley (1992)
24. Ho Yu, Ch., Jannasch-Pennell, A., DiGangi, S.: Compatibility between text mining and qualitative research in the perspectives of grounded theory, content analysis, and reliability. Qual. Rep. **16**(3), 730–744 (2011). http://www.nova.edu/ssss/QR/QR16-3/yu.pdf
25. Hopkins, D.J., King, G.: A method of automated nonparametric content analysis for social science. Am. J. Polit. Sci. **54**(1), 229–247 (2010)
26. Ignatow, G., Michalcea, R.: Text Mining. Guidebook for Social Sciences. SAGE, London (2016)
27. Kelle, U.: Computer-Aided Qualitative Data Analysis: Theory, Methods and Practice. SAGE, London (1995)
28. Krippendorf, K.: Content Analysis. An Introduction to its Methodology. SAGE, Thousand Oaks (2004)
29. Kuckartz, U.: Qualitative Text Analysis: A Guide to Methods, Practice and Using Software. SAGE, London (2014)
30. Larose, D.T.: Odkrywanie wiedzy z danych: wprowadzenie do eksploracji danych. Wydawnictwo Naukowe PWN, Warszawa (2006)
31. Leetaru, K.: Data Mining Methods for the Content Analyst: An Introduction to the Computational Analysis of Content. Routledge Communication Series. Taylor and Francis, Routledge (2012)
32. Lewins, A., Silver, Ch.: Using Software in Qualitative Research: A Step-by-Step Guide. SAGE, London (2007)
33. Manning, Ch.D., Raghavan, P., Schütze, H.: Introduction to Information Retrieval. Cambridge University Press (2008). http://wwwnlp.stanford.edu/IR-book/
34. Meyer-Schonberger, V., Cukier, K.: Big Data. Rewolucja, która zmieni nasze myślenie. MT Biznes, Warszawa (2014)
35. Miles, M.B., Huberman, M.A.: Analiza danych jakościowych, translated by Stanisław Zabielski. Trans Humana, Białystok (2000)
36. Munn, K., Smith, B. (eds.): Applied Ontology. An Introduction. Transaction Books Rutgers University, Piscataway (2008)
37. Ogden, C.K., Richards, I.A.: The Meaning of Meaning. Harcourt: Brace & World Inc., New York (1923). http://courses.media.mit.edu/2004spring/mas966/Ogden%20Richards%201923.pdf
38. Piatetsky-Shapiro, G., Frawley, W.: Knowledge Discovery in Databases. AAAI Press, Menlo Park (1991)
39. Sandelowski, M., Barroso, J.: Classifying the findings in qualitative studies. Qual. Health Res. **13**(7), 905–923 (2003)
40. Schreier, M.: Qualitative Content Analysis in Practice. SAGE, London (2012)
41. Tashakkori, A., Teddlie, Ch.: Mixed Methodology: Combining Qualitative and Quantitative Approaches. SAGE, Thousand Oaks (1998)
42. Tomanek, K., Bryda, G.: Odkrywanie wiedzy w wypowiedziach tekstowych. Metoda budowy słownika klasyfikacyjnego. In: Niedbalski J, Metody i techniki odkrywania wiedzy. Narzędzia CAQDAS w procesie analizy danych jakościowych. Łódź: Wydawnictwo Uniwersyetu Łódzkiego, s. 219–248 (2014)
43. Weber, R.P.: Basic Content Analysis. SAGE, Newbury Park (1990)
44. Wiedemann, G.: Opening up to big data: computer-assisted analysis of textual data in social sciences. Forum Qual. Sozialforschung **14**(2) (2013)
45. Xiaowei, Y.: Ontologies and how to build them (2004). https://users.cs.duke.edu/~xwy/publications/ontologies.pdfg

Automatic Content Analysis of Social Media Short Texts: Scoping Review of Methods and Tools

Judita Kasperiuniene[1]([envelope]) [ORCID], Monika Briediene[1,2],
and Vilma Zydziunaite[1]

[1] Vytautas Magnus University, Kaunas, Lithuania
{judita.kasperiuniene,monika.briediene,
vilma.zydziunaite}@vdu.lt
[2] Baltic Institute of Advanced Technologies, Vilnius, Lithuania

Abstract. Content analysis is a widely applied method, applicable to qualitative and quantitative data. In content analysis, computer programs could be used not only for manual typing of codes and categories but for automatic screening of texts, software-assisted identifying and coding of words, phrases, paragraphs or events. Methods for analyzing social media textual data are usually associated with computer science, social media and communication scholar articles. These empirical sources focus on optimizing the goals of computer sciences and mostly evaluate the percentage of documents, phrases or words correctly categorized into a set of themes. Our study aimed to scope the nature and extent of empirical research articles and online materials around automatic content analysis not limited to computer science, media, and communication. This article analyzed and systematized the empirical articles of the last five years examining the categories of automatic content analysis and the related research questions focusing on social sciences and provided practical examples of methods, tools and empirical applications. Research gaps were identified and recommendations for the application of automated content analysis are provided.

Keywords: Automatic content analysis · Machine learning ·
Natural language processing · Social media · Online texts · Software tools ·
Text classification · Text segmentation

1 Introduction

In the continuously growing methodological pool of modern social research, the content analysis remains widely applied method [24] and potentially one of the most important research techniques [35, 41] applicable to qualitative and quantitative data. With this method, verbal, written or visual information could be researched - systematically examined, studied and grouped. Today, when social media is gaining increasing importance, content analysis can be applied to examine networked short texts. Many scientists, who apply content analysis, manually read and analyze texts. Initial manual analysis of social media texts "eats" researcher's time. Rarely are scholars able to manually read all the social media short texts, and such type of studies

A. P. Costa et al. (Eds.): WCQR 2019, AISC 1068, pp. 89–101, 2020.
https://doi.org/10.1007/978-3-030-31787-4_7

become very human resource consuming. Automated content analysis can be used to save scholars' time, providing a wide range of tools to measure diverse quantities of interest. In this article, we supported [21] stating, that automatic content analysis could be a solution, although these methods never replace carefully and close manual reading and analyzing of online texts.

Computer programs could be applied at two phases of content analysis. Firstly, for storing, analyzing and reporting research data, and secondly – for automatic screening of texts, identifying and coding of words and phrases [39]. A six-step algorithmic solution could be used: (i) separating texts to segments; (ii) calculating similarities, (iii) clustering; (iv) cluster labeling; (v) analyzing facets (categories); (vi) mapping results [10]. Automatic classification could be supervised or unsupervised. The researcher needs to adapt and prepare online short texts for automatic content analysis. This preparation consists of extracting data from social media, networks or virtual communication channels; preprocessing and setting research objectives. In a preprocessing phase, the researcher prepares raw data for future processing procedures. In many cases, two different research objectives for automatic text analysis are set. First – to classify given data to known and unknown categories (e.g. [2, 33]) and second – to scale the data (e.g. [7, 25, 53]) to *wordscores* (dimensional information) and *wordfishes* (positional information) with supervised or unsupervised learning. Wordscores is a widely used procedure to extract scores for new text parts or segments on the basis of scores for words derived from documents with known scores (e.g. [9, 19, 38]), and wordfishes is a statistics algorithm that determines the position of a text part or segment repetition in a document [32]. If the categories of classification are known, dictionary-based or supervised methods could be applicable [23]. Using supervised methods, it is possible to make an individual classification of data or measure proportions classifying texts to individual data or ensembles [34]. If the categories are unknown, fully automated or computer-assisted clustering give a better probability. With these methods, single or mixed membership models could be applied in document, data or author level [47]. Although different methods of automated text analysis exist, anyone method's performance is context-specific [21].

The aim of this scoping review is to identify the nature and extent of methods and tools for the automatic content analysis of social media short texts. While focusing on qualitative content analysis, the following research questions were addressed: *(i) how the automatic content analysis tools are used for the empirical analysis of social media short texts? (ii) what is the scope and categories of tools and technologies used in empirical automatic content analysis research?*

2   Methodology

Scoping review is a method to comprehensively synthesize evidence across a range of study designs [45]. In this research, scoping review methodology [4, 30] was applied for data collection from empirical articles and web portals, systematically mapping the scientific literature and newest knowledge available on the topic; identifying automatic content analysis categories and popular techniques, sources of empirical evidence and gaps in scholar research. Articles included using these criteria: (i) only empirical open accessed articles (2015–2019) were researched; (ii) articles were freely accessible from Sage, Springer, and Emerald portfolios; (iii) the methods and/or tools on automatic content analysis were researched in these articles; (iv) articles in which social media short texts were used for empirical data analysis were studied. Additionally, web portals, which provided info not accessible from researched articles, were included.

The literature and web portal search process involved two stages. First, a broad systematic search using keywords was conducted in *Web of Science, Scopus, Emerging Sources Citations Index,* and *Google Scholar* databases. We selected these databases because they are the most commonly cited for social sciences, technologies and multidisciplinary research. Also, a hand-searching of key journals, existing networks, and conference sites was done. Reflecting time and technology development constraints, we included scholar articles only of the last five years. We searched for the latest information from the online portals, but in some cases, it was not possible to recognize the date of online text appearance. We read the abstracts evaluating how much they respond to our research questions. This stage resulted in charting the key items of information obtained from the primary search process. We recorded information as follows – authors; year of publication; aims of the study; study field; methods and techniques used for automated text preprocessing and analysis; type of data; probabilities and errors, resulting from data analysis and conclusions (their suitability for analogous research) or important results. Second, the literature was organized thematically, and associated automatic content analysis research questions formulated.

To inform the search strategy, we developed an inclusive list of keyword combinations related to our topic. Keywords *"automatic content analysis"*, *"automatic classification of texts"*, *"automatic text segmentation"*, *"automatic text clustering"*, *"automatic text retrieval"*, *"social media short texts"* were used. The exploration combined a logical operator AND among the keyword inside, to make the search more accurate and the results more relevant to the criteria. To ensure intercoder reliability, the first researcher developed a representative set of units to test the reliability. The coding was done by the first and second researcher separately. Coding decisions were made independently by each researcher. Quality control was done manually.

The number of articles researched during the first stage of analysis exceeded a few tens of thousands. The article search was done in *Sage, Emerald* and *Springer* portfolios. Firstly, the search was done in keywords, but the further selection of relevant articles was based on the titles and abstracts. Then, over 300 article abstracts were manually reviewed. The abstracts raised different research questions, solved different tasks, analyzed different social media spaces, indicated different programming languages used. Distribution of article abstracts among institutions and countries was very diverse. The geographic scope of social media data analysis in these articles was not investigated. 106 articles were selected as fulfilling the search criteria. The relevance of the Sage to our research area was 0.60%, Emerald – 0.25%, and Springer – 0.73%. The most relevant keywords were: "automatic text segmentation", "automatic classification of texts" and "social media short texts". Additionally, we identified research gaps and draw recommendations for the application of automated content analysis.

3 Findings

Social media data is clearly the largest, richest and most dynamic evidence base of human behavior, bringing new opportunities to understand individuals, groups and society (e.g. [5, 31]). Working with text data from social networks becomes difficult due to the following indications: interactivity, decentralized and hyperlinked structures, non-normative language, foreign language insertions, markings, emoticons, symbols and many more. Due to a large amount of data, it is very difficult to process it manually. Therefore, analysis work requires technologies that can help optimize the process. While online data is available in text, graphics, animation, video, and audio formats, we only targeted social media short text analysis. In this scoping review, we started from *natural language processing* methods and techniques but we're not limiting ourselves only with these fields.

3.1 Methods and Tools for Short Text Retrieval and Classification

Technically, text classification could be described as the automatic process of assigning codes and categories to interview transcriptions or social media text data. It's one of the fundamental tasks of computer-assisted *natural language processing* (NLP). Text classification is crucial for *automatic content analysis* (ACA) as well as text retrieval, knowledge management, and decision making as it converts text from raw data to a real knowledge [18]. For text classification, a common expression of a task is used:

$$Y = f(X, \theta) + \epsilon$$

In the preceding formula, X is a suitably chosen text representation (e.g., a vector), θ is the set of unknown parameters associated with the function f (also known as the classifier or classification model) that need to be estimated using the training data, and ϵ is the error of the classification [33]. The short text has been the prevalent format for networked information, especially with the development of online social media [57]. Although sophisticated signals delivered by the short texts make it a promising source for knowledge retrieval, the automatic content analysis is not easily organized. Usually, short texts are full of noise, has no metadata, therefore, prior to performing automated processes, such data must be further processed, and methods applied. *Machine learning* tasks must deal with a very high dimensional and sparse feature space, in which most features have low frequencies [52]. Performing natural language processing, it turns out to be clear that sentiments or opinions from social media provide the most up-to-date and inclusive information [56], therefore *sentiments analysis* and *opinion mining* become extremely important. Two ideal types may be distinguished in automatic content analysis. *First* is the general, 'a-priori' type, which relates to the automatic coding based on a categorization scheme (dictionary) constructed by the analyst according to the theoretical framework. *Second*, the empirical attempt concerns the automatic (inductive) extraction of categories which are indicative of topics or themes dealt within a specific body of text data, and which may then comprise the scheme categories to be used for coding [41]. In the past, content analysis was mostly conducted manually, with investigators interpreting a text by classification, categorization, and subjective interpretation. Recently, the automatic content analysis could be developed using different methodological approaches. These methods vary from deductive "top-down" techniques such as visibility analysis, sentiment analysis, subjectivity analysis and string comparisons counting to inductive "bottom-up" approaches such as frames, topics, principal component analysis, cluster analysis, semantic network analysis [8]. The most widely used methods for ACA in empirical studies are provided in Table 1.

Table 1. Studies on popular ACA methods and tools for text retrieval and classification.

References	Topic	Tool(s)	Method(s)	Language	NLP usage
Huang [25]	Wordfishing for unsupervised learning of textual data	R	Wordfish	N/A	Yes
Elkink [19]	Wordscoring and wordfishing for social science	–	Wordscore, Wordfish	N/A	Yes
Kobayashi et al. [33]	Text classification for organizational research	R	Random forest, Support vector machines, Naïve Bayes	EN	Yes
Batrinca and Treleaven [5]	Techniques, tools, and platforms for social media analytics	APIs, Gnip, DataSift, Google Refine, R, Matlab, Python, RapidMiner	MapReduce, Computational statistics, Machine learning, Naïve Bayes	N/A	Yes
Allahyari et al. [2]	Text mining techniques for classification, clustering and text extraction	BOW toolkit, Mallet, WEKA	VectorSpace Mode, Naïve Bayes, Nearest neighbor classifier, Decision tree, Support Vector Machines, k-means clustering, Hidden Markov model	N/A	Yes
Crossley, Kyle and McNamara [13]	Automatic tools for sentiment, social cognition, and social-order analysis	SÉANCE, LIWC	MANOVA, DFA	EN	Yes
Ding and Pan [16]	Sentiment analysis tools for opinion mining	Text Processing, Semantria, SANN	Naïve Bayes	EN	Yes
Kaefer, Roper and Sinha [29]	Qualitative content analysis of newspaper articles	NVivo, QDAS, CAQDAS, MaxQDA, Dedoose	–	N/A	No
Trilling and Jonkman [53]	Scaling up ACA	R, Python, AmCAT	Principal Component Analyses	N/A	Yes

With lexical and semantic software and statistical tools, qualitative information could be extracted identifying underlying key attributes, factors, and themes. Each method has specific strengths and/or weaknesses, depending on research analysis goals. Choosing the right approach is important in order to maximize efficiency and the relevance value of the insights (e.g. [2, 26, 29]). All the researched papers testified to the need to process the text first for proper text analysis. Therefore, various text processing methods were used: grammar induction of *Instagram* texts [15], lemmatization [42], morphological and word segmentation [4], part-of-speech tagging [20], parsing of *Twitter* or *Instagram* texts (e.g. [6, 48, 49]), sentence breaking in advertisements, coming from *Facebook* [37], and stemming non-normal language texts with slang [40]. When processing social media short texts, one or more of the methods used in their clustering were chosen (see column *Method(s)* in Table 1). The choice of method depended not only on the length of the text characters and the additional characters represented emotions but also on the language of empirical texts. However, publications do not emphasize a single language survey, often using tools or methods that work with data in any language (in Table 1, *Language* marked as N/A). Research equipment was categorized into programming languages and tools with graphical user interfaces (when programming knowledge is not required) (see [2]). Tools were divided into paid and free, created by researchers or developed by commercial enterprises (see [5, 33]). The choice of the tool was dependent on the user or scholar's skills and field of investigation. Additionally, scholars stressed the importance to consider the available sample and the type of data [33]. The choice of method for automatic content analysis depended on the selected tool and the type of data. Moreover, the use of the method was of great importance for the use of the available computer memory. Due to the lack of resources, the researchers were forced to choose not so complex methods (see [5]).

3.2   Stages and Features of Automated Content Analysis

The fundamental process of ACA can be divided into three stages: concept identification, concept definition and text classification [43]. In the previous section, we described tools and methods focusing on social media short text retrieval and classification. Now we go deeper, explaining features of ACA methodologies to identify, collect, preprocess, and analyze networked texts (Table 2).

Table 2. Categories of automatic content analysis identified of scoping review (Researched articles N = 106).

Identified categories	Articles		Associated research questions
	No.	%	
Automatic scanning	17	16	How to identify text?
Automatic classifying	26	24	
Automatic text segmenting	18	17	
Automatic clustering	13	12	

(*continued*)

Table 2. (*continued*)

Identified categories	Articles		Associated research
	No.	%	questions
Automatic text retrieving	12	11	**How to collect text?**
Gathering networked data (texts from *Facebook*, *Instagram*, *Twitter*, etc.)	32	**30**	
Text extracting	17	16	
Natural language processing	65	**61**	**How to preprocess texts for analysis?**
Machine learning (including Deep learning)	55	**51**	**How to analyze text?**
Working with special text-analysis tools	26	24	
Analyzing non-regular language texts	28	26	How to research specific language texts?

Three of the most frequently asked questions in empirical ACA studies were related to *preprocessing* the networked texts (61% of researched cases); *analyzing* social media short texts (51% of researched cases) and *collecting* texts (30% of research cases). The topics of text identification or studies of specific language texts were less analyzed. Scientists are investigating the problem of gathering online text data. In most researched cases, three stages of social media short text retrieval were applied: *tokenization* (generation of respective tokens from given documents), *stemming* (removing prefixes and suffixes from the words, thereby seems to be imperative in the information retrieval process) and *removing stop words* (e.g. [46]). Natural language processing is the application of computer-based and mathematical methods for analyzing social media short texts. NLP lies in the fusion of artificial intelligence, linguistics, mathematics, and computer science (e.g. [51]) and is studied in more than 60% of researched articles. Text processing for analysis usually starts after collecting data and is done in these steps: (i) punctuating (choosing the classes of characters for further analysis and removing non-informative symbols); (ii) choosing important domains of information; (iii) lowercasing letters; (iv) stemming or reducing the word to its basic form; (v) removing stop words; (vi) removing frequently used terms (e.g. [14]).

Authors often come across with shorts and full noises texts. Almeida et al. [3] are dealing with data with normally rife of slangs, idioms, symbols, and acronyms that make even tokenization a difficult task. Scholars apply methods that normalize terms and create new attributes in order to change and expand original text samples aiming to alleviate factors that can degrade the performance of the algorithm, such as redundancies and inconsistencies [3].

While pre-processing texts of Twitter data, [28] replaced negative text mentions; removed URL links; reverted words that contained repeated letters to their original language form; removed numbers; removed stop words and expanded acronyms and slang to their original words by using an acronym dictionary. The graphical visualization of the text normalization process is provided and empirically tested by [50]. It is worth mentioning that the quality of the data is also important. In [11] study, the best results and follow-up were performed only after a specific data set was collected. In this case, qualitative and quantitative research comes into play. Machine learning use algorithms and mathematical models in order to perform specific tasks and instructions. These methods could be applicable to social media short text analysis. Scoping review showed that more than 50% or researched articles answered the question, how the automated content analysis could be done. Conroy, Rubin and Chen [12] used machine learning techniques (linguistic cue approaches) for fake online news detection and proposed operational guidelines for a feasible fake news detecting system. Yousefi-Azar and Hamey [55] presented a query-based single-document summarization scheme with an unsupervised deep neural network and deep auto-encoder to learn features.

Ali, Kwak and Kim [1] obtained results with mixing several machine learning approaches, as *supper vector machine (SVM)* and *fuzzy domain ontology (FDO)*. The proposed system retrieves a collection of reviews about hotel and hotel features. The SVM identifies hotel feature reviews and filter out irrelevant reviews (noises) and the FDO is then used to compute the polarity term of each feature [1]. Other machine learning models such as *artificial neural networks* for spam detection in social media (e.g. [27]), *decision trees* for semantic text analysis and spam content detection in online instant messaging environments (e.g. [3]), *support vector machines* for classification of online reviews (e.g. [1]), automatic detection of cyberbullying (e.g. [54]), detection of depressive moods (e.g. [22]) or emotional distress (e.g. [11]) from *Twitter* texts and *Bayesian networks* for opinion mining (e.g. [1]), predicting and benchmarking trolling and aggressive networked behavior (e.g. [36]) were also explained in researched articles. While identifying the nature and extent of the ACA methods and tools, we have found that the associated research questions, challenges that scientists face and the problems they solve, overlap (Fig. 1).

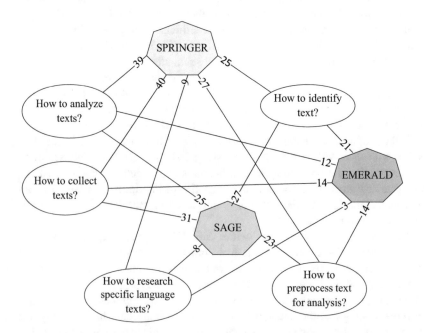

Fig. 1. Concept map representing the distribution of scoping review of publications by sources, associated research questions (ellipses) and scientific publishers (heptagons).

The largest variety of scientific empirical articles describing the use of ACA and its application was found in the scholar journals of the *Sage* and *Emerald* publishing houses. The topics of identifying, collecting, preprocessing, analyzing social media data and working on non-normal language short texts were covered. Although, Sage provided more articles on this topic. In *Springer*, the topics of identifying, collecting, and analyzing social media short texts were covered in detail. By providing this information, we want to show the prevailing trends and the topics of scientific journals that are continuously changing.

4 Discussion

The scoping review showed that automatic content analysis becomes more and more popular. However, in many empirical cases studied, ACA is perceived as a mathematical or computer algorithm, or set of procedures that allowed conditionally easy extraction and handling of text. Scientists solved the tasks of text classification, segmentation, and clustering. The researches argued how to identify different type of texts (e.g. [25, 57]) and what mathematical tools and methods could be applicable in a specific situation while categorizing, predicting, and exploring social media short texts (see Table 1). In qualitative research, this discussion is in line with the solving of the operationalization task – defining categories and units of measurement or accessing the face and content validity [31, 41]. While developing a content analysis research, it is recommended to create a codebook containing the categories and determine sample [21]. In ACA, this step is replaced with text retrieving, extracting and pre-processing procedures [2, 5]. Data coding and developing themes and categories are understood as text analysis in ACA [44].

Limitations. Apart from the fact that automated content analysis is an innovative and rapidly developing method, in this scoping review of methods and tools, we found some limitations. The key limitations of ACA applications were: (i) ACA works best when social media short text data are trained and applied to a specific situation or context. In other mediated contexts (even similar ones), the reliability and validity of automatic content analysis are questionable. For data validity, machine learning must be applied to a potentially larger sample of online texts. Additionally, a researcher must review and revise the findings of ACA; (ii) NLP tools require clear and consistent definitions of the language to be defined. In many cases, automatic content analysis is subjective because of the moderation of networked content. The pre-processing of networked texts is necessary; (iii) the relatively low accuracy and reliability of the intermediate links in natural language processing studies warn that tools are widely applied to the moderation of social media content; (iv) even the latest NLP tools do not equate to the researcher's ability to analyze texts and extract codes, categories and their meaning [17].

5 Conclusions

The scoping review showed that ACA has a great potential to fill an important methodological gap in developing the content analysis of social media texts. This technique provides the methods and tools for automatic text extraction from mediated environments, scanning, classification, segmentation, clustering, and analysis. Natural language processing techniques could be used for preprocessing of texts. Machine learning is popular for social media text analysis. Automatic content analysis methods could be used by researchers who do not have programming skills.

This article aims to explain the variety of ACA methods and tools. Scoping review showed that in data gathering and analysis of big amounts of social media short texts automatic methods could be faster than manual work. In many cases, networked texts must be rearranged before qualitative analysis and in these situations, automatic pre-processing techniques help. Although, automated coding and categorizing is not that objective and valid as a researcher itself. Additionally, ACA is context sensitive.

In this paper, we have discussed challenges related to automated content analysis. Following critical realism, we have positioned automated content analysis as an interface between computer-assisted abstraction and man-made interpretation. Furthermore, on a big data sample, automatic content analysis is an irreplaceable qualitative research technique.

Acknowledgments. This research was supported by the European Social Fund according to the activity "Development of Competencies of Scientists, other Researchers, and Students through Practical Research Activities" of Measure No. 09.3.3-LMT-K-712 (Project No. 09.3.3-LMT-K-712-02-0079).

References

1. Ali, F., Kwak, K.S., Kim, Y.G.: Opinion mining based on fuzzy domain ontology and Support Vector Machine: a proposal to automate online review classification. Appl. Soft Comput. **47**, 235–250 (2016)
2. Allahyari, M., Pouriyeh, S., Assefi, M., Safaei, S., Trippe, E.D., Gutierrez, J.B., Kochut, K.: A brief survey of text mining: classification, clustering and extraction techniques. arXiv preprint arXiv:1707.02919 (2017)
3. Almeida, T.A., Silva, T.P., Santos, I., Hidalgo, J.M.G.: Text normalization and semantic indexing to enhance instant messaging and SMS spam filtering. Knowl.-Based Syst. **108**, 25–32 (2016)
4. Arksey, H., O'Malley, L.: Scoping studies: towards a methodological framework. Int. J. Soc. Res. Methodol. **8**(1), 19–32 (2005)
5. Batrinca, B., Treleaven, P.C.: Social media analytics: a survey of techniques, tools and platforms. AI Soc. **30**(1), 89–116 (2015)
6. Bharti, S.K., Babu, K.S., Jena, S.K.: Parsing-based sarcasm sentiment recognition in Twitter data. In: Proceedings of the 2015 IEEE/ACM International Conference on Advances in Social Networks Analysis and Mining, Paris, pp. 1373–1380. IEEE (2015)
7. Borg, I., Groenen, P.: Modern multidimensional scaling: theory and applications. J. Educ. Meas. **40**(3), 277–280 (2003)

8. Boumans, J.W., Trilling, D.: Taking stock of the toolkit: an overview of relevant automated content analysis approaches and techniques for digital journalism scholars. Digit. J. **4**(1), 8–23 (2016)

9. Bruinsma, B., Gemenis, K.: Validating Wordscores: the promises and pitfalls of computational text scaling. Commun. Methods Measures **13**(3), 212–227 (2019)

10. Chang, Y.H., Chang, C.Y., Tseng, Y.H.: Trends of science education research: an automatic content analysis. J. Sci. Educ. Technol. **19**(4), 315–331 (2010)

11. Cheng, Q., Li, T.M., Kwok, C.L., Zhu, T., Yip, P.S.: Assessing suicide risk and emotional distress in Chinese social media: a text mining and machine learning study. J. Med. Internet Res. **19**(7), e243 (2017)

12. Conroy, N.J., Rubin, V.L., Chen, Y.: Automatic deception detection: methods for finding fake news. Proc. Assoc. Inf. Sci. Technol. **52**(1), 1–4 (2015)

13. Crossley, S.A., Kyle, K., McNamara, D.S.: Sentiment Analysis and Social Cognition Engine (SEANCE): an automatic tool for sentiment, social cognition, and social-order analysis. Behav. Res. Methods **49**(3), 803–821 (2017)

14. Denny, M.J., Spirling, A.: Text preprocessing for unsupervised learning: why it matters, when it misleads, and what to do about it. Polit. Anal. **26**(2), 168–189 (2018)

15. Desai, S., Han, M.: Social media content analytics beyond the text: a case study of university branding in Instagram. In: Proceedings of the 2019 ACM Southeast Conference, pp. 94–101. ACM (2019)

16. Ding, T., Pan, S.: An empirical study of the effectiveness of using sentiment analysis tools for opinion mining. In: Proceedings of the 12th International Conference on Web Information Systems and Technologies, Rome, vol. 2, pp. 53–62. SciTePress (2016)

17. Duarte, N., Llanso, E., Loup, A.: Mixed messages? The limits of automated social media content analysis. https://cdt.org/files/2017/11/Mixed-Messages-Paper.pdf. Accessed 11 July 2019

18. ElGhazaly, T.: Automatic text classification using neural network and statistical approaches. In: Shaalan, K., Hassanien, A., Tolba, F. (eds.) Intelligent Natural Language Processing: Trends and Applications, vol. 740, pp. 351–369. Springer, Cham (2018)

19. Elkink, J.A.: Data analytics for social science wordscores & wordfish. http://www.joselkink.net/files/POL30430_Spring_2017_11_wordscores.pdf. Accessed 11 July 2019

20. Ghosh, S., Ghosh, S., Das, D.: Part-of-speech tagging of code-mixed social media text. In: Proceedings of the Second Workshop on Computational Approaches to Code Switching. Association for Computational Linguistics, Austin, Texas, pp. 90–97 (2016)

21. Grimmer, J., Stewart, B.M.: Text as data: the promise and pitfalls of automatic content analysis methods for political texts. Polit. Anal. **21**(3), 267–297 (2013)

22. Guntuku, S.C., Yaden, D.B., Kern, M.L., Ungar, L.H., Eichstaedt, J.C.: Detecting depression and mental illness on social media: an integrative review. Curr. Opin. Behav. Sci. **18**, 43–49 (2017)

23. Guo, L., Vargo, C.J., Pan, Z., Ding, W., Ishwar, P.: Big social data analytics in journalism and mass communication: comparing dictionary-based text analysis and unsupervised topic modeling. J. Mass Commun. Q. **93**(2), 332–359 (2016)

24. Hsieh, H.F., Shannon, S.E.: Three approaches to qualitative content analysis. Qual. Health Res. **15**(9), 1277–1288 (2005)

25. Huang, L.: Use wordfish for ideological scaling: unsupervised learning of textual data part I. https://sites.temple.edu/tudsc/2017/11/09/use-wordfish-for-ideological-scaling/. Accessed 11 July 2019

26. Huddy, G.: How text analytics works for social media. https://www.brandwatch.com/blog/social-media-text-analytics/. Accessed 11 July 2019

27. Jain, G., Sharma, M., Agarwal, B.: Spam detection in social media using convolutional and long short-term memory neural network. Ann. Math. Artif. Intell. **85**(1), 21–44 (2019)
28. Jianqiang, Z., Xiaolin, G.: Comparison research on text pre-processing methods on Twitter sentiment analysis. IEEE Access **5**, 2870–2879 (2017)
29. Kaefer, F., Roper, J., Sinha, P.: A software-assisted qualitative content analysis of news articles: example and reflections. Forum Qual. Soc. Res. **16**(2), 1–20 (2015)
30. Kastner, M., Tricco, A.C., Soobiah, C., Lillie, E., Perrier, L., Horsley, T., Welch, V., Cogo, E., Antony, J., Straus, S.E.: What is the most appropriate knowledge synthesis method to conduct a review? Protocol for a scoping review. BMC Med. Res. Methodol. **12**(1), 1–10 (2012)
31. Kim, I., Kuljis, J.: Applying content analysis to web-based content. J. Comput. Inf. Technol. **18**(4), 369–375 (2010)
32. Kluver, H.: The promises of quantitative text analysis in interest group research: a reply to Bunea and Ibenskas. Eur. Union Polit. **16**(3), 456–466 (2015)
33. Kobayashi, V.B., Mol, S.T., Berkers, H.A., Kismihók, G., Den Hartog, D.N.: Text classification for organizational researchers: a tutorial. Organ. research Methods **21**(3), 766–799 (2018)
34. Kotzias, D., Denil, M., De Freitas, N., Smyth, P.: From group to individual labels using deep features. In: Proceedings of the 21st ACM SIGKDD International Conference on Knowledge Discovery and Data Mining, Sydney, pp. 597–606. ACM (2015)
35. Krippendorff, K.: Content Analysis: An Introduction to Its Methodology. Sage, California (2018)
36. Kumar, R., Ojha, A.K., Malmasi, S., Zampieri, M.: Benchmarking aggression identification in social media. In: Proceedings of the First Workshop on Trolling, Aggression and Cyberbullying, Santa Fe, pp. 1–11. ACL (2018)
37. Lee, D., Hosanagar, K., Nair, H.S.: Advertising content and consumer engagement on social media: evidence from Facebook. Manage. Sci. **64**(11), 5105–5131 (2018)
38. Lowe, W.: Understanding wordscores. Polit. Anal. **16**(4), 356–371 (2008)
39. Macnamara, J.R.: Media content analysis: its uses, benefits and best practice methodology. Asia Pac. Public Relat. J. **6**(1), 1–34 (2005)
40. Maylawati, D.S.A., Zulfikar, W.B., Slamet, C., Ramdhani, M.A., Gerhana, Y.A.: An improved of stemming algorithm for mining indonesian text with slang on social media. In: 6th International Conference on Cyber and IT Service Management, Parapat, Indonesia, pp. 1–6. IEEE (2018)
41. Neuendorf, K.A.: The Content Analysis Guidebook, 2nd edn. Sage, London (2016)
42. Nguyen, T.H., Shirai, K., Velcin, J.: Sentiment analysis on social media for stock movement prediction. Expert Syst. Appl. **42**(24), 9603–9611 (2015)
43. Nunez-Mir, G.C.: How to synthesize 100 articles in under 10 minutes: reviewing big literature using ACA. https://methodsblog.com/2017/01/12/big-literature-aca/. Accessed 11 July 2019
44. Nunez-Mir, G.C., Iannone III, B.V., Pijanowski, B.C., Kong, N., Fei, S.: Automated content analysis: addressing the big literature challenge in ecology and evolution. Methods Ecol. Evol. **7**(11), 1262–1272 (2016)
45. O'Brien, K.K., Colquhoun, H., Levac, D., Baxter, L., Tricco, A.C., Straus, S., O'Malley, L.: Advancing scoping study methodology: a web-based survey and consultation of perceptions on terminology, definition and methodological steps. BMC Health Serv. Res. **16**(1), 1–12 (2016)
46. Panda, M.: Developing an efficient text pre-processing method with sparse generative Naive Bayes for text mining. Int. J. Mod. Educ. Comput. Sci. **10**(9), 11–19 (2018)

47. Roiger, R.J.: Data Mining: a Tutorial-Based Primer, 2nd edn. CRC Press, Boca Raton (2017)
48. Sanguinetti, M., Bosco, C., Mazzei, A., Lavelli, A., Tamburini, F.: Annotating Italian social media texts in universal dependencies. In: Fourth International Conference on Dependency Linguistics, pp. 229–239. Linköping University Electronic Press, Linköping (2017)
49. Sharma, A., Gupta, S., Motlani, R., Bansal, P., Srivastava, M., Mamidi, R., Sharma, D.M.: Shallow parsing pipeline for Hindi-English code-mixed social media text. arXiv preprint arXiv:1604.03136, pp. 1–6 (2016)
50. Singh, T., Kumari, M.: Role of text pre-processing in Twitter sentiment analysis. Procedia Comput. Sci. **89**, 549–554 (2016)
51. Tixier, A.J.P., Hallowell, M.R., Rajagopalan, B., Bowman, D.: Automated content analysis for construction safety: a natural language processing system to extract precursors and outcomes from unstructured injury reports. Autom. Constr. **62**, 45–56 (2016)
52. Tommasel, A., Godoy, D.: Short-text feature construction and selection in social media data: a survey. Artif. Intell. Rev. **49**(3), 301–338 (2018)
53. Trilling, D., Jonkman, J.G.: Scaling up content analysis. Commun. Methods Measures **12**(2–3), 158–174 (2018)
54. Van Hee, C., Jacobs, G., Emmery, C., Desmet, B., Lefever, E., Verhoeven, B., Hoste, V.: Automatic detection of cyberbullying in social media text. PLoS ONE **13**(10), 1–22 (2018)
55. Yousefi-Azar, M., Hamey, L.: Text summarization using unsupervised deep learning. Expert Syst. Appl. **68**, 93–105 (2017)
56. Yue, L., Chen, W., Li, X., Zuo, W., Yin, M.: A survey of sentiment analysis in social media. Knowl. Inf. Syst. **60**(2), 617–663 (2019)
57. Zuo, Y., Zhao, J., Xu, K.: Word network topic model: a simple but general solution for short and imbalanced texts. Knowl. Inf. Syst. **48**(2), 379–398 (2015)

Biomechanical Analysis of Nurses Students of Midwifery in Vertical Deliveries

Mário Cardoso[1] , Maria Helena Presado[1] ,
Armando David Sousa[1,2] , Ana Leonor Mineiro[1,3] ,
Fátima Mendes Marques[1] , Cristina Lavareda Baixinho[1(✉)] ,
and Luís Miguel Moreira Pinto[1,4]

[1] Nursing Research & Development Unit,
Nursing School of Lisbon, Lisbon, Portugal
{mhpresado, crbaixinho}@esel.pt
[2] Hospital Center of Funchal, Madeira, Portugal
[3] Hospital Garcia de Orta, Almada, Portugal
[4] University Beira Interior, Covilhã, Portugal

Abstract. The musculoskeletal injury related to work is a health problem of nurse-midwives, due to the specificity of the activities uncovered in the birthing block. The reinforcement of the importance of the adoption of the principles of biomechanics at the time of delivery is a fundamental aspect to improve the quality of life of nurse-midwives. This work aims to analyse the effect of inappropriate equipment on biomechanical of nurses students of midwifery in vertical deliveries during simulated practice. A qualitative study was performed using videograms of the simulated clinical practices of two types of vertical deliveries (parturient in the seated position and "four supports"), which were later analyzed using webQDA® software. Results demonstrated difficulty in the length of the principles of biomechanics when faced with inappropriate equipment, different approaches were adopted and adapted for the same type of delivery, allowing the analysis and evaluation of the same.

Keywords: Nurse-midwife · Musculoskeletal injuries · Biomechanics · Simulation

1 Introduction

Nursing has an extensive history of using simulation as a teaching/learning strategy, which is recognized as an essential method in transferring skills from the simulated context to the reality of clinical practice [1–3].

The realism of a simulated clinical experience reaches its maximum exponent when associated with a set of materials and equipment that recreate an environment similar to the one of the clinical practice. If we associate the environment with robotic technology and computer science, we get a hi-fidelity simulation, which enables the re-creation of scenarios as close to the real as possible. The simulated high-fidelity practice (SHFP) is a practical training methodology, enabling learning, renewal of skills and professional skills, promoting safety, education, innovation, research, quality and confidence, reducing the risk for the users [4].

© Springer Nature Switzerland AG 2020
A. P. Costa et al. (Eds.): WCQR 2019, AISC 1068, pp. 102–113, 2020.
https://doi.org/10.1007/978-3-030-31787-4_8

The SHFP provides the ability, in a controlled, repeatable and readable environment, to reproduce and amplify in a fully interactive and real-time manner, gestures, procedures, postures and clinical acts in a variety of situations [5]. Being the epicentre of learning, SHFP promotes, through a participatory and interactive environment, the acquisition, extension and deepening of clinical and technical knowledge (technical and non-technical), critical and reflexive thinking and decision-making [6–8].

Through SHFP, it is possible to obtain objective data on the performance of students/professionals, insofar as it increases the awareness of individual difficulties and perceives the positive and lesser aspects, thus encouraging students to take an interventional attitude in their learning. Conclusively, students show satisfaction and motivation with the SHFP, in that they can objectively perceive their evolution, being more aware of their real abilities [3, 9, 10].

The specificity of nurse-midwives activity leads to the appearance and aggravation of the musculoskeletal injuries related to work (MIRW), which constitute the major occupational health problem in nurses [11–14]. The complexity of their activities, in block of births, involves the adoption of postures with body misalignment and consequent postural instability; the rapid movements in situations of stress, require the transition from a static position to a dynamic position, with the application of forces, often overloading, that exceed their capacities [13, 15–21].

This fact is particularly true in assisting women in childbirth in an upright position. Nurse-midwives perceive delivery during upright delivery, as presenting an increased risk for MIRW, for having to adopt uncomfortable postures (e.g. squatting, sitting on the floor or a bench with pronounced bending of the trunk) for a long time, thus implying the adoption of non-ergonomic postures [15, 18, 22, 23].

The assistance to the woman during birth in an upright position presents numerous biomechanical requirements and if we combine it with the lack of adapted equipment, increase the difficulty in complying with the principles of biomechanics increases. This often causes inadequate postures, evidenced: by the majority adoption of static positions; postures that require unnatural alignments of the body, exposing it to maximum ranges of movement, with greater overload in the cervical and lumbar regions; changes in the base of support, with a base of support little broad, contributing to the decrease of the balance; performing repetitive movements; misuse of the levers to force [13, 14, 20].

There are several risk factors that hinder the adoption of the principles of biomechanics by nurse-midwives in the delivery of vertical positions [13, 14, 23] being associated with the physical environment - small delivery rooms and physical spaces with inadequate organizations; the lack of maintenance of the functional equipment, with little ergonomic characteristics; to the parturient - in relation to their position and behaviour; the specificity of the task - complexity and diversity of activities, with different focus of attention simultaneously, requiring frequent biomechanical adjustment; to the characteristics of the professional - anthropometric characteristics, inexperience in performing this type of deliveries. Health professionals should have experience in attending births in vertical positions, generating confidence and safety in the conduct of this type of delivery, often being able to anticipate which positions to adopt in the face of a certain situation, in order to maintain safe practices [21, 24, 25].

As mentioned in the study of Baixinho et al. [18], the intuitive knowledge - based on experience, allows nurse-midwives to anticipate and control the process more effectively and safely. One of the factors that also hamper the adoption of the principles of biomechanics by nurse-midwives is their ignorance about the principles of biomechanics, having no ergonomic training, to the detriment of those who already have some training [12]. Although vertical birth care translates into an exponential risk of MIRW, it is observed that nurse-midwives prioritize the safety and comfort of the parturient, to the detriment of their satisfaction and the adoption of biomechanically safe postures during the delivery in vertical positions [13–15, 18, 21, 26, 27].

We believe that delivering safe and quality care through evidence-based practice is imperative, as well as ensuring the safety, health, and comfort of nurse-midwives [13, 14, 21]. The use of the principles of biomechanics protects the musculoskeletal system, prevents the adoption of incorrect postures and abnormal movements, reduces local mechanical stress in muscles, ligaments and joints, fatigue, errors, accidents and the risk of MIRW [13].

Nurse-midwives demonstrate concerns about MIRW prevention concerning physical environment control, parturient preparation, and the adoption of behaviours and practices that maintain the principles of biomechanics [17, 18]. Although this awareness of MIRW prevention exists, the rapid and unpredictable change in the behaviour of women in the situation, complications resulting from childbirth, environmental conditions and the diversity of activities make it difficult to adopt the principles of biomechanics [13, 15, 17, 18, 28].

In this context, SHFP constitutes an important training methodology for the prevention of MIRW in nurse-midwives, allowing a safe and comfortable practice in their professional performance [17, 19, 21, 29, 30].

Through SHFP, it is possible to work with nurse-midwives, safely and in a controlled environment, the postures they adopt during birth in vertical positions, planning the activity with the organization of the workspace and intervene on the factors that hinder adoption principles of biomechanics, [13]. With SHFP, it is possible to anticipate the problems/consequences that occur in clinical practice and which pose an increased risk of MIRW and thus minimize their impact. The SHFP allows theoretical knowledge to find expression in practice, allowing professionals to raise awareness and qualification for the application of the principles of biomechanics, to be able to apply them consistently and automatically in daily activity. By reducing the biomechanical requirements, the adoption of appropriate postures by the nurse-midwives is promoted, to perform the task with comfort, safety and efficiency [30–33].

It is essential that nurse-midwives reflect on their clinical practice, analysing not only the care they provide but also their biomechanics in the course of their execution. Biomechanical knowledge must be present in clinical practice, leading to a safe, efficient, quality and productive care system for both professionals and users [12].

The present study aims to analyse the effect of inappropriate equipment on biomechanical of nurses students of midwifery in vertical deliveries during simulated practice.

2   Method

We chose a qualitative descriptive and exploratory study, with a qualitative method, using video recording of the students of the Maternal Health and Obstetrics Post-graduate Course to understand how the use of inappropriate equipment at the time of delivery may favour the adoption of incorrect postures, and identify contributions of the simulated practice in the transfer of the biomechanical knowledge to nurse-midwives clinic.

Firstly, the objectives of the SHFP sessions were explained, and their contribution to the study was highlighted. The principles of biomechanics were presented, demonstrating their applicability and considerations in their adoption. Subsequently, the students were distributed in groups and proceeded to the simulated practice and training of skills in assisting women in childbirth in an upright position. For this study, we only analyzed the videos with the parturient in the seated position and four supports, corresponding to seven videos, (three in the four-position position and four in the seated position). The participants in this study were seven female licensed nurses, students Maternal Health and Obstetrics Postgraduate Course who performed deliveries in a simulation context. The images were obtained using a video camera positioned strategically in the corner of the delivery room, with fixed support, performed by one of the researchers, making sure of its operability and the quality of the image. The film record allows us to study complex human actions, challenging to be described in detail by a single observer [34]. The repeated visualization of the images gives more rigour and credibility to the study. The use of videos has been a new reality in the analysis of posture and body movement [35, 36], gestures and facial expression [37], or emotions [38, 39].

Given the complexity of the data processing and analysis process, the six phases for video analysis proposed by Lima [40]: watch videos, select critical events, describe critical events, transcribe them, discuss data with peers, and delete videos [39].

After viewing and selecting the frames (24 frames), the seven researchers analysed it and categorized the images. When obtained the consensus of all the team, we transcribing the videos and respective frames, a task performed with the software webQDA®, they were coded and analysed based on the grid developed by Presado et al. [36]. Moreover, according to the categories following the principles of biomechanics: (1) body movement; (2) body alignment; (3) balance; (4) strength; (5) friction and friction, this grid was still applied in other studies [14, 35]. The assignment of a colour code facilitated the coding and definition of categories, in a structured and interconnected way, using webQDA® software. In the constitution of the corpus and the definition of the categories, representatively, completeness, homogeneity and pertinence of the same were guaranteed for the object of study.

Video recordings took place on March 2, 2018, in the Obstetrics simulation laboratory of the Nursing School of Lisbon, after being approved by the Ethics Commission Process n° 02/2017/CE/ESEL of 24/6/2017, and having been guaranteed voluntariness, confidentiality and anonymity of participants and findings.

3   Results

The actor parturitions were performed by a student in the bed in '*four*-position' position and the sitting position in the birthing cage. There were different approaches to bed birth in a four-position position, a posterior approach, and two lateral approaches to the right of the woman patient. It should be noted that the bed used was not adapted to the childbirth, not allowing the articulation of the same in height, making it impossible for the students to adjust. For the approach to seated childbirth, the use of two banks was verified, and one student did not resort to any support equipment. In the analysis of the videos were taken into consideration the extreme body posture of the students at the time of approach to childbirth. Five categories emerged for each type of delivery, related to the five principles of body mechanics (Fig. 1).

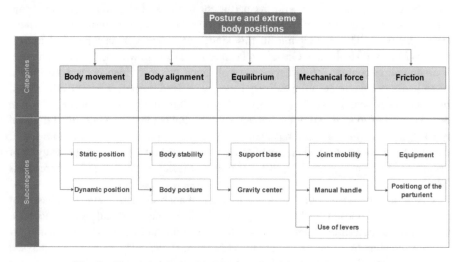

Fig. 1. Categories and subcategories of activity-related risk factors

Subcategories were created to synthesize the action. The use of the videos made it possible to observe in detail the body postures adopted by the students (Table 1), allowing a rigour in obtaining the data. Three moments of the postural evaluation were defined in the approach to the perineum; in the new-born's delivery and delivery to the mother.

In the analysis of Table 1, we found that there are a total of 184 references regarding the principles of body mechanics, from the seven analysed sources (seven students). We found that the body alignment obtained a higher number of recording units (57 FI), both in the parturient with four supports (34 FI) and in the seated position (23 FI). It should be noted that a more significant number of instability observation units were observed regarding body alignment in four-arm delivery (7 IF) compared to sitting (6 IF). It was possible to find that in all births, the students remained in static positions throughout the delivery (23 FI).

Table 1. Results of postural extreme body evaluation of students during the simulated practice of childbirth: observation units (FI) by category and subcategory

Evaluation	Category	Subcategory	FI Pregnant sitting pose	FI Pregnant in four supports	Total FI
Posture and extreme body positions	Body movement	Static position	8	15	23
		Dynamic position	0	4	4
		Subtotal	8	19	27
	Body alignment	Body stability	10	11	21
		Body posture	13	23	36
		Subtotal	23	34	57
	Equilibrium	Support base	9	12	21
		Gravity centre	9	14	230
		Subtotal	18	26	44
	Mechanical force	Joint mobility	4	4	8
		Manual handle	6	8	14
		Use of levers	9	14	23
		Subtotal	19	26	45
	Friction	Equipment	5	3	8
		Position of the parturient	3	4	7
		Subtotal	8	7	15
Total			76	108	184

Regarding the adoption of the body posture in the approach to the seated childbirth, it was verified that the students adopted different positions of strategies, being that the position of seated knee - seize, presented greater dominance of the centre of gravity and maintained the balance, although with total knee flexion (tibiofemoral flexion of 160°), while in the sitting position on a small bench with lower limbs in W, we obtained a deviation from the centre of gravity, with no support base in both feet, causing a slight imbalance in the student. Another position that revealed instability was the squatting approach and the sitting with the left foot of full knee support and flexion.

It should be noted that in all approaches to childbirth, students had to perform trunk flexion and a slight lateral tilt to gain access to the field of action at the time of expulsion and receipt of the new-born since this was at the lower level to the student. There were only four situations of static position change for dynamics, having occurred in the squatting position in which after a few minutes the student feels the need to change her position and in the delivery of the new-born to the mother. We did not simulate deliveries with complications, so it was not possible to identify the body postures adopted in these situations.

With regard to labour in four supports, was verified that two types of approaches were made to the field of action, one with student positioned herself with right knee in total flexion on the bed of the woman in the parturient and with the left leg in extension with support on the floor, and a lateral approach to the parturient, forcing the bending, inclination and rotation of the trunk right, using the sides of the bed as support lever. In this approach it was possible to identify that the moment of delivery of the new-born to the mother, requires the change of the static position to dynamics, not guaranteeing the corporal alignment, being the passage effected in imbalance.

In more careful observation of the videos, we found in lateral reception of the new-born with the parturient in four supports, the female students perform an extension of the hands and with the weight of the new-born increase their instability.

In the design of the body posture adopted by students in approach to the seated delivery, we verified the use of the middle bench as an important factor to maintain the balance and the need to perform abduction of the knees and flexion of the trunk superior to 60°, to access the field of action. The use of mirrors facilitated the visualization of the foetal coronation by the student and the parturient, reducing the flexion movements of the trunk and being a stimulating factor for the parturient herself.

4 Discussion

The analysis of the videos made it possible to analyse the body postures assumed by the specialists (that is, the students) in performing the vertical deliveries. Although this type of delivery has advantages for the mother and the baby, it poses complex challenges for the professional and their 'biomechanical' decision making because it requires the nurse to have to squat or kneel to work [19], impaired body posture and alignment, prolonged sitting and standing periods, pronounced and prolonged neck flexion, dorsiflexion under load, repetitive and/or strong movements, upper limb elevation higher than 90° [19].

The findings of our study are in agreement with other investigations that observe that these professionals assume inadequate postures, using excessive muscle strength [19, 41], due to professional requirements, inadequate working environment and depending on the type of delivery [41, 42].

However, births with water births are considered the most painful by postures, mobilization of marquees, beds and other heavy equipment that must be mobilized, as well as manual manipulation of the woman and the new-born [19, 41], we find in this investigation that deliveries in vertical position have specific risks associated to the type of delivery; to the position adopted by the parturient and its implications in the adjustment of the position of the specialist nurse, because the approach may be anterior or posterior or lateral. Besides, the parturient may be in bed, on a stool, or even on foot, and may also change the position several times because she is uncomfortable, and the nurse must quickly adjust to a new location.

In addition, of these specificities associated with the nature of the activity, the duration of labour itself, which may take hours, implies the maintenance of static positions for long periods in the period of labour expulsion, and can reach 4 min, and despite not being in a positioning position, it changes the principles of biomechanics, contributing to misalignment and incorrect posture in the standing position [35].

This static position can be maintained in the orthotic position, with the different body segments aligned, but most of the time, the specialist is with the knee and hip joint in flexion. Further studies are needed to explore the impact of this position on biomechanics and the genesis of musculoskeletal injury.

In some of these deliveries, professional kneels to receive the new-born, supporting the weight of the body itself in the knees and adopting dorsiflexion with arms extended, which increases pressure in the shoulder's articulation. It is in this position that they make the rapid movement of supination to receive the new-born. This movement, in this precise activity, has been associated with an increase in shoulder tendonitis [36, 43]. In the observed cases in which the specialist chooses to be seated the adoption of similarly painful positions is verified, which is reinforced by the findings of other studies that consider that even when in a position, the professional is subject to postures and extreme positions associated with body movement, mechanical strength and friction [35].

Birth in vertical positions makes it difficult to organize the physical space of the delivery room, which was not designed for the specifics of the same, which implies a reorganization of the same, often, at the moment in which labour already occurs. It should be noted that the organization of space, anticipation and planning of the movements by the nurse are behaviours to reduce the risk of injury but are challenging to adopt in emergencies [35].

Although it is not the objective of this study to evaluate the potential of the simulated practice in a laboratory context, we reinforce the contribution to the training of these professionals for their biomechanical safety. The use of this methodology promotes critical thinking, also enables the active development of transferable and transformational leadership skills and, therefore, improves students' critical thinking and clinical reasoning in more complex care situations and assists in the development of skills techniques, through the improvement of reflexes, perception, physical abilities [4].

We reinforce the opinion of other authors who consider that the studies on the biomechanical safety of these professionals are exasperated, especially those that use qualitative methodology [36, 44].

5 Conclusions

The analysis of the 24 frames of labour in a vertical position allows observing that in the practices and behaviours of the nurse-midwives, although the biomechanics principles are present, the body posture is influenced by the type of delivery and equipment and stands out the permanence in the static position and with uncomfortable poses. Although it is possible to predict the specific risks of musculoskeletal injury and to anticipate the risk, the particular nature of this type of delivery, with the parturient changing positions, presents difficulties in the planning of the body postures to be adopted by the professional and in the organization of space physical and equipment.

The rapid and sometimes unexpected change in the woman's behaviour, as well as the emergence of unforeseen and emergencies, make it challenging to adopt and maintain body alignment, balance and proper posture.

The focus on the process of childbirth seems to be conditional on giving specific guidance to the woman who could make the application of force and the use of the levers less painful. The difficulties in maintaining the principles of biomechanics are increased by the meagreness of space, materials and equipment damaged.

The scarcity of studies on the principles of biomechanics in this type of birth guides us to the need for a more comprehensive understanding of this phenomenon, with a more significant number of participants and diverse contexts.

This study presents limitations associated with the method, number of participants and intentionality of the participants' choice. Visual analysis of the findings is complex and cannot be taken out of context, the fact that participants know they are being filmed may have altered the 'usual biomechanical behaviour'. The limited number and intentional selection of participants does not allow generalization of findings.

It is common for biomechanics studies to be inserted in a quantitative paradigm, results obtained in this qualitative research emphasize the importance of videos and simulated practice for the study of musculoskeletal injuries, as well as training on biomechanics and practice change.

WebQDA software, was an important contribution in the evidence of qualitative research, because it facilitates the transcription, coding and categorization and analysis of the findings.

References

1. Sanford, P.G.: Simulation in Nursing Education: a review of the research, the qualitative report. Florida **15**(4), 1006–1011 (2010)
2. Martins, J.C.A., Mazzo, A., Baptista, R.C.N., Coutinho, V.R.D., Godoy, S., Mendes, I.A.C., Trevizan, M.A.: A Experiência Clínica Simulada no Ensino de Enfermagem: retrospectiva histórica. Acta Paul Enferm. **25**(4), 619–625 (2012)
3. Baptista, R.C.N., Martins, J.C.A., Pereira, M.F.C.R., Mazzo, A.: Simulação de Alta-fidelidade no Curso de Enfermagem: Ganhos percebidos pelos estudantes. Rev Enf Referência **IV**(1), 135–144 (2014)
4. Presado, H., Colaço, S., Rafael, H., Baixinho, C.L., Felix, I., Saraiva, C., Rebelo, I.: Learning with high fidelity simulation. Cien Saud Coletiva **23**(1), 51–55 (2018)
5. Martins, J.C.A., Mazzo, A., Mendes, I.A.C., Rodrigues, M.A.: A simulação no ensino de enfermagem. In: Série monográfica 10, ESEnfC. Unidade de Investigação em Ciências da Saúde: Enfermagem (ed.), p. 310 (2014)
6. Fonseca, A., Mendonça, C., Gentil, G., Gonçalves, M.: Centro de Simulação Realística: Estrutura, Funcionamento e Gestão. In: Série Monográfica Educação e Investigação em Saúde: A Simulação no Ensino de Enfermagem, pp. 207–226. Unidade de Investigação em Ciências da Saúde – Escola Superior de Enfermagem de Coimbra, Coimbra (2014)
7. Ventura, C.: Enquadramento e Justificação Ética: Ética e Simulação em Enfermagem. In: Série Monográfica Educação e Investigação em Saúde: A Simulação no Ensino de Enfermagem, pp. 29–38. Unidade de Investigação em Ciências da Saúde – Escola Superior de Enfermagem de Coimbra, Coimbra (2014)

8. Mills, J., West, C., Langtree, T., Usher, K., Henry, R., Chamberlain-Salaun, J., Mason, M.: "Putting it together": unfolding case studies and high-fidelity simulation in the first-year of an undergraduate nursing curriculum. Nurse Educ. Pract. J. **14**, 12–17 (2014)

9. Najjar, R.H., Lyman, B., Mihel, N.: Nursing students experiences with high-fidelity simulation. Int. J. Nurs. Educ. Sch. **12**(1), 1–9 (2015)

10. Martins, J.C.: A Aprendizagem e desenvolvimento em contexto de prática simulada. Rev Enf Referência **IV**(12), 155–162 (2017)

11. Serranheira, F., Cotrim, T., Rodrigues, V., Uva, A.: Estudo nacional de caracterização da sintomatologia músculo-esquelética em enfermeiros: resultados preliminares. In: Colóquio Internacional de Segurança e Higiene Ocupacionais, pp. 608–610. Guimarães, SPSHO (2011)

12. Sousa, A.D.: Lesões músculo-esqueléticas ligadas ao trabalho nos enfermeiros especialistas em saúde materna e obstetrícia no decorrer do parto. Dissertação de Mestrado em Enfermagem de Saúde Materna e Obstetrícia. Escola Superior de Enfermagem de Lisboa, Lisboa (2018)

13. Mineiro, A.L.S.: Prevenção de Prevenção de Lesões Músculo Esqueléticas no Enfermeiro Especialista durante o parto em posições verticais. Dissertação de Mestrado em Enfermagem de Saúde Materna e Obstetrícia. Escola Superior de Enfermagem de Lisboa, Lisboa (2018)

14. Mineiro, A.L., Presado, M.H., Cardoso, M.: Posturas do enfermeiro obstetra na assistência ao parto em posições verticais. In: Costa, et al. Atas do 8° Congresso Ibero-Americano em Investigação Qualitativa, vol. 2, pp. 807–816 (2019)

15. Nowotny-Czupryna, O., Naworska, B., Brzek, A., Nowotny, J., Famula, A., Kmita, B.: Professional experience and ergonomic aspects of midwives' work. Int. J. Occup. Med. Environ. Health **25**(3), 265–274 (2012)

16. Long, M.H., Johnston, V., Bogossian, F.E.: Helping women but hurting ourselves? Neck and upper back musculoskeletal symptoms in a cohort of Australian Midwives. Midwifery **29**, 359–367 (2013)

17. Baixinho, C.L., Presado, M.H., Marques, F.M., Cardoso, M.: Prevenção de lesões músculo-esqueléticas: relatos dos enfermeiros especialistas em saúde materna e obstetrícia. Atas - Investigação Qualitativa em Saúde **2**, 488–497 (2016)

18. Baixinho, C.L., Presado, M.H., Marques, F.M., Cardoso, M.: A segurança Biomecânica na prática clinica dos Enfermeiros Especialistas em Saúde Materna e Obstetrícia. Revista Brasileira em Promoção da Saúde **29**, 36–43 (2016)

19. Baixinho, C.L., Presado, M.H., Marques, F.M., Cardoso, M.: Posturas dos estudantes durante o trabalho de parto: análise de filmes de prática simulada. Atas - Investigação Qualitativa em Saúde **2**, 513–522 (2017)

20. Wang, J., Cui, Y., Xu, X., Yuan, Z., Jin, X., Li, Z.: Work-related musculoskeletal disorders and risk factors among Chinese medical staff of obstetrics and gynecology. Int. J. Environ. Res. Public Health **14**(562), 1 (2017)

21. HSE Manual handling risks to midwives associated with birthing pools: literature review and incident analysis. HSE Report (2018)

22. Waldenstrom, U., Gottvall, K.: A randomized trial of birthing stool or conventional semirecumbent position for second-stage labor. Birth Issues Perinat. Care **18**(1), 5–10 (1991)

23. Baixinho, C.L., Presado, M.H., Marques, F.M., Cardoso, M.: Lesões músculo-esqueléticas nos enfermeiros especialistas em saúde materna: autopceção dos fatores de risco. In: Costa et al. Atas do 4° Congresso Ibero-Americano em Investigação Qualitativa e do 6° Simpósio Internacional De Educação e Comunicação: Investigação Qualitativa na saúde, vol. 1, pp. 193–198 (2015)

24. APEO & FAME: Iniciativa parto normal - Documento de consenso. Lusodidacta, Loures (2009)
25. WHO: WHO recommendations: intrapartum care for a positive childbirth experience. WHO, Geneva (2018)
26. Coppen, R.: Midwives' views on birthing positions. In: Birthing Positions: Do Midwives Know Best?, pp. 61–78. MA Healthcare Limited, London (2005)
27. Jonge, A., Teunissen, D.A.M., Van Diem, M.T.H., Scheepers, P.L.H., Lagro-Janssen, A.L. M.: Women's positions during the second stage of labour: views of primary care midwives. J. Adv. Nurs. **63**(4), 347–356 (2008)
28. Tsekoura, M., Koufogianni, A., Billis, E., Tsepis, E.: Work-related musculoskeletal disorders among female and male nursing personnel in Greece. WJRR **3**(1), 08–15 (2017)
29. Boulton, R.: The risk of musculoskeletal injury to midwives during routine hospital-based work activities. HSE Report (2011)
30. Thinkhamrop, W., Sawaengdee, K., Tangcharoensathien, V., Theerawit, T., Laohasiriwong, W., Saengsuwan, J., Hurst, C.P.: Burden of musculoskeletal disorders among registered nurses: evidence from the Thai nurse cohort study. BMC Nurs. **16**(68), 1–9 (2017)
31. Vélez, A.G.B., Aguiar, K.C., Santos, A.L.: Análise ergonômica das posturas que envolvem a coluna vertebral no trabalho da equipe de enfermagem. Texto & Contexto Enfermagem **13** (1), 115–123 (2004)
32. OE Cuidados à pessoa com alterações da mobilidade - posicionamentos, transferências e treino de deambulação. Guia orientador de boas práticas. Ordem dos Enfermeiros, [S.l] (2013)
33. Costa, R., Azevedo, C.G., Silvestre, S.: Biomecânica: Terapêutica de posição – contributo para um cuidado de saúde seguro. In: Lourenço, M.J., Ferreira, O., Baixinho, C.L. (coord.) Lusodidacta, pp. 67–73 (2016)
34. Loizos, P.: Vídeo, filme e fotografias como documentos de pesquisa. In: Bauer, M.W., Gaskell, G. (orgs.) Pesquisa qualitativa com texto, imagem e som, 2ª ed., pp. 137–155. Vozes, Petrópolis (2008)
35. Sousa, A.D., Baixinho, C.L., Marques, F.M., Cardoso, M., Presado, M.H.: Biomechanics of nurse-midwives in the delivery: contribution of qualitative research. In: Costa, A., Reis, L., Moreira, A. (eds.) Computer Supported Qualitative Research, WCQR 2018. Advances in Intelligent Systems and Computing, vol. 861. Springer, Cham (2019)
36. Presado, M.H., Cardoso, M., Marques, F.M., Baixinho, C.L: Posturas dos estudantes durante o trabalho de parto: análise de filmes de prática simulada. In: Investigação Qualitativa em Saúde. Editado por 6º Congresso Ibero-Americano en Investigação Qualitativa (CIAIQ 2017), vol. 2, pp. 488–497 (2017)
37. Streeck, J. (ed.): Embodied Interaction (Learning in Doing: Social, Cognitive and Computational Perspectives) - Language and Body in the Material World. Cambridge University Press, Cambridge (2014)
38. Ritchie, S.M., Newlands, J.B.: Emotional events in learning science. In: Bellocchi, A., Quigley, C., Otrel-Cass, K. (eds.) Exploring Emotions, Aesthetics and Wellbeing in Science Education Research, pp. 107–111. Springer, Basel (2016)
39. Sousa, A., Cardoso, M., Presado, M.: Metodologia adotada na analise de vídeos em investigação: revisão sistemática. In: CIAIQ 2019 (2019)
40. Lima, F.H.: Um método de transcrição e análise de vídeos: a evolução de uma estratégia. In: VII Encontro Mineiro de educação matemática, Universidade Federal de São João del Rei, 9–12 outubro 2015, pp. 1–11 (2015)

41. Nevala, N., Ketola, R.: Birthing support for midwives and mothers - ergonomic testing and product development. Ergon. Open J. **5**, 28–34 (2012)
42. Jellad, A., Lajili, H., Boudokhane, S., Migaou, H., Maatallah, S., Frih, J.B.S.: Musculoskeletal disorders among Tunisian hospital staff: prevalence and risk factors. Egypt. Rheumatol. **35**, 59–63 (2013)
43. Taghinejad, H., Azadi, A., Suhrabi, Z., Sayedinia, M.: Musculoskeletal disorders and their related risk factors among iranian nurses. Biotech Health Sci. **3**(1), e34473 (2016)
44. Long, M.H., Bogossian, F.E., Johnston, V.: The prevalence of work-related neck, shoulder, and upper back musculoskeletal disorders among midwives, nurses, and physicians: a systematic review. Workplace Health Saf. **61**, 223–229 (2013)

Conflict Mediation at School: Literature Review with webQDA®

Elisabete Pinto da Costa[1,2(✉)] ⓘ and Susana Oliveira Sá[1,2(✉)] ⓘ

[1] Lusófona University of Porto and Research Centre for Education
and Development (CeIED), Universidade Lusófona de Humanidades
e Tecnologias (ULHT), IMULP, Lisbon, Portugal
elisabete.pinto.costa@gmail.com, susana.sa@iesfafe.pt
[2] Institute of Higher Studies de Fafe (IESF) and Interdisciplinary Research
Centre for Education and Development (CeIED), Universidade Lusófona
de Humanidades e Tecnologias (ULHT), Lisbon, Portugal

Abstract. School mediation, as a socio-educational intervention strategy, has deserved the attention of schools, as well as academia that seeks to recognize its advantages and potentials at the personal, interpersonal and organizational levels. With this article, one intends to present a study that starts from the following research question: How does conflict mediation in the school context emerge in studies in Educational Sciences published in the EBSCOhost and Scopus databases? This research followed the Cronin, Ryan and Coughlan Protocol (2008) and was performed using a systematic literature review supported by the webQDA® software, which allowed the construction of a more systematized conceptual framework on this theme. It has been found that mediation has been affirmed in the studies that analyze and evaluate, with scientific accuracy, the social phenomenon of conflict, the intervention methodology of mediation and the construction of a citizen coexistence, as socio-educational constructs of the school, that (still) seeks answers to deal with the challenges of "mass schooling" in increasingly multi-intercultural societies.

Keywords: Systematic literature review · Conflict mediation · School mediation · webQDA®

1 Introduction

Experiences of conflict mediation in school context are found, internationally, in the 70s in the United States and in the 80s in Europe [1]. In Portugal, the experiences of school mediation date back to the year 2000 [2]. The school mediation devices are pointed out as good practice, according to the confirmed results, by foreign studies, which have determined the research agenda, and in the national context in recent scientific articles. From a literature review for another study [3], one determined that the contribution to list advantages and virtues of school mediation outnumber those which share typologies and results of intervention, research or evaluation. One also determined that the research on the topic has been increasing. In the path of the statement of [4], contributions are required to convince the scientific community and the School of the social utility and the educational relevance of mediation.

© Springer Nature Switzerland AG 2020
A. P. Costa et al. (Eds.): WCQR 2019, AISC 1068, pp. 114–123, 2020.
https://doi.org/10.1007/978-3-030-31787-4_9

In this article, one proposes to present a more detailed research about what has been produced in the research setting in the last decade and published in the EBSCOhost and a Scopus databases about school mediation as a social technology [1].

2 Conflict Mediation

Mediation is presented as a methodology that allows to manage conflicts constructively and improve social interaction. Its educational potential has been highlighted in several studies, both empirical and theoretical [5–16].

The empirical reality reveals that educational institutions seek to deal with conflicts in the early stages or even before they occur, so as not to trigger dynamics of violence or bullying episodes [17]. These advantages justify that different mediation initiatives become part of the educational reality, especially those in which students are mediators of their own colleagues. Several researches demonstrate that when one questions those youngsters who had a conflict, they refer that their colleagues acted before teachers or their families [14]. Because they are on equal terms with their peers, mediator students use a language and a dialogue that is closer to their peers, which may be more useful than the perspective provided by an adult. It also creates a climate of freedom, whereas teachers or other adults may be seen as the authority who normally performs punitive roles.

Although much of the literature focuses on peer mediation, there is a dissemination of studies that analyze this intervention methodology from the point of view of educational advantages in interpersonal and social terms, of procedural requirements and of organizational implications [3]. In the first case, a constructivist perspective of promoting students' skills prevails, and in the second case, it is advocated that conflict mediation lacks sustainability to assert itself as a school culture.

3 Methodological Procedures

The theoretical argumentation of this study is based on a literature review proposal, supported by the qualitative methodological perspective [18, 19] and supported by the fact that conflict mediation at school emerges as an object of research in Educational Science.

Thus, in the context of the literature review of scientific work obtained through the EBSCOhost and Scopus databases and the reference manager Mendeley, referring to a decade, more specifically between 2009 and 2018, one sought to identify the *outputs* of conflict mediation in school context.

The systematic literature review followed the Cronin, Ryan and Coughlan research Protocol [20] and its procedures (step by step) are summarized in Table 1:

Table 1. Research steps of systematic literature review according [20]

1st step	Research question	How does conflict mediation in the school context emerge in studies in Educational Sciences published in the Scopus and EBSCOhost database?
2nd step	Criteria for data collection (inclusion of metadata consulted)	Metadata in the EBSCOhost and Scopus database in the field of education
3rd step	Select the metadata that make up the corpus of analysis	Paper's indexed with the keywords: Mediation; Conflict; School Mediation and coexistence
4th step	Analyze the research corpora by constructing a conceptual map	Generate classification categories with the webQDA® qualitative analysis software
5th step	Disseminate search results	Write and publish this study

The analysis *corpus* of this work was composed initially of an exploratory search in the two databases for articles that addressed the theme of School Mediation, which allowed the identification of the keywords and which later supported the search for the articles that would be analyzed.

It was proposed that the research be limited to thesis and dissertations, but only one response was obtained in due time for the authorizations made for the authors [21]. This study was excluded as it focused on parental conflict.

One started (2st step on the Table 1) the search through the EBSCOhost database, using the following terms to narrow the search: (a) Mediation; (b) Conflict; (c) School Mediation; (d) Peer Review; (e) Open Access; (f) Articles and (g) Last 10 years. Eight metadata were obtained. Then, the search in the ERIC database was performed, and 75 metadata emerged. However, one could not import them into webQDA® [22] due to the incompatible format with the software (BibTeX (*.bib); XML (*.xml) or RIS (*.ris)). Finally, based on the 7 terms (mentioned above), one searched the Scopus database and obtained 102 metadata. Then, using one of the operations allowed by the software, the metadata were saved in BibTex format (i.e.: authors, title, abstract, keywords, magazine identification, year of publication and DOI). In the end, one took into consideration a total of 82 metadata collected from the EBSCO/host and Scopus databases (3rd step on Table 1).

The identified data *corpus* was organized into three categories (4th step on Table 1): (1) Included; (2) Excluded and (3) Duplicated which, through the use of webQDA® software, contributed to the construction of the conceptual map. The latter became essential for data extraction and analysis considering the categories [23] proposed for the research.

Thus, the keywords of the 82 identified papers were extracted, and 44% of these were not repeated in the papers. The most frequent keywords were: Mediation (n = 32); Conflict (n = 19); Resolution (n = 8); Students (n = 5); Family (n = 3);

Social (n = 3); Training (n = 3); Education (n = 3); Violence (n = 3); Process (n = 3); Dispute (n = 2); Program (n = 2); Communication (n = 2); Mediator (n = 2); Solving (n = 2); Negotiation (n = 1) and Peace (n = 1). One proceeded to the metadata analysis, from which the items of the analysis corpus were extracted, specifically the indexed articles, referring to mediation in the educational area. Finally, in a third stage, the indexed metadata were selected according to the keywords identified in the selected articles. At the end of this procedure, 82 metadata were identified and analyzed using webQDA® software.

An initial reading of this data *corpus* allowed us to eliminate those works without direct relation to School Mediation. The reason for the ineligible exclusion of several studies (n = 62) was due to the fact that the metadata associated "conflict" with other areas of intervention, such as: parental area or school performance, and not exactly conflict mediation at school. It should be noted that one metadata came in duplicate.

After this exhaustive selection procedure, 14 metadata were identified, which were subjected to a thorough reading, constituting the definitive analysis *corpus* of the study.

3.1   Conceptual Development with webQDA®

The descriptors defined in the webQDA® were journal article, year of publication, authors, keywords and state.

Descriptors act as classification attributes, but can be applied to only part of the document, that is, from the source under analysis, [21].

Once this assignment was accomplished, one started to develop the conceptual map built from the webQDA® software.

When the content analysis was carried out, the ideas were cut out into reference units, words or phrases, texts contained in the information material produced, which corresponded to clear, objective and meaningful ideas in the context of the research. Subsequently, after deep reading, the reference units were grouped into indicators, which later allowed us to clarify the definition of each of the categories. As frequency unit, one took the reference unit, which was counted as many times as present in the discourse. The data analysis and treatment were carried out with the support of content analysis software in qualitative research, webQDA®, through open procedures, corresponding to a permanent process of progressive creation, in which the reflection and analysis of data are rigorously and constantly triangulated, which makes the methodological process reliable.

One proceeded, in an early stage (5th step on Table 1), to the analysis of the results of the 14 metadata. Afterwards, in order to validate categorization, to guarantee a greater reliability of the study, the 2 researchers did the categorization [21], through online collaborative work, supported by the software webQDA®. These validated the three emerging categories: Conflict, Mediation and Coexistence, (see Table 2).

Table 2. Categories and indicators in conflict mediation and coexistence

Categories	Indicators
Conflict	Democratic management
	Strengthens relationships
	Promotion of social skills
Mediation	Planning
	Implementation
	Assessment
Coexistence	Dialogue
	Participation
	Decision-making
	Assumption of responsibilities

4   Analysis and Discussions

During the period under analysis, 2008–2018, the research in the referred databases (with the previously announced imposed restrictions) contributed to recognize the field and revisit the literature on the subject in focus. According to the literature review technique using the software webQDA®, it was possible to move forward with various data analysis, whose discussion is developed below.

In 11 of the 14 selected studies, authors from more than one institution were identified, thus allowing the identification of interinstitutional partnerships. The production on the subject is still limited, although growing, and most publications seem to derive from authors who prioritize Conflict Mediation in the school context as their objects of study.

In the national context, it was found that the scientific production on the theme follows the trend of international scientific production. This evolution is pointed out in a study by [24] in which several publications are identified where conflict mediation in school context is the main object of study.

4.1   Data Analysis

It was found that the analyzed papers were predominantly theoretical-empirical, with qualitative approach (5 papers) and some with quantitative approach (6 papers). The most commonly used research methods were case studies (mixed) and the data collection instruments were predominantly interviews and also observation guides of student meetings. The most used data analysis technique was content analysis (35%), followed by statistical analysis, either descriptive (21%) or inferential (3.5%) or both (2.1%). No experimental or quasi-experimental work was identified and only two presented a correlational design. Of the total studies, 6 are quantitative [25, 26, 32, 33, 35, 37] in nature, with inferential approaches.

4.2   Research Focuses on Conflict Mediation at School

In the period under review, one found a diversity of research focuses on conflict mediation at school: intervention as a model of public policy; type of conflicts dealt with in mediations and results of mediation sessions; skills learned and enhanced through mediation; mediation; results of coexistence and assessment.

The first work soon emerged at the beginning of the research period, in 2008, [24, p. 597], refer to mediation as "a model of public politics to reduce the scholar violence and to promote the peace culture".

The following year, 2008, [25, p. 639] study was found, which focuses on the typology of mediated conflicts: "the majority of the conflicts referred to mediation were physical, verbal, and non-verbal violence, relationship and communication conflicts, and conflicts of interest" and on the positive results of mediations: "98.9% resulted in agreement and 1.1% in no-agreement".

In 2010 and 2012 no publications on this theme were identified, with this search filter.

In 2011, [26, p. 2324] published a study on peer mediation and on the promotion of "empathic skills" and [27, p. 57] refer to the "possibilities that mediation offers when resolving the conflictive situations that people with special needs and their families live. It is necessary to create a cooperative spirit in order to transform into satisfaction the frustration and sorrow that people with special needs and their families experience, from the acknowledgement of their rights with the purpose to increase their participation in the society and their access to the decision-making".

In 2013, [28, p. 163] points out the "possibilities that mediation offers when resolving the conflictive situations that people with special needs and their families live".

With 2 references in 2014, the studies focused on school mediation as an integrated intervention in institutional plans and programmes of mediation in school conflict. In Spain, [29, p. 271] carried out "a comparative study to find out which autonomous regions support school mediation, including it in the social agreements that they have signed, as well as in its plans and coexistence institutional programmes". In [30, p. 375] identified several "difficulties of different nature (...) (lack of involvement, internal and external coordination difficulties, among others aspects), on the other hand, the participants also expressed their satisfaction with the Mediation Program (as it facilitated a better coexistence in their school or the development of personal skills)".

The study (2015), presented a discussion of relationship asymmetry in the academic setting is provided. In [31, p. 29] performed a study "whose objective is verifying from a structural perspective the influence of a set of measures to reduce such problems (unexcused absences punctuality and attendance, disrespect, damage to facilities and property, bullying and relationships between members of the school community)".

In 2016, 2 papers were found. In [32, p. 650], whose study had the purpose "to investigate the effectiveness of Polatli Negotiator Mediator Leader Students project". And finally, [33, p. 29] who points out Mediation "as an innovative procedure for the consensual settlement of conflicts". This study starts from the "premise that, in the institutionalized self-image of mediation, the position of the mediating third party is, in fact, empty".

Two studies were identified in 2017. In [34, p. 537] highlighted that "school mediation has proven its educational potential, as it influences the improvement of the individual, of the interpersonal relationships as well as of the school climate". In [35, p. 165] analyzed students' "changing perceptions of conflict after training them in mediation skills".

In 2018, two studies were selected. In [36, p. 233] "Mediation is not used and even known conflict resolution strategy in Lithuania's schools. Therefore, taking into account the lack of research on school based conflicts this research sets as its object conflict resolution education" and [37, p. 79], defend: "The social reality is in constant change; therefore an effort to upgrade the educational responses to improve coexistence is required from the educational system. In this context, school mediation has proven its educational potential, as it influences the improvement of the individual, of the interpersonal relationships as well as of the school climate".

There is no visible division on separation dates between descriptive and experiential studies, they mingle.

In a brief summary, one can bring together the most outstanding aspects for an evaluation of the metadata meta-analysis.

The term coexistence comes from a little broadly.

4.3 Development of the Conceptual Map of Conflict Mediation in School Context

Still based on the literature review using webQDA®, one intended to present a more systematic conceptual map about this topic, that has been emerging in studies on socio-educational constructs of School improvement (see Fig. 1).

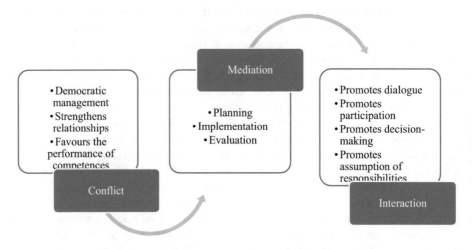

Fig. 1. Conceptual map of conflict mediation in school context

Mediation boosts the participation of the students in the democratic management of conflicts, strengthens their interpersonal relations and favors the performance of competences necessary for a true learning of the civic values of democratic citizenship. Finally, it promotes dialogue, participation, decision-making and responsibility, among other objectives necessary for interaction [8, 13, 23].

This action proposes the introduction of the conflict mediation theme in the school curriculum, aiming at the opportunity to verbalize the issue and to clarify what is expected of children and youth in the set of social behaviors. Otherwise, such action implies telling the youth and the child that their differences can become antagonisms and that if they are not understood, they develop into conflict, which may end in violence. It is noteworthy that this learning and this social perception, when developed in the student, tend to constitute lifelong learning. Thus, it is necessary the qualification/training of teachers and school professionals on the subject, covering the subjects related to the policy of reducing violence and the promotion of positive coexistence in the school space, towards responsible interaction.

5 Conclusions

To answer the research question (1st step on Table 1), the literature review highlights the theoretical contributions on the conflict around the concept, theories, typologies, causes, consequences, approaches and diagnosis. Regarding mediation and conflict mediation, the following contributions are pointed out: concept, principles, models, types, advantages and empirical studies in order to: (a) strengthen school management through clarifications resulting from the periodic evaluation of episodes of violence; (b) offer solutions through integrated networks of public agents; (c) create routines for the implementation and implementation of conflict mediation intervention methodologies, promotion of good living environments and physical restructuring [1, 7, 8, 10].

For the last reference in analysis, interaction, which arose in a smaller number, the following stand out: concept, articulation between coexistence and discipline, regulatory models, intervention proposals and empirical studies. Using webQDA®, one intended to present a more systematic conceptual map on this topic, that has been emerging in studies, which analyze and evaluate, with scientific accuracy, the social phenomenon of conflict, the intervention methodology of mediation and the interaction as socio-educational constructs in School, that (still) seeks answers to deal with the challenges of the "mass schooling", the "school for all" and (more recently) an increasingly multi-intercultural society [2, 23].

There are several gaps in the literature, for example: a clearer need to address conflict mediation steps in the school context for a promotion of positive coexistence in the school space, towards responsible interaction, the proposal of new models with successful examples of experience, as they refer [4, 24]. To be explored in the future in this field of research [1, 7, 8, 11, 12]

The limitations of the study are quite wide ranging from those inherent to the analysis software to those associated with the search mode: (1) limitation of the databases; (2) search keywords (Boolean AND or Boolean OR); (3) reference manager (incomplete metadata transport); (4) associated with limitations in database searches (limited to keywords).

References

1. Cunha, P., Monteiro, A.: Gestão e Mediação de Conflitos na Escola. Pactor, Lisboa (2019)
2. Ferreira, E.: A mediação na escola. In: Vasconcelos-Sousa, J. (Coord.) Mediação, pp. 144–153. Quimera, S/l (2002)
3. Pinto da Costa, E.: Mediação de Conflitos na Escola: da Teoria à Prática. Edições Universitárias Lusófonas, Lisboa (2019)
4. Torremorell, C.: Guía de mediación escolar. Octaedro, Barcelona (2002)
5. Cowie, H., Wallace, P.: Peer Support in Action. From Standing to Standing By. Sage Publications, Londres (2000)
6. García-Longoria, M.P.: La mediación escolar, una forma de enfocar la violencia en las escuelas. Alternativas. Cuadernos de Trabajo Social 10, 319–327 (2002)
7. Grande, M.J.C.: Convivencia escolar. Un estudio sobre buenas prácticas. Revista de Paz y Conflictos 3, 154–169 (2010)
8. Moral, A., Pérez, M.D.: La evaluación del "Programa de prevención de la violencia estructural en la familia y en los centros escolares". Revista Española de Orientación y Psicopedagogía (REOP) 21, 25–36 (2010)
9. Ibarrola-García, S., Iriarte, C.: Desarrollo de las competencias emocional y sociomoral a través de la mediación escolar entre iguales en educación secundaria. Revista Qurriculum 27, 9–27 (2014)
10. Paulero, R.: Alumnos mediadores. Construyendo la paz. Revista Chilena de Derecho y Ciencia Política 1, 89–104 (2011)
11. Torrego, J.C., Galán, A.: Investigación evaluativa sobre el programa de mediación de conflictos en centros escolares. Revista de Educación 347, 369–394 (2008)
12. Bonafé-Schmitt, J.P.: La médiation scolaire par les élèves. ESF, París (2000)
13. Boqué, M.C.: Cultura de mediación y cambio social. Gedisa, Barcelona (2003)
14. García-Raga, L., Martínez, M.J., Sahuquillo, P.: Hacia una cultura de convivencia. La mediación como herramienta socioeducativa. Cultura y educación 24, 207–217 (2012)
15. López, R.: Las múltiples caras de la mediación. Y llegó para quedarse. Universitat de València, Valencia (2007)
16. Pulido, R., Martín-Seoane, G., Lucas-Molina, B.: Orígenes de los Programas de Mediación Escolar: Distintos enfoques que influyen en esta práctica restaurativa. Anales de Psicología 29, 385–392 (2013)
17. Torrego, J.C., Villaoslada, E.: El Modelo Integrado de regulación de la Convivencia y el Tratamiento de los Conflictos. Un proyecto que se desarrolla en centros de la comunidad de Madrid. Tabarque 18, 31–48 (2004)
18. Flick, U.: An Introduction To Qualitative Research, 5th edn. Sage Publications, London (2014)
19. Flick, U.: Doing Triangulation and Mixed Methods. Sage Publications, London (2018)
20. Cronin, P., Ryan, F., Coughlan, M.: Undertaking a literature review. In: Doing Postgraduate Research, vol. 17, pp. 411–429 (2008)
21. Soares, C., Fornari, L., Pinho, I., Costa, A.P.: Revisão da Literatura com apoio de software: contribuição da pesquisa qualitativa. Ludomedia, Oliveira de Azeméis (2019)
22. Costa, A.P., Moreira, A., Souza, F.N.: webQDA (version 3.1) - Qualitative Data Analysis. Aveiro University and MicroIO, Aveiro (2019)
23. Bardin, L.: Análise de conteúdo. Edições 70, São Paulo (2011)
24. da Costa, E.P., Sá, S.: Teacher narratives on the practice of conflict mediation. In: Costa, A.P., Reis, L.P., Moreira, A. (eds.) WCQR 2018. AISC, vol. 861, pp. 156–169. Springer, Cham (2019). https://doi.org/10.1007/978-3-030-01406-3_14

25. Chrispino, A., Dusi, M.L.H.M.: A proposal to sharpen the public politics in order to reduce the scholar violence and to promote the peace culture. Ensaio **61**, 597–624 (2008)
26. Turnuklu, A., Kacmaz, T., Turk, F., Kalender, A., Sevkin, B., Zengin, F.: Helping students resolve their conflicts through conflict resolution and peer mediation training. Procedia Soc. Behav. Sci. **1**, 639–647 (2009)
27. Sahin, F., Sülen, S., Nergüz, B., Serin, O.: Effect of conflict resolution and peer mediation training on empathy skills. Procedia Soc. Behav. Sci. **15**, 2324–2328 (2011)
28. Chrispino, A., Santos, T.: Teaching policy to violence prevention: teaching techniques that can contribute to school violence reduction. Ensaio **19**, 57–80 (2011)
29. Munuera Gómez, M.: Mediation with special needs people: equal opportunities and access to justice. Politica y Sociedad **50**, 163–178 (2013)
30. Viana-Orta, M.: School mediation in institutional plans and programmes of school coexistence in Spain. Revista Complutense de Educacion **25**, 271–291 (2014)
31. Pulido Valero, R., Calderón-López, S., Martín-Seoane, G., Lucas-Molina, B.: Implementation of a school mediation program: analyzing the perceived difficulties and ways to improve it. Revista Complutense de Educacion **25**, 375–392 (2014)
32. Seeman, J.I., House, M.C.: Authorship issues and conflict in the U.S. academic chemical community. Account. Res. **22**, 346–383 (2015)
33. Morueta, R.T., Vélez, S.C.: Relations between some preventive actions on peaceful coexistence at school in centers of good practices. Estudios Sobre Educacion **29**, 29–59 (2015)
34. Yildiz, D.G., Çetin, H., Türnüklü, A., Tercan, M., Çetin, C., Kaçmaz, T.: Evaluation of polatli negotiator mediator leader students project. Elementary Educ. Online **15**, 650–670 (2016)
35. Münte, P.: Mediation practice in need of professionalisation or social engineering? Zeitschrift fur Rechtssoziologie **36**, 29–57 (2016)
36. Leonov, N.I., Glavatskikh, M.M.: Changing the image of a conflict situation while training school students in mediation skills. Psychol. Russ. State Art **10**, 165–178 (2017)
37. Iuladiene, G.: Mediation at school in Lithuania (Case study). Pedagogika **129**, 220–233 (2018)
38. García-Raga, L., Bonet, R.M.B., Lasagabaster, J.M.: Meaning and sense of school mediation based on high school mediators' views. Revista Espanola de Orientacion y Psicopedagogia **29**, 79–93 (2018)

Meaning of Cervical Cancer Screening to Women

Susana Almeida[1] , Emília Coutinho[2(✉)] , Vitória Parreira[3] ,
Paula Nelas[2] , Cláudia Chaves[2] , and João Duarte[2]

[1] Unidade de Cuidados de Saúde Personalizados de Seia,
Unidade Local de Saúde da Guarda, Guarda, Portugal
susana.de.almeida@sapo.pt
[2] Viseu Higher School of Health, Viseu, Portugal
ecoutinhoessv@gmail.com, pnelas@gmail.com,
claudiachaves21@gmail.com, duarte.johnny@gmail.com
[3] Nursing School of Porto, Porto, Portugal
vitvik@gmail.com

Abstract. The crucial role played by cervical cancer screening (CCS) in women's health, originated a qualitative, cross-sectional, exploratory and descriptive study aiming to understand what cervical cancer screening and nursing interventions carried out during medical appointments really mean to women. The sample consisted of female users of a healthcare unit, among which 20 women who had missed their cervical cancer screening sessions were contacted and selected. The semi-structured interview was the instrument selected to collect data. Data analysis was carried out using content analysis techniques, with semantic categorization, and an inductive approach using version 11 of the NVivo program. Data analysis gave rise to a set of arguments that shed some light on why women choose not to attend CCS sessions and on which nursing interventions conducted in those CCS sessions are really meaningful to them.

This research was useful to increase understanding and provide room for a reorientation of nurses' professional practices that will help create optimal conditions for the promotion of CCS adherence and for the health care and welfare of women, families and the community as a whole.

Keywords: Meaning · Screening · Cervical cancer ·
Reasons for non-adherence · Nursing interventions

1 Introduction

Cervical cancer (CC) has a significant impact on global public health, especially in developing countries. Research currently carried out around the world highlights the importance of cancer screening programs to identify three oncologic pathologies. Cervical cancer [1] is one of those pathologies. The Portuguese Society of Gynecology [2] stresses that the decrease in the occurrence and mortality caused by this condition will only be achieved if preventive measures are implemented.

© Springer Nature Switzerland AG 2020
A. P. Costa et al. (Eds.): WCQR 2019, AISC 1068, pp. 124–135, 2020.
https://doi.org/10.1007/978-3-030-31787-4_10

Cervical cancer Screening (CCS) sessions are secondary prevention measures and, therefore, play an extremely important role in the prevention of cervical cancer. CCS allows the early detection of premalignant lesions using organized or opportunistic screening programs. A cervical screening test– a conventional cervical Pap smear- is recommended for all sexually active women aged between 25 and 64 and this test should be performed every 3 years when the results obtained are considered normal. Should the results be abnormal, they will have to be handed over so other medical procedures may be applied in accordance with Norm n° 018/2012 of 21/12/2012 [3]. According to the Portuguese Society of Gynecology [2] the higher the percentage of target population screened, the more effective the organized screening will be. Ideally, at least 70% of the target population should take part in CCS programs. Some authors [4] disclosed that, despite the influence played by age, a personal invitation and the setting of a clear date for the screening session will considerably increase screening adherence; others [5] added that an invitation letter, an informative leaflet or a phone call to remind the users of the session will significantly increase adherence. The €2.78 spent with this demarche would increase by 1% cervical cancer screening coverage resulting only in €490 per 1000 women screened for cervical cancer.

While the benefits of CCS are widely supported, the number of women who are actually invited to undergo cervical cancer screening far outnumbers those who are effectively screened. This proves the existence of a real CCS adherence problem [1]. There is much to be gained from the implementation of measures designed to encourage women to willingly join CCS. As far as CCS is concerned, the first measure would be to channel nursing interventions to benefit the general public. On the other hand, a specific, properly planned, structured and enlightening nursing consultation covering a wide range of problems that affect women and that are directed to the particular life cycle they are going through is of paramount importance.

In order to develop such competences, nurses have to understand what cervical cancer screening really means to each and every woman and, only then, will they be able to define, assess and analyze the needs of the community vis a vis education and health care and act accordingly [6].

2 Methods

The starting point was the concern caused by a low cervical cancer screening adherence among 25–64 year-old women, given the impact that such decision has on their health. Data was provided by a PHUC (Personalized Healthcare Unit) of a selected LHU (Local Healthcare Unit) located in the centre of Portugal and the sample consisted of women who were listed in the Sistema de Informação Nacional dos Cuidados de Saúde Primários- SINUS (Portuguese National Primary Healthcare Information System). The aim was to understand the significance for the women, of cervical cancer screening and nursing interventions carried out during consultation, to accurately identify areas related to the performance of nurses specialized in maternal and obstetric care that would help understand such low adherence and, subsequently, outlining strategies to increase the demand for that kind of health care. Consequently, the research question was defined: -what do CCS and nursing interventions conducted during consultation mean to women? Based on the objective and on the research questions set, a qualitative, cross-sectional, exploratory and descriptive study was conducted. All the ethical principles inherent to the development of the study were upheld. Participants were invited and accepted that they were voluntarily taking part in the study by signing an informed consent form. Authorization to conduct the study was previously required and the Ethical Committee of the LHU where the study was to be conducted gave a favorable opinion. Once data saturation was achieved, we had a sample composed of 20 women. Non-probability convenience sampling was used to select the participants and the inclusion criteria were that participants had to be users of that healthcare unit who had missed one or more CCS sessions and who were willing to take part in the study. To identify the women who were eligible to participate in the study, the institution computer and paper records were consulted. Information concerning the number of women listed in the PHCU whose age ranged between 25 and 64 was obtained from the National Primary Healthcare Information System and provided by the LHU Statistics, Planning and Management Support Office. The users' presence in the cervical cancer screening sessions was confirmed by the results of the cytology smear tests. Since there were no computer records to back up this information, the eligible participants were identified using the hard copy files available at the health care unit.

Once this first stage was completed, a phone call was made to each candidate. Each one was briefly informed of the objectives of the study and the procedures to be followed. Ethical principles for medical research were also ensured in accordance with the principles laid down in the Declaration of Helsinki. In order to interview the women who had agreed to participate in the study, a face-to-face appointment was arranged according to the availability of each participant.

The data collection instrument selected was the semi-structured interview. The design of the interview was based on an existing guide that took into account the questions previously prepared.

The interviews took place in a private office in the Personalized Health Care Unit (PHCU). Each session began with some brief questions related to the participants' socio-demographic background. The verbatim transcription of the interview (which was recorded with the consent of each participant) was then obtained.

During the interview each woman answered the questions asked and that were part of the guide. Among other things, they talked about their feelings, their thoughts and knowledge to express the emotions and meanings triggered by different aspects or situations portrayed in the study. Each interview lasted on average 25 min. The interviews were recorded and were then fully transcribed as recommended by Bardin [7]. The assumptions presented by this author were followed for the content analysis methodology.

During the pre-analysis stage, a quick first reading of the content of the interviews was performed, regardless of the topics most frequently mentioned by the participants in their answers. Instead, the aim was to observe each new topic, idea or issue to organize the ideas that were similar and those that were brand new.

During the analysis of the data gathered, focus was particularly placed on the choice of registration and coding units, and then the coding of the data into categories and subcategories, according to the registration units previously defined, was performed.

Throughout the whole process of data organization: pre-analysis, data analysis (coding), processing of the results (categorization) and inference [7], the use of NVivo computer program (version 11) was required. This program allowed for a greater accuracy and control of the analysis of the data obtained from the interviews. During the study, precautions were taken to ensure data confidentiality and the participants' anonymity. To achieve these objectives identification codes were used, in which the letter "E" corresponds to the interview and the numbers 1 to 20 correspond to the sequence in which the interviews were conducted.

The study was carried out by six researchers but only the first three were involved in the coding process. Initially the process was conducted by the first two authors who discussed the coding criteria, especially issues related to the definition of the registration and context units. They were both responsible for the definition of the categories according to the subcategories that emerged from the research. Subsequently, these criteria were shared with the third researcher who ultimately validated the coding process. The last three authors were involved in the design of the theoretical framework and in the final review of the article.

In order to answer the aforementioned research questions and to implement the study objectives, several categories were taken into account (10). Because of their extension, only three of those categories will be considered in this article: the meaning of CCS consultations to women; the reasons why women don't go to CCS consultations and the nursing interventions women value during their CCS consultation.

Data regarding the participants' background show that the 20 participants interviewed are all recorded users of the PHCU of a Local Healthcare Unit located in the centre of Portugal; 19 of them are Portuguese and one of them is Brazilian. Their age ranges from 33 to 64 years, with a median age of 47 years.

As for their marital status, fifteen are married, three are divorced, one of them is a widow and another one is currently in a non-marital relationship. As far as education is concerned, data show that six of the participants have concluded lower secondary education, five have completed basic education, five others have a higher education degree and one of them has a master's degree.

As for professions, data tells us that five of the participants are nurses, three are retirement home assistants, three are operation assistants, two are technical assistants, and two are factory workers. There was also an occupational therapist, a cook and a retirement home caregiver. Two women were unemployed by the time the interview was conducted.

3   Results and Discussion

The objectives and the research questions were the structuring axis of the study and influenced the different phases of data analysis. Three categories arose from those initial questions: the meaning of CCS consultations to women; the reasons why women don't attend CCS consultations and the nursing interventions women value during their CCS consultation. Those categories are presented below. Tables will be used to present data where the letter "n" represents the number of women who mentioned a given subcategory; the "ur" abbreviation refers to the number of registration units.

3.1   Meaning of CCS Consultations to Women

Based on the analysis of the interviews' verbatim, we are able to identify different meanings women attached to CCS consultations. One of the most frequently mentioned was "checking for abnormal cells" which was referred by 18 women; 13 of the women referred to "disease prevention". The following answers express the meanings attached by the participants to CCS consultations:

> "Cervical cancer screening consultations are important to detect situations that may subsequently lead to cervical cancer" (E9)
> "Going to these consultations is important to achieve early prevention. Situations that will later develop into cancer might be detected prematurely. If women had participated in those screenings and had taken their cytology smear tests, their medical condition would not have been so serious... they would have been treated earlier (...) This consultation is useful to prevent cervical cancer" (E13)
> "This consultation is important to prevent cervical cancer" (E20)
> "I went to that sort of consultation once and I discovered I had a wound on my cervix. I underwent early treatment and everything turned out all right" (E9)

Data concerning the meaning attributed by the participants to CCS are exhibited in Table 1.

Table 1. Meaning attributed to CSS

Category	Subcategory	n	ur
Meaning attributed to CCS	Disease detection	18	45
	Disease prevention	13	26
	Information about the disease	8	16
	Confirmation that everything is fine	7	11
	Treatment of the disease	6	8
	Counseling and sharing	1	1

When asked about what cervical cancer screening consultations mean to them, 18 of the 20 women interviewed stated that this screening is an important way to detect cancer. This general position shows how significant disease detection is for women, a position that is clearly supported by some authors [8] whose studies highlight the role played by Pap smears in the early detection of cellular lesions before they could turn into cancer. According to the HPV Institute [9], when women become infected with HPV and the infection is not treated on time, may lead to the appearance of abnormal cells in the cervix. If they are not discovered and treated on time, those abnormal cells may evolve from a precancerous situation into cancer. This conception supports a position that prizes diagnosis or the early information about cellular alterations and is in clear contrast with the meaning attached to CCS found in the fourth subcategory which states that women go to those screening predominantly to confirm that everything is fine. This position was selected by seven of the participants. It is as if two paradigms are in clear contrast: the former that values the identification/detection of the disease, reported by eighteen of the women, and the role it plays in the treatment of the disease, reported by six of the participants, and the latter that prizes the health promotion paradigm that no one seems to value.

Such concept is deeply rooted in a society where the main focus is placed on curative medicine neglecting preventive medicine and health promotion, with all the consequences that it may entail. If the CCS consultation is no more than the act of detecting the disease, women can unwittingly assume that if they don't go to these consultations the disease won't be detected and, therefore, they won't get sick. It's sort of escape strategy from an unpleasant possibility.

This point of view is also shared by other authors [10] who state that preventive behaviors, such as Pap smear, may foster negative feelings, which are experienced by women in a very specific way reflecting their individuality and their socio-cultural background.

Disease prevention was the answer given by 13 participants and became, therefore, the second most mentioned subcategory. This assumption is supported by some researchers [11] who consider prevention to be the most effective way to control oncologic disease. On the other hand, when a woman adopts a preventive attitude, this behavior is determined by her own beliefs and perceptions of what health and disease really mean, by her perception of the prevention test, by her own experiences or by those lived and shared by others [12]. The fact is that, according to the Portuguese Gynecology Society [2], women have to adopt primary and secondary preventive measures to reduce the occurrence of the disease and the mortality it normally causes.

3.2   The Reasons Why Women Don't Attend CCS Consultations

Cervical cancer screening is an important tool for early cervical cancer detection, however there's a significant number of women who do not take it periodically, as it is recommended. Following the analysis of the interviews, we could identify different reasons why women don't go to screening consultations. The most common reasons were the doctors' poor availability and the users' work schedule. Each one of those reasons was identified by 6 of participants.

The following are participants' answers that describe some of the meanings attached to CCS by women.

"Years and years went by, one, two, three, four and they never called me... The appointment never happened... When I wanted to take the test, the only thing Doctor X would do was require the tests. Then he would leave them on his desk and that was it... Later, he would take a look at the tests and that was it...Actually, I have never been examined by the doctor.
"I missed this last consultation because I was working, I couldn't make it to the appointment"
(E5)
"The reason why I didn't show for CCS all these years was the lack of information. It was not because I didn't want to go, but simply because I didn't know these consultations existed" (E4)
"I was too scared and too embarrassed to go to the consultation... (...) I don't feel comfortable being examined by a male doctor... That's the reason why I missed the examination" (E1)

Data about the reasons why women don't go to CCS consultations are exhibited in Table 2.

Table 2. Reasons why women don't go to CCS consultations

Category	Subcategory	n	ur
Reasons why women don't go to CCS consultations	Unavailability of doctors to schedule	6	14
	Patients' Work Schedule	6	9
	Being unaware of the existence of CCS	5	17
	Feeling embarrassed to be examined by a male doctor	5	9
	Carelessness	2	7
	Having lived abroad	2	2
	Living in difficult conditions	1	5
	Unavailability of doctors to examine the patients	1	4
	Menstrual cycle	1	2

As we analyze the category "Reasons why woman don't go to cervical cancer screening consultations", the most frequently mentioned causes are extrinsic in nature and are related to the way health care service and health care workers schedules are organized. The poor availability of the doctors to call the patients for a CCS appointment and the fact that the patients' work schedule does not allow them to go to the consultations are the reasons that have been referred most insistently by the participants (both reasons were referred by 6 of the women surveyed).

The poor availability of the doctor to examine the patients is mentioned by one woman only. In this particular, other authors [8] have suggested the existence of several reasons that have led to the interruption of treatments or to the inexistence of CCS consultations: the lack of healthcare personnel in the unit or the fact that the healthcare provider responsible for these kind of tests was absent the day the consultation had been scheduled, the long waiting lines, the fact that the system is unable to meet the users' needs, the fact that it is too difficult to find the time to make a new appointment, the endless waiting period people have to go through in order to get the right consultation date. Some of the participants also complain that sometimes healthcare providers don't even order the tests.

With the current scientific and technological progress, it is hard to believe that healthcare services are not sufficiently developed to ensure an electronic system that will control the doctors' appointments and that will simultaneously inform healthcare personnel that they are failing to do their job or to ensure extended consultation hours outside normal working hours that would be made available on some days of the week. Users would be informed in due time of those services, of course.

Since Portuguese Healthcare Administration Services have been taking the centrality of healthcare information systems into its own hands (SClinico is a good example of that) it is really hard to accept that these women have to experience situations in which a doctor is not available to make an appointment or to see his patients or that they have to struggle with the impossibility of getting a CCS consultation because of their work schedule. However, some of the reasons why they don't take CCS consultations are intrinsic to women: not being aware of the existence of such consultations or feeling embarrassed because they had to be examined by a male doctor are reasons referred by five women respectively.

These two reasons are based on a set of beliefs and feelings experienced vis a vis CCS consultation, based on psychosocial or cultural factors and that will undoubtedly influence decisions, behaviors and attitudes, such as the patients' decision not to take those consultations.

These factors are emphasized by some authors [13] who claim that this is due to the lack of awareness of uterine cancer, of the Pap smear and of the way it is performed. Among the causes identified, there are also different types of access barriers and some other personal reasons. There are those [14] who suggest the existence of intrinsic reasons - like the patients' embarrassment and concern and their fear of the diagnosis - and claim that those are the motives why women between 40 and 65 don't go to CCS consultations. There are also those who associate these feelings of embarrassment, nervousness and fear with the sexual constraints resulting from the way these women were brought up and with the lack of information [15]. Other authors [12] also consider embarrassment to be one of the obstacles that prevent women from taking this medical exam.

It should be stressed that a user-driven health system would seek to ensure contexts that would help its users overcome the embarrassment and discomfort they might feel and would make its best efforts to ensure users' privacy and to offer them extended consultation hours that wouldn't collide with their work schedules.

3.3   Nursing Interventions Women Value During Their CCS Consultation

The women interviewed identify a set of nursing interventions they value when they attend the CCS consultation. The information nurses provide to the patients, reported by 16 of the 20 participants, how it helps them feel confident, referred by 9 of the participants, and how it helps build a closer relationship, referred by 6 of the participants.

The participants' answers gave rise to some registration units that express the type of nursing interventions women value during their CCS consultations:

"The nurse's role is to teach (...) to inform us" (E3)
"The nurse's role in this consultation is to inform us. There are many people like me who do not know (...) Their role is to inform us (...) It is up to the nurses to explain how things work (...) The work carried out by the nurse in this consultation is important. They teach us and tell us how to prevent the disease" (E6)
"The nurse's role in this consultation is to help us feel good" (E15)
"They [the nurses] are closer to us... we spend more time together (...)" E16

Data related to the nursing interventions valued by women during their CCS consultations are shown in Table 3.

Table 3. Nursing interventions valued by women during CCS consultations

Category	Subcategory	n	ur
Nursing interventions valued by women during CCS consultations	To Inform	16	40
	To make them feel confident	9	13
	To build a closer relationship	6	14
	To clarify patients' doubts	5	7
	To help them relax	5	6
	To cooperate with the doctors during the medical exam	5	6
	To provide support	4	8
	To raise awareness about the importance of CCS	4	5
	To advise	3	9
	To help achieve a good understanding	2	2

As healthcare providers, it is essential that health professionals, especially nurses, are aware of the factors that women value during their CCS consultations.

This study shows that women want nurses to be more available and closer to their patients, they want them to be able to provide better and clearer explanation about prevention, diagnosis, treatment and access to CCS consultation. Some studies [16] also highlight the important role nurses play in CCS, because they are healthcare professionals whose action is deeply educational and who can provide women with the right information about the advantages of taking the Pap smear test and about the importance of safe sex. Other studies [17] add that nurses should carry out CCS control actions and give priority to cases in which risk, vulnerability and inequality criteria are evident.

So, in addition to their importance as healthcare educator, nurses must also implement programs that will encourage cervical cancer screening as a way to prevent the development of cancer through early surveillance. This notion is also suggested by some other studies [18] that claim that the quality of the service provided to the community has to improve and that every available resource has to be properly used. They claim that it is imperative to find governance and organizing strategies that will ensure the success of such program.

As an empowerment tool, nursing professionals should adopt strategies meant to increase women and general population's health literacy by seizing every opportunity to promote clarifications and teaching. Training actions aiming at specific groups and awareness-raising campaigns on cervical cancer prevention meant for the entire population are also some of women's expectations.

Every citizen is responsible for promoting his/her health, but healthcare professionals, particularly those who are working in primary health care, are responsible for developing awareness-raising strategies that will be aimed at the entire population.

The Portuguese Ministry of Health [19] claims that the empowerment of the population is a priority, and the two fundamental cornerstones of such action are health promotion and disease prevention which are cross-cutting issues that exist in all levels of health care provision.

4   Conclusions

The main conclusions of the study presented below are based on the research questions designed.

One of the challenges health care system has to face – to achieve a greater presence of women in CCS consultation – is huge and requires systematic and concerted health-promoting policies. To this end, awareness-raising actions and community-based initiatives aimed at different generations of women and that will meet their expectations and needs have to be efficiently coordinated. The introduction of such changes will require a massive effort on the part of all health professionals but especially on the part of nurses. But we recognize that achieving a greater presence of women in CCS consultations is an objective that largely depends on their own capacity to overcome personal and cultural barriers strongly rooted in healthcare services. That is why it is so important to understand and value the experiences of each one.

When they attend a CCS, women focus mainly on the detection of the disease. Disease prevention is only their next concern. The reasons why woman don't go to CCS consultation that stem from this study are fundamentally related to the health services they are offered – they have mostly to do with the unavailability of the doctor to call the patients to a CCS appointment and the fact that patients can't seem to balance their work schedule and the CCS consultation schedule. However, intrinsic conditions affecting women's behavior were observed as well: the embarrassment caused by the presence of a male doctor, a common feeling of discomfort and the fact that many women don't know that CCS even exists. These intrinsic or endogenous constraints are naturally associated with women's beliefs and taboos, as previously mentioned, and are intimately related to the life experience of each one of those women.

As for the nurses' interventions women value most when they decide to take this kind of consultation, are the informing and inspiring confidence.

The limitations of the present study are related with its dimension which render it impossible to present the totality of the data that would otherwise provide a more complete understanding of the object of study.

Because of the results achieved, this study is particularly relevant to boost the awareness of women who are at risk due to their lack of information. In view of these results, this is a path that healthcare services and professionals have to follow if they truly want to implement a number of significant changes, interventions and programs that need to be carried out as soon as possible.

Now that we know the meanings women attribute to CCS consultations, the reasons why they refuse go to those consultations and the type of nursing interventions they expect, we can surely say that this study can significantly contribute to a clearer understanding of these women's perceptions vis a vis a public health service that has to excel in the care it provides to its users and that will have to empower nurses with the authority to come forward with new proposals involving the implementation of screening programs and other conditions to ensure the pursuit of excellence in health care provision.

On the other hand, those results can lead to an increasing demand for CCS since better informed women will finally accept it as one of the most effective preventive measures to avoid cervical cancer. It is clear that the results obtained in this study will have to be shared with health care professionals who are specialized in gynecology or oncology so they might become aware of all the changes to be made and that will help them modify their professional practices in favor of women's health and of the well-being of the whole society.

Funding and Acknowledgment. This study is financed by national funds by FCT - Foundation for Science and Technology, I.P., as part of the project UID/Multi/04016/2016. Funding is supported by FCT and CIDETS - Center for Studies in Technology and Health Education, Portugal. We would like to thank the following entities with whom we have worked: IPV - Polytechnic Institute of Viseu, CI & DETS - Center for Studies in Education, Technologies and Health, RESMI - Higher Education Network for Intercultural Mediation, Sigma Theta Tau International, Phi Xi Chapter, UICISA: E - Health Sciences Research Unit: Nursing Core UICISA: E/ESEnfC Higher School of Health of the Polytechnic Institute of Viseu.

References

1. Miranda, N., Gonçalves, M.B., Santos, C.A.e.G.: Programa Nacional para as Doenças Oncológicas 2017. Direção-Geral da Saúde, Lisboa (2017)
2. Sociedade Portuguesa de Ginecologia: Consenso sobre infecção por HPV e neoplasia intraepitelial do colo vulva e vagina (2014)
3. Direção-Geral da Saúde: Norma nº 018/2012 de 21/12/2012 (2012)
4. Acera, A., Manresa, J.M., Rodriguez, D., Rodriguez, A., Bonet, J.M., Trapero-Bertran, M., Hidalgo, P., Sanchez, N., de Sanjose, S.: Increasing cervical cancer screening coverage: a randomised, community-based clinical trial. PLoS ONE **12**, e0170371 (2017)
5. Trapero-Bertran, M., Acera Perez, A., de Sanjose, S., Manresa Dominguez, J.M., Rodriguez Capriles, D., Rodriguez Martinez, A., Bonet Simo, J.M., Sanchez Sanchez, N., Hidalgo Valls, P., Diaz Sanchis, M.: Cost-effectiveness of strategies to increase screening coverage for cervical cancer in Spain: the CRIVERVA study. BMC Public Health **17**, 194 (2017)
6. Amann, G.P.v., Monteiro, H., Lea, P.: Programa Nacional de Saúde Escolar 2015. In: Saúde, D.-G.d. (ed.) pp. 105 p. Direção-Geral da Saúde, Lisboa (2015)
7. Bardin, L.: Análise de Conteúdo (2016)
8. Santos, U.M., Souza, S.E.B.d.: Papanicolau: Diagnóstico Precoce ou Prevenção do Câncer Cervical Uterino? Revista Baiana de Saúde Pública **37**, 941 (2014)
9. Instituto do HPV: Guia do HPV (2013)
10. Rafael, R.d.M.R., Moura, A.T.M.S.d.: Modelo de Crenças em Saúde e o rastreio do câncer do colo uterino: avaliando vulnerabilidades [Health Belief Model and cervical cancer screening: assessing vulnerabilities]. Revista Enfermagem UERJ **25**, e26436 (2017)
11. Macip, S.: Cancro: Conhecer, Confortar, Vencer (2013)
12. Ferreira, M.d.L.d.S.M.: Motivos que influenciam a não-realização do exame do Papanicolau segundo a percepção de mulheres. Esc Anna Nery Rev Enferm **13**, 378–384 (2009)
13. Santos, A.C.S., Varela, C.D.d.S.: Prevenção do Câncer de Colo Uterino: motivos que influenciam a não realização do exame de papanicolaou. Revista Enfermagem Contemporânea **4**(2), 179–188 (2015)

14. Leite, K.N.S., da Silva, J.P., de Sousa, K.M., Rodrigues, S.d.C., de Souza, T.A., Alves, J.P., de Souza, A.R.D., Rodrigues, A.R.d.S.: Exame Papanicolau: fatores que influenciam a não realização do exame em mulheres de 40 a 65 anos. Arquivos de Ciências da Saúde **25**, 15–19 (2018)
15. Peretto, M., Redivo, D.L.B., Reckziegel, B.H.M.: O não comparecimento ao exame preventivo do câncer de colo uterino: razões declaradas e sentimentos envolvidos. Cogitare Enfermagem 17 (2012)
16. Barbosa, L.R.: Intervenvenções de Enfermagem Utilizadas Enfermagem Utilizadas no Rastreamento Precoce do Câncer Cervico Uterino: Revisão Integrativa Rev. de Atenção à Saúde 13(44), 94–99 (2015). 13 n44.2530
17. Junior, J.C.O., Oliveira, L.D.d., Sá, R.M.d.: Fatores de adesão e não adesão das mulheres ao exame colpacitológico. Revista Eletrônica Gestão & Saúde **6**,184–200 (2015)
18. Moutinho, J.M.: Manual de Procedimentos do Rastreio do Cancro do colo do Útero - Unidades de Patologia Cervical. Portugal, Administração Regional de Saúde do Norte, IP, Departamento de Estudos e Planeamento, Coordenação Regional dos Rastreios Oncológicos Porto: ARSN (2009)
19. Ministério da Saúde: Retrato da Sáude 2018. In: SNS (ed.) Ministério da Saúde, Lisboa (2018)

Measurement of the Difference Between Students and Teachers' Average Networks Using Associative Pathfinder Networks

Juan Arias-Masa[1]([⊠]) (ID), Juan Ángel Contreras-Vas[1] (ID),
Violeta Hidalgo Izquierdo[1] (ID), Rafael Martín Espada[1] (ID),
and Juan Arias-Abelaira[2] (ID)

[1] University of Extremadura, Mérida, Badajoz, Spain
{jarias, jaconvas, vhidalgo, rmmartin}@unex.es
[2] University of Alcalá, Madrid, Spain
juan.ariasa@edu.uah.es

Abstract. This article includes an experimental study of research in higher education. The experience has been carried out with the Systems Interconnection students of the Degree in Telematics Engineering of the University of Extremadura and describes how the teacher is able, with the help of the Associative Pathfinder Networks and their similarity test, to obtain, in real time, reliable information of the cognitive structure of his students according to a certain subject of study that is imparting. It is a practical application of the Theory of Nuclear Concepts of doctors Casas y Luengo. The basic objective is to be able to determine if it is possible to obtain information in real time from the assimilation that students have about the contents taught by the teacher during the instruction. For this, 6 concepts of the subject were chosen and with them an evaluation of similarity of concepts was made in three instants of the teaching-learning process, for each one of those instants of the instruction the average networks were calculated and these were compared with the network of science, obtaining a high similarity especially in the instant after the exam where you can see how the student group has assimilated the concepts under study.

Keywords: Theory of nuclear concepts · Average network of students · Higher education

1 Introduction

The actors of university education are increasingly aware that the teaching-taking process is as effective and efficient as possible. Thus, since the establishment and creation of the European Higher Education Area (EHEA) in Spain [1], based on Bologna's model [2] there are many teachers and research groups that are aware that the university should not only be open to public debate about its purpose and meaning in this 21st century but it should be converted into an engine of economic development. Thus, it has to improve the connection with the working market, it has to worry about its graduates, because all of this is a recurrent theme in recent times of educational policy of higher education.

© Springer Nature Switzerland AG 2020
A. P. Costa et al. (Eds.): WCQR 2019, AISC 1068, pp. 136–144, 2020.
https://doi.org/10.1007/978-3-030-31787-4_11

The educational process, or teaching-learning, is a cycle where feedback is very necessary from the whole process to the teacher. Thus, if he quickly receives information about whether the students are assimilating the content and skills that the teacher is trying to convey, there will be more possibilities to vary the course in this teaching-learning process.

The knowledge of the cognitive structure of the students by the teaching teams through knowledge representation techniques such as the Pathfinder Associative Networks (PAN, hereinafter) [3] will be very useful to make those decisions about the teaching-learning process.

There are many works developed by this research team about the use of Izquierdo [4] and Contreras [5], as well as the theses developed in the Ciberdidact research group [6], from the first one Casas Garcia [7] to Corcho [8], and others several such as Arias [9] Contreras Vas [10], etc. In all of them, we have worked extensively with the PAN and they are based on the Theory of Nuclear Concepts developed by García [11]. A summary of this theory can be found Bizarro Torres [12], namely:

- Concepts are not learned in isolated form, they are associated with others in the form of structure and they are forming networks.
- The networks that are created throughout the evolutionary development have fewer concepts, but they are better linked.
- The most important concepts of cognitive structure are not only the most significant because of their degree of generality or abstraction, it also has to do with the teacher's examples.
- For this, the core concepts around which networks are formed are those that are more significant to the student.

To obtain data of the cognitive structure of the students and later these data can be represented graphically with computer programs for their visual evaluation by the teaching team, as well as, to be able to make comparisons of similarities between them we have the PAN that can obtain their data based on "Similarity Scores" of concepts described in Schvaneveldt [3].

Fig. 1. Access to MeBa server with registered user

This research team obtains similarity scores, in real time and through the internet, by using the Meba server [13]. To do this, it uses the collection of distance information between concepts. The server Meba allows access by registered user with login and password as we can see in Fig. 1, so each student will be reflected in the server database the information provided when the test similarity of concepts.

2 Objective

The objective of this research is to be able to determine the viability of the teacher having information, in real time, about the approximation of the students' networks to the teacher's one as the subject progresses. The students' networks represent the cognitive structures of these in relation to a certain topic. We will represent these structures with Pathfinder Associative Networks based on distance matrices.

3 Experimentation

This study has been carried out in the first topic of the subject of Systems Interconnection [14] in the course 2017–18. This subject belongs to the degree in Telematics Engineering that is taught at the University of Extremadura. They are students in the second year and it is the first time they have done this experience. This first topic of the subject is called "Introduction to the link layer", and within this topic the concepts under study are:

- **Protocol.** It is the set of rules or specifications (syntactic, semantic and procedural) that govern the exchange of correct information and efficiently between entities (devices or elements) of level (N) located in different systems {even entities (N)} in an open communications environment.
- **Primitives (N).** It is an order or invocation of a particular service by which an entity (N) requests an entity (N–1) a certain communication service (connection, transfer or release) with a pair, or the answer to that request.
- **Even entities (N).** Entities of the same level (N) of different systems that cooperate each other.
- **Levels or layers (N).** It is a set of entities of a certain rank that belong to different systems, they cooperate each other and with the entities of the rank immediately inferior in their respective systems to communicate the entities of the immediately superior rank of those systems.
- **Functions (N).** They are the tasks, activities or group of tasks of level (N) with own meaning that the entities realize to give a certain service.
- **Point of access to the service.** It is the address where the exchange of primitives and parameters between two levels takes place.

In addition to those concepts by which they are asked for and that can be studied in [15] there are several basic concepts of study such as: Level (N) services, Parameter, Interface, System, etc. But in the research studies that the group Ciberdidact [6] has carried out using the data collection for the PAN, it is recommended that no more than

6 or 7 concepts were used in each sample or take of information. This is because the number of comparisons between pairs of concepts increases and makes the individuals who perform the test may not take the interest that the test deserves. Therefore, if the that happens, the coherence of their networks weakens, and consequently, the data will not be valid. Therefore, this study has been limited to those 6 concepts previously cited and described.

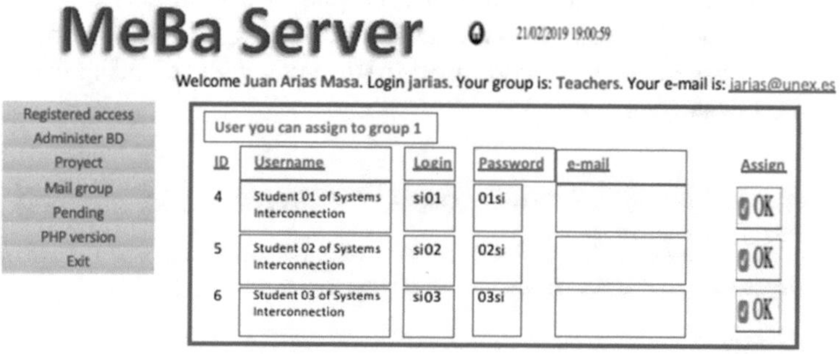

Fig. 2. Example of assigning users to groups

Once the concepts that are the subject of the study have been determined, the next step is to register all the necessary information in the Meba server in order that the students can perform this evaluation in real time. To do this, each student is registered with their login and password, and they are also grouped in the same working group. Subsequently, an assignment is made between groups and topics, so that students, when they enter the system, can go directly to perform the requested evaluation. In Fig. 2 you can see an example of how this assignment is made. In addition, it is necessary to assign concepts to a specific topic. This is also done very easily with

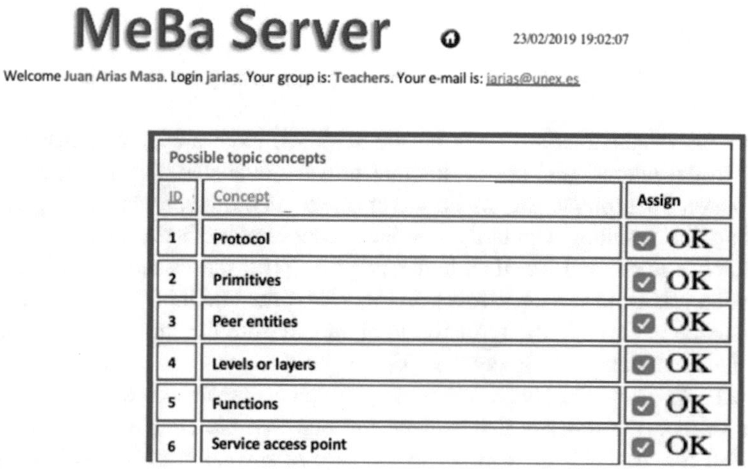

Fig. 3. Example of assignment of concepts to topics

Meba, see Fig. 3. After having the assignments, it is only necessary for students to enter as registered users with their login and personal password and perform the Pathfinder evaluation.

For the evaluation of similarity of concepts, these will be shown two by two to each student as it can be seen in Fig. 4. In that window of information capture, where in addition to presenting the instructions in both Spanish and Portuguese, the student must "click with the mouse" in the lower triangle. This triangle will change color as "clicked" the student, to present two new concepts, so that all possible pairs will be shown. Thus, in our case we have worked with 6 concepts, 15 comparisons have to be made, that is, in general they are performed (n*(n–1))/2, with "n" being the number of concepts to be evaluated.

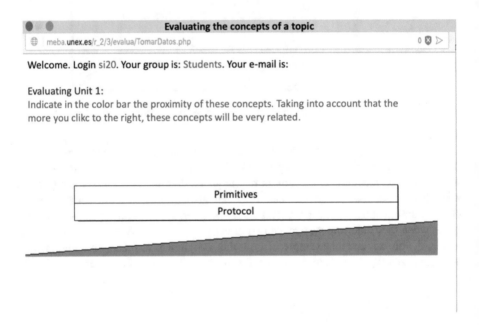

Fig. 4. Example of evaluation of similar concepts with Meba.

After the evaluations are done by the students, from any device that is able to connect to the internet and has an Internet browser available, that is, from any computer, Tablet, Smartphone, etc. Meba server offers the professor who has organized the evaluation the resulting similarity matrices associated with the evaluation of each student. With them, and the JPathfinder program [16], the teacher can obtain all the information related to these networks. In this study, the first thing that is done is to see the coherence of each similarity matrix to eliminate from the study those matrices that have a degree of coherence lower than 0.15 in the given answers. Starting, to eliminate these matrices from the study, we must proceed to create the average network or average matrix of similarity that will be the one that we compare with the science network or the teacher's network, or in the case of several teachers, with the average network of the entire teaching team.

```
Data is13
Similar
6 Nodes
0 decimal places
        0 minimum weight
        768 maximum weight
Lower triangular:
560
580   349
219   426   570
750   624   612   546
274   326   549   547   178
```

Fig. 5. Example of triangular matrix of similarity.

An example of a similarity matrix that Meba offers and that is then introduced in JPathfinder is shown in Fig. 5. In this figure we can see that the user who has performed the evaluation has been "is13", which is a similarity matrix, that have been worked with 6 nodes and in addition to being decimal values the range of values is between 0 and 768 (in Meba, what is really done is to read the pixel where the student "click" with his mouse).

4   Results and Discussion

In this experiment, data has been taken in three instants, namely:

- Before the instruction.
- During the Instruction.
- After the theoretical examination of topic 1.

Of these three tests, in this document we only present the results of the second and third data collection. This is done because their results are more valid because all the students' networks have valid coherence and the average network is close to science's one.

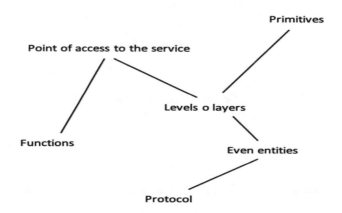

Fig. 6. Average Pathfinder Associative Network of the students during the instruction.

However, this research team has analyzed and compared data from the data collection "before the instruction" to the other two data points. But this research team believes that they are not scientifically valid, and we hope to be able to contrast them with new academic courses where the same data collection could be done.

The average network of the second data collection that took place during the teaching of the topic, is shown in Fig. 6. When we say during the teaching we mean that the complete subject has not already been explained, only the concepts on which the evaluation of similarity is concerned.

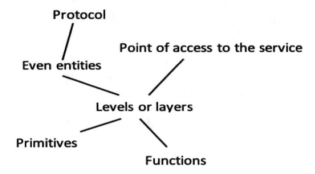

Fig. 7. Average Pathfinder Association network of students after the exam

The third data collection is shown in Fig. 7. When we compare both networks with each other, in front of the science network, Fig. 8, we find that both have a high similarity of 0.49 and 0.667 respectively. These values of similarity, whose range can vary between 0 and 1 are very high for being an average network, it is the average of the students not the similarity value of any student.

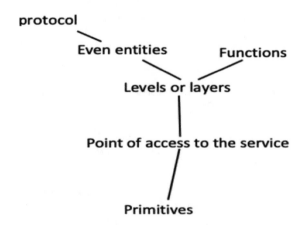

Fig. 8. Associative Network Pathfinder of the science network

In any case, in the three networks, the central node is the one of "levels or layers" that is a nuclear node, as it is defined by [17] where it is indicated that *"Nuclear nodes. A node is considered to be a nuclear node the one which has three or more than three links.* " It is also said by [18] that "a nuclear concept is an important concept that users have more anchored in their mind and use to organize and structure their cognitive network." This nuclear node that coincides in the three networks is the basis of the definition of the rest of the basic concepts, all of them are based or are defined based on the concept of level N or layer N.

The visual difference that is seen in the average networks that students have after they have done the theoretical exam (Fig. 7), that is, they have studied the subject in depth, and the science network, is the node of "Primitive" that in the case of the science network is linked to the "Service access point", while in the student network it is linked to the "levels or layers" node. From the theoretical point of view of the subject, both networks are possible, because "the primitives" is the way in which a "level or layer" requests a service at a "lower level", and the site where "the primitive" is passed it is in the "access point to the service" for that both branches are valid. Therefore, we consider that the average network of students after they have studied the subject is very close to the network of science.

The comparison between the network of science and the average network of students during the instruction allows us to obtain real-time data of the perception that students are having of the contents that are being taught, and also in a non-invasively way [7]. In addition, in the results we have contrasted that both networks are quite close, the average of the students and the one of the sciences, this will serve as feedback in real time to the professor to be able to advance in his explanation.

5   Conclusions

The main purpose of this research was to determine if a teacher can have information in real time of the information that their students are receiving while a certain topic is being taught. We have chosen a study of 6 specific concepts of a theme and students have been tested on similarity to these 6 concepts. These tests have been done before the instruction (which in this study we have discarded), during the instruction and after the theoretical exam. Both the average network during the instruction and after the exam coincide with the network of science, which is the teacher's network in our case.

Therefore, we have achieved an easy method to be used by teachers, who quickly and in real time, they can see if the cognitive structures of their students start to approach the network of science, which represents the information that the teacher wants to transmit to their students.

References

1. Pallisera, M., Fullana Noell, J., Planas Lladó, A., del Valle Gómez, A.: La adaptación al Espacio Europeo de Educación Superior en España: los cambios/retos que implica la enseñanza basada en competencias y orientaciones para responder a ellos. Rev. Iberoam. Educ. **52**(4) (2010)
2. Curiel, M.L.M.: El proceso de Bolonia y las nuevas competencias. Tejuelo Didáctica la Leng. y la Lit. Educ. no. 9, pp. 19–37 (2010)
3. Schvaneveldt, R.: Pathfinder Associative Networks. Ablex Publishing, Norwood (1990)
4. Hidalgo Izquierdo, V., Arias Masa, J., Casas García, L.M., Luengo González, R., Castillo Martínez, A.: Redes cognitivas de alumnos y Red de la Ciencia: similaridad (2008)
5. Contreras, J.A., Arias Masa, J., Luengo Gonzalez, R., Casas Garcia, L.M.: Nuclearity indexes (full and reduced), as a contribution to the Theory of Nuclear Concepts/Indices de nuclearidad (completo y reducido), como aportacion a la Teoria de Conceptos Nucleares. RISTI (Revista Iber. Sist. e Tecnol. Inf.) no. E4, pp. 16–35 (2015)
6. Grupo Ciberdidact. http://www.unex.es/investigacion/grupos/ciberdidact
7. Garcia, C., Manuel, L.: El estudio de la estructura cognitiva de los alumnos a traves de redes asociativas Pathfinder. Aplicación y posibilidades en Geometría, Badajoz (2002)
8. Corcho, P.: "Enseñanza de los Elementos Notables del Triángulo utilizando Objetos de Aprendizaje y LMS", Tesis Doctoral. Universidad de Extremadura, Cáceres (2016)
9. Arias Masa, J.: Evaluación de la calidad de Cursos Virtuales: Indicadores de Calidad y construcción de un cuestionario a medida. Aplicación al ámbito de asignaturas de Ingeniería Telemática. Universidad de Extremadura (2008)
10. Vas, C., Angel, J.: Enseñanza por Competencias: Conceptos propios, requisitos previos e influencia en el rendimiento académico de los alumnos, para la asignatura de Bases de Datos, en los estudios universitarios de Grado en Informática., Badajoz (2016)
11. Casas, L.M., Luengo, R.: Teoría de los Conceptos Nucleares. Aplicación en Didáctica de las Matemáticas. Líneas Investig. en Educ. matemática. Badajoz, Spain Serv. Publicaciones FESPM (2004)
12. Bizarro Torres, N., Luengo González, R., Casas García, L.M., Torres Carvalho, J.L.: Aplicación de las Redes Asociativas Pathfinder al análisis de los conceptos forma, tamaño y color en alumnos con Discapacidad Intelectual (2015)
13. Arias-Masa, J.: Web del Servidor MeBa. Proyecto II (2017). http://meba.unex.es/. Accessed 22 Feb 2017
14. Arias-Masa, J.: Ficha 12a de Interconexión de Sistemas curso 2017-18 (2018)
15. Diaz-Diaz, M.: Técnicas y redes de comunicación de datos (2001)
16. Schvaneveldt, R.: JPathfinder. Las Cruces, NM USA (2017)
17. Contreras Vas, J.Á.: Enseñanza por Competencias: Conceptos propios, requisitos previos e influencia en el rendimiento académico de los alumnos, para la asignatura de Bases de Datos, en los estudios universitarios de Grado en Informática (2016)
18. Luengo, R.: La Teoría de los Conceptos Nucleares y su aplicación en la investigación en Didáctica de las Matemáticas. UNIÓN-Revista Iberoam. Educ. Matemática **34**, 9–36 (2013)

Indigenous People in the Brazilian Context: An Analysis of the Social Representations

Sílvia Barbosa Correia⬭, Luciana Maria Maia(✉)⬭,
Luana Elayne Souza⬭, Tiago Jessé Lima⬭,
and Samuel Figueredo Maia⬭

University of Fortaleza, Fortaleza, Brazil
silviapsi.barbosa@gmail.com, lumariamaia@hotmail.com

Abstract. This study aims to find out about the social representations of the indigenous peoples in Brazil, taking into consideration the images and meanings shared by the non-indigenous population. The findings are based on the application of the free association test and interviews with 38 participants. The material was submitted to an analysis of content supported by the theory of social representations, with the help of Atlas.ti software. The analysis presented 475 excerpts from narratives, grouped into categories: indigenous conceptions; origin of knowledge; relationship between indigenous peoples and society. The free association resulted in 190 evocations, classified into: distinct subject, subject with rights, excluded subject, and valued subject. The representations about the indigenous peoples are anchored in previous knowledge that places them in a condition of the primitive subject. The distancing of the population reinforces stereotypes that subjugate the indigenous peoples. Knowledge demystifies stereotypes, making the indigenous peoples respected in their own culture and, at the same time, recognized as a subject with rights, just like any other citizen.

Keywords: Social representations · Indigenous peoples · Brazil

1 Introduction

The figure of the Indian present in the history of Brazil has been constructed by the colonizers who came across the differences between themselves and the native peoples. The constructed image that they were inferior, backward people considered to be devoid of culture, precisely by associating them with wild animals, has established itself in the imaginary of the Brazilian people throughout history. The process of colonization, the contact between indigenous peoples and the colonizers, has changed the way of life and many of the characteristics of the ethnic groups scattered throughout the Brazilian nation. The consequences of this process of acculturation and economic and social power relations between non-indigenous society, state and the Brazilian Indians have strengthened the process of exclusion and the distance between indigenous and non-indigenous peoples [1].

Studies indicate that the image of the indigenous person is marked by stigmas that put them in an unfavorable condition vis-à-vis society in general. Lazy, savage, backward and opportunistic are some representations of the indigenous people, which disqualify and dehumanize the indigenous peoples in the face of society. The lack of knowledge and proximity to these groups, the influence of media that reinforces stereotypes, and economic interest's contrary to the way of life of these populations are some elements present in the social representations of the Brazilian Indian [2–5].

It is recognized that indigenous people have been suffering from various forms of exclusion and transformation in their condition vis-à-vis non-indigenous society: from savages to being protected and, only recently, being the subject of undeniable rights [6]. These transformations were permeated by denial, exclusion and extermination of indigenous groups. In Ceará, north-east Brazil, the denial of indigenous identity for more than a hundred years legitimized the idea that in that state the presence of these peoples no longer existed. Only in the 1980s did ethnic groups re-emerge demanding the recognition of their ethnic identity and the guarantee of their rights as an indigenous people [7].

Based on these preliminary considerations, this paper intends to discover the social representations of the indigenous peoples in Brazil, from a context that is characterized by the recognition of the indigenous person as a Brazilian citizen, while, at the same time, as someone who is distinguished by a violation of rights and manifestations of discriminatory attitudes [8]. The theoretical argument originates from the Theory of Social Representations, which allows for the study of the relationships between social and cognitive phenomena through communication and thought. The social representations have the function of making the strange into something familiar, from two constitutive processes: anchoring and objectification. The anchoring consists of the process from which the representation in the social space is rooted, allowing for the understanding of how the construction of a social representation takes place, from values, beliefs and knowledge. Objectification, however, consists of the process by which social groups construct a common knowledge based on shared exchanges and opinions [9]. As a specific theoretical contribution to social representations, it is based on the societal approach, which conceives of social representations as "organizing principles of symbolic relations between individuals and groups" [10]. In this approach, members of a population share certain beliefs about a given social relationship.

In this way, social representations are established from the relations of communication, which are possible as a result of reference points common to the individuals or groups involved in these symbolic exchanges. In relation to individual positions, the societal approach presupposes that individuals differentiate between themselves in the relationships they maintain with these representations. Finally, this approach recognizes that the anchoring of position statements is based on collective symbolic contexts, involving hierarchies of values, perceptions about relationships between groups, categories, and shared social experiences in groups [10, 11].

In this sense, social representations contribute to the definition of identity and guarantee the specificities of the groups, allowing its members to construct a positive social and personal identity in the light of prevailing and historically constructed norms and social values. It can thus be affirmed that social representations contribute to the process of social differentiation insofar as the specificities of social groups generate their own representations that can favor the establishment of attitudes and practices of social discrimination [12].

2   Method

2.1   Participants

The participants were a total of 38 people from the general population, considered as "non-indigenous", aged between 18 and 50 years, the majority of them female (N = 27, 71,%), single (N = 29, 76.3%), and with incomplete higher education (N = 17, 44.7%), living in the city of Fortaleza, state of Ceará, north-eastern Brazil. The number of participants was considered enough according to the criteria of saturation of the data [13]. Table 1 shows the main sociodemographic characteristics of the participants.

Table 1. Characterization of participants.

Gender	f	%
Feminine	27	71,1
Masculine	10	26,3
Other	01	2,6
Marital status	f	%
Single	28	73,7
Married	08	21,1
Widowed	01	2,6
Other	01	2,6
Ethnicity	f	%
Caucasian	10	26,3
Mixed	23	60,6
Afro-Brazilian	04	10,5
Asian	01	2,6
Religion	f	%
Catholic	16	42,2
Protestant or Evangelical	05	13,2
Spiritualist	06	15,8
None	09	23,7
Other	02	5,1
Education	f	%
Elementary	01	2,6
Middle	10	26,3
Higher incomplete	16	42,2
Higher completed	11	28,9

2.2   Instruments

For the data collection, all 38 participants answered three instruments: (a) a sociode-mographic questionnaire, with the following items: age, gender, birthplace, marital status, level of education, religion, color/ethnicity perception and monthly income; (b) Free Word Association Test (FWAT) - participants were asked to say the first five words they thought of when they heard the word "indigenous"; and (c) semi-structured interview script, composed of 10 questions about conceptions about this ethnic group, knowledge and ways of accessing this information.

In the case of this research, because the object of investigation is far from the investigated population, the preliminary use of the FWAT facilitated the involvement of the participants in the interview. The FWAT is considered a technique easy to under-stand, administrate and adapt to the interests of a given study [14]. In this research, the semi-structured interview was used as the main instrument and the FWAT as a com-plementary instrument. Both techniques have been used in qualitative studies [15].

2.3   Procedures for Collecting and Analyzing Data

The instruments were applied in April 2018, in public places in the city of Fortaleza. On the occasion, after being given information about the objective of the research, the Informed Consent Form was presented and signed, which contained information about the research and the rights of the participants.

The terms evoked from the FWAT were recorded, later typed into a spreadsheet, grouped into five categories, and for each category the rates of evocation were cal-culated. The categorization was made according to adjectives that translate the way in which the non-indigenous person understands and defines the indigenous person and that reflects the way she/he relates to them. The proposed categories emerged from the reading of the terms evoked by the participants, with reference to previous studies on the theme [1–5].

The interviews were recorded and transcribed and later added into the software Atlas.ti, version 8.0, composing the textual body of analysis. Atlas.ti is a software of the CAQDAS (Computer Assisted Qualitative Data Analysis Software) category and has been increasingly used for the purpose of analyzing large amounts of data, con-tributing to the rigor and scientific accuracy of qualitative data analysis. Atlas.ti can be used with different types of theoretical-methodological approaches, starting from the analysis of a hermeneutic unit, which allows us to make conceptual relations and have a view of the whole of the object investigated [16, 17].

The treatment of the content from Atlas.ti was initially based on the preliminary analysis of the interviews, considered as the primary analysis documents. The data were organized according to the precepts of content analysis which is divided into three phases: pre-analysis, material exploration and interpretation. The analysis was based on the selection of quotations, codes, code groups, elaboration of memos and outputs.

3   Results

The FWAT for the inductive stimulus "indigenous" enabled the evocation of 190 words, which were organized into five categories, and were then indicated to be about this ethnic group. The categories and the frequency with which the terms were evoked were: distinct subject (86), primitive subject (58), subject with rights (22), excluded subject (15) and valued subject (09). As explained, these data were taken in a preliminary way, in order to bring the participants closer to the research object. The FWAT results can be seen in Table 2.

Table 2. Categories of the Free Word Association Test for the term "Indigenous"

Categories	f	%
Distinct subject	86	45,3
Primitive subject	58	30,5
Subject with rights	22	11,6
Excluded subject	15	7,9
Valued subject	09	4,7
Total	190	100

The analysis of the interviews, with the help of the Atlas.ti, indicated the existence of 475 excerpts of narratives, which were grouped into three broad categories: Conceptions about the indigenous people, Origin of knowledge and Relationship between indigenous and non-indigenous peoples, and then divided into 14 codes. The categories were created from elements that could correlate and, in this way, contribute to the understanding of the structuring of social representations about the indigenous people in the Brazilian context. Specifically, the elements that were used as possibilities to analyze the constitution of the social representations about the indigenous people were: knowledge about the indigenous people; origin of this knowledge; idea about the indigenous people; and how society in general relates to this population. Besides these elements, it was considered in the analyzes the social representations of social groups that can interfere in the way the relationship with and between its members is established. In the context of indigenous populations, the intergroup relationship occurs between general society and indigenous peoples of different ethnicities and contexts. The codes created give consistency to the categories [11]. The structure of content analysis of interviews, with categories and codes, can be visualized in Fig. 1.

The first category: Conception about the Indigenous person, with 288 excerpts from narratives, was subdivided into three subcategories: (i) characterization of the Indigenous person, which has the following codes and frequencies: physical characteristic (20); psychological characteristic (13); sociocultural characteristic (92), totaling 125 excerpts from narratives; (ii) definition of indigenous person (54); and (iii) differentiation, which has the following codes: differences between indigenous and non-indigenous persons (71), differences between primitive and civilized indigenous persons (38), totaling 109 excerpts from narratives.

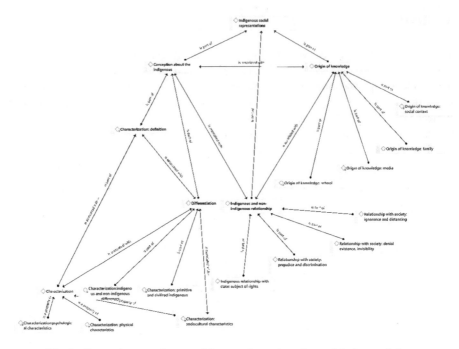

Fig. 1. Semantic network created from codes, categories and their correlations.

In relation to this first category, it is possible to argue that the conceptions about the indigenous people are permeated by aspects ranging from the physical characterization, linked to stereotypes shared in the imaginary of the non-indigenous person; passing through psychological aspects, in which elements of behavior appear - the way of being of the indigenous people; reaching the most evident element for the non-indigenous person, that is the socio-cultural differentiation, in which there are the characteristic aspects of the indigenous culture that distinguish the difference between these ethnic groups and the society in general. The narratives of the following participants illustrate this discussion: "we identify the Indian by his features, such as straight hair, narrow eyes, brown skin and big mouth" (physical characterization); "They speak little, they observe much" (psychological characterization); "They worship their gods, have religions, herbs, I don't know much", "I believe they have rituals, shamans … something more with nature and spirits", "They are more collaborative, like living in a large family" (sociocultural).

It is possible to understand the predominance of certain socio-cultural characteristics over others from the evidence that these aspects favor the creation of a symbology about an ethnic group that, at first, is not familiar. Social representations also emerge as a way of giving a symbolic value, understanding a particular object, defining it, and giving it identity [9]. In this way, conceiving and explaining what is "being indigenous", from elements of culture, enables the understanding of the non-indigenous person.

The results of the FWAT seem to reinforce the evidence of socio-cultural aspects from the higher frequency of words evoked in the categories "distinct subject" (45.3%) and "primitive subject" (30.5%), which bring together certain terms (culture, headdress, dance, rituals, painting, bow, arrow, hut) that symbolize how striking this aspect is in constructing the conception of the indigenous individual from the non-indigenous. In the theory of social representations, anchoring is classifying, naming something [9]. The anchoring is identified in symbolic aspects of the "primitive Indian" who had survival habits as a savage.

The second category: Origin of knowledge about the indigenous people, presented the following codes: school (31), media (18), social context (16) and family (1). These codes totaled 66 excerpts of narratives.

In this category narratives emerged in which the participants highlighted different contexts in which they had access to knowledge about the indigenous peoples. The predominant source of knowledge is school, where one acquires knowledge of the history of Brazil and, consequently, the history of the indigenous peoples in the period of colonization of the country. The following narratives illustrate this form of access to knowledge: "I know through what was done in school", "I know through books, actually, when I was still studying ... just this", "I remember a little from the time in school, it was glossed over, but it's been so long". The second source of knowledge is the media in general. News, reports, documentaries are some examples of publications that favor the construction of a social representation regarding a certain phenomenon or object. For example, "what I know is what we see on television", "I know only what I saw in school and what I see in the newspapers", "in the reports they say they are more modern, that they are no longer naked". The third source of knowledge is the social context, that is, the contact or experience that the participant has experienced is able to modify his or her understanding of the indigenous people and the issues that pervade their reality. Some narratives indicate this aspect: "I have already studied at FUNAI (National Foundation of the Indian), so I dealt a lot with the natives and I saw them practically every day ... I saw their struggle with the lands, and I started to study, read the indigenous legislation, and saw how history was distorted", "Today studying anthropology, they are nothing like we think they are, that vision of the primitive Indian. Actually, they are civilized". The last source of knowledge was the family. The identification of a participant whose family is of indigenous origin reveals an approximation with the reality of this ethnic grouping, providing further knowledge. "I know through history, because I studied ... my family is also of indigenous origin".

The evocations of the FWAT reinforce the idea that the school, the media and the social context contribute to forming the idea of the indigenous people as a primitive subject: "forest", "nature", "wild", as an excluded subject: "discriminated", "intolerance", or as a subject with rights: "resistance", "struggle", and "land".

The third category: the relationship between indigenous and non-indigenous people, generated the following codes: recognition of prejudice and discrimination of society (76), ignorance and detachment from society (24), denial of existence/invisibility (4) and, finally, recognition of the indigenous people as a subject with rights (17), totaling 121 excerpts from narratives.

In this category, narratives emerged that explain how society and the State relate to indigenous peoples in the Brazilian context. The recognition of prejudice and discrimination is predominant in the narratives of the participants, even if it does not present itself as a personal positioning. Some excerpts from the narratives explain this aspect of the relationship: "There are many prejudiced people, who consider them layabouts, who do not work. They are not able to perceive the warriors they are!", "someone ignorant, uncivilized", "someone who is too outdated or something that no longer exists". Another position in relation to the indigenous population is the ignorance and distancing, verified in some narratives: "my knowledge is very superficial in this question, really very little", "I have never had any contact, to be very sincere. I do not know if they live well", "I do not know very much, but I know they are very connected with nature". Another form of relationship is the denial of existence, exemplified in the following narrative: "Actually, I do not think there is an Indian living in our social environment". Finally, narratives emerge that indicate how indigenous people relate to the State. There is recognition from the participants that the indigenous people have the same rights as the whole of society, as well as the same treatment given by the State. The narratives portray the idea of rights that are not very accessible: "Unfortunately they live in a precarious way. The government does not pay attention to them", "I think the Indians have achieved some rights, but they are still very discriminated against today".

The evocations of the FWAT that come closest to these narratives are those that were categorized as "excluded subject" (prejudice, discrimination) and "subject with rights (struggle, equality). In this category it can be identified that the approximation of indigenous groups by society in general, whether through knowledge or through contact or lived experience, allows different possibilities in relation to how to see, understand and relate to indigenous peoples.

4 Discussion

This paper presents the results of an empirical study that aimed to discover the social representations of the indigenous peoples in Brazil. Based on studies in the area, this work assumed as a presupposition that the indigenous people are a social minority that strongly experiences a reality of prejudice and discrimination in Brazilian society and that this process often occurs in a distinguishable way, through a stereotyped view, from the denial of rights and existence of these ethnic groups.

In order to reach the proposed objective, the use of the FWAT and the interview were used as instruments of collection, and then combined with the use of the Atlas.ti software for analysis and presentation of the findings. It was observed that the combination of these two collection strategies - FWAT and interview, was positive in the sense that they complemented the information and analyzes. The use of Atlas.ti helped in the process of coding, that is, naming the excerpts of interviews in order to allow grouping from the similarities of meaning, and then confirming the categories conceived from the interview script. The possibility of creating memos, records of previous analyzes, as well as establishing the relationship between codes and, subsequently, between categories, creating so-called networks, allowed a clearer and more consistent presentation of the research. Regarding the resource of the calculation of frequency of code registration, such as the magnitude and density, it was possible to verify the higher incidence of some codes over the others.

With this, it was possible to verify that the use of technological resources can help in the work of data analysis and interpretation of the results of a qualitative research, allowing more rigor and reliability of the analyzes. It is worth noting that the use of technological resources, as well as the combination of data collection instruments, does not diminish the role of the researcher. On the contrary, it increases the need for prior knowledge about the researched subject and the ability to establish possible relationships between the categories created and named by the researcher.

The semantic network constructed from the analyzed categories and codes from the textual body of interviews allowed the visualization of the relationship between the three main categories - Conception about the indigenous people, Origin of the knowledge about the indigenous people and the relationship established between indigenous and non–indigenous peoples. These three categories complement and justify the social representations formed regarding the indigenous people, from the perspective of the non-indigenous, insofar as they legitimize the constructed image anchored in ideas, often stereotyped, and that are objectified in the way the relations between the social groups in question are given.

Regarding interpretation, the results suggest that the representations about the indigenous people are still strongly anchored in the image of the indigenous person as a primitive subject. The distancing between these ethnic groups and the general population reinforces stereotypes that put the indigenous person in a situation of social disadvantage. These representations reflect a historical image, built by the indigenous people in the past centuries, which acts as an impediment or hinders their acceptance as part of contemporary society.

From a societal perspective, the representations depend on the relationship between the groups involved, justifying the chain of relations, while maintaining the specificity and identity of each group [10, 11]. In this way, the identity of a group, the result of categorization and comparison processes, is based on the construction of stereotypes, highlighting the differences between groups and similarities in the group of belonging. In relation to the construction of stereotypes, there is a tendency to maintain positive characteristics in the group of belonging and negative characteristics for the external group [18].

The expression of discrimination and, consequently, social exclusion, portrays the misconception that is still around today, regarding the indigenous peoples. The demystification of stereotypes, so rooted in the social representations hitherto constructed, is possible from the approach and recognition and respect to the difference. Distancing and stagnant knowledge in the past reinforce the ideas of undeveloped and backward people, as one participant's speech suggests: "… ignorant, uncivilized." The constitution of the identity of individuals or groups depends on how they recognize themselves and are recognized by others and, consequently, determines the practices in relation to them, whether in the sense of guaranteeing rights, meeting demands, or even fighting prejudice [4].

The results also suggest, although in a discreet way, the existence of elements anchored in the defense of respect for difference and equity between distinct social groups, recognizing this group as subjects with rights. These changes can be associated with the greater visibility of this group, since information and knowledge make it possible to glimpse new images and, consequently, new possibilities of social participation, without this representing a threat to the maintenance of the cultural identity of these ethnic groups. According to the perspective of Doise [10], the social representations of the indigenous people are influenced by the relationship established with society, while this relationship interferes with the constitution of new representations.

5   Conclusions

The data presented in this article come from an exploratory research that investigated the social representations of the indigenous peoples in the Brazilian context. In this way, the present study, based on the theory of Social Representations, allowed for an understanding of the constructed image of the indigenous people in the present day.

These representations involve elements of a social construction based on prior knowledge, as well as elements that are established from the relationship between these different groups, considering political and social issues involved in the historical context. It is noteworthy that Brazil is experiencing a very significant historical moment regarding the struggle for the affirmation of the rights of different social minorities, making the context favorable to the investigation about the relationship between the indigenous population as a minority group and Brazilian society.

This research is not without limitations and, therefore, it is considered that the size of the sample and its homogeneous characteristic made the generalization of these results possible. In this sense, future studies should investigate these social representations in larger and more diverse samples of the general population. It is also pertinent to investigate whether these representations vary according to the different group belongings to which the participants are part of, as well as to consider the degree of proximity of the participants with these ethnic groups. In addition, it is recommended that future studies include the indigenous people themselves as participants in the research, considering the relationship between representations and the construction of social identity. It is also believed that it is relevant to investigate if there are contents that were not brought by the participants, which means, in other words, to analyze representations that may be hidden as a result of the social norm.

Finally, it is considered that this research contributed to reflection on the (im)-possibilities of social participation of this minority, in view of the shared images about the indigenous people, the norms and the achievements that characterize their social relations.

Acknowledgements. This research was supported by the University of Fortaleza through the Call for Research 30/2017 of Support for Teams of Research and the bestowing of a scientific initiation grant to Samuel Figueredo Maia; there was also support from the Cearense Foundation for Support to Scientific and Technological Development (Funcap) through the granting of a doctoral scholarship to Sílvia Barbosa Correia.

References

1. Baniwa, G.S.L.: O índio brasileiro: o que você precisa saber sobre os povos indígenas no Brasil de hoje. Ministério da Educação, Brasília (2006)
2. Braga, C.F., Campos, P.H.F.: Invisíveis e subalternos: as representações sociais do indígena. Psicologia Sociedade **24**(3), 499–506 (2012)
3. Lima, M.E.A., Faro, A., Santos, M.R.: A desumanização presente nos estereótipos de índios e ciganos. Psicologia Teoria e Pesquisa **32**(1), 219–228 (2016)
4. Lima, M.E.A., Almeida, A.M.M.: Representações sociais construídas sobre os índios em Sergipe: ausência e invisibilidade. Paidéia **20**(45), 17–27 (2010)
5. Stock, B.S., Fonseca, T.M.G.: Para desacostumar o olhar sobre a presença indígena no urbano. Psicologia Sociedade **25**(2), 282–287 (2013)
6. Souza, M.N., Barbosa, E.M.: Direitos indígenas fundamentais e sua tutela na ordem jurídica brasileira. Âmbito Jurídico **14**(85), 1–4 (2011)
7. Pinheiro, J.: Ceará: terra da luz, terra dos índios: história, presença e perspectiva. MPF/FUNAI/IPHAN, Fortaleza (2002)
8. Conselho Indigenista Missionário [CIMI]: Relatório violência contra povos indígenas no Brasil: dados 2017. CIMI (2017)
9. Moscovici, S.: Representações Sociais: investigações em Psicologia Social. Vozes, Petrópolis (2010)
10. Doise, W.: Da psicologia social à psicologia societal. Psicologia: Teoria e Pesquisa, **18**(1), 27–35 (2002)
11. Almeida, A.M.O.: Abordagem societal das representações sociais. Sociedade e Estado **24**(3), 713–737 (2009)
12. Vala, J., Castro, P.: Pensamento social e representações sociais. In: Vala, J., Monteiro, M.B. (eds.) Psicologia social. 9ª edição, pp. 569–602. Fundação Calouste Gulbenkian, Lisboa (2013)
13. Minayo, M.C.S.: Amostragem e saturação em pesquisa qualitativa: consensos e controvérsias. Revista Pesquisa Qualitativa **5**(7), 1–12 (2017)
14. Palacios-Espinosa, X., González, M.I., Zani, B.: Las representaciones sociales del cáncer y de la quimioterapia en la familia del paciente oncológico. Avances en Psicología Latinoamericana **33**(3), 497–515 (2015)
15. Fernandes, F.S., Ferraz, F., Salvaro, G.I.J., Castro, A., Soratto, J.: Representações Sociais dos profissionais de saúde sobre a terminalidade infanto-juvenil. Revista CEFAC **20**(6), 742–752 (2018)
16. Forte, E.C.N., Pires, D.E.P., Trigo, S.V.V.P., Martins, M.M.F.: A hermenêutica e o software Atlas TI: união promissora. Texto Contexto Enfermagem **26**(4), 1–8 (2017)
17. Silva, A.M.T.B., Constantino, G.D., Premaor, V.B.: A contribuição da teoria das representações sociais para análise de um fórum de discussão virtual. Temas em Psicologia **19**(1), 233–242 (2011)
18. Tajfel, H.: Social psychology of intergroup relations. Ann. Rev. Psychol. **33**(1), 1–39 (1982)

Qualitative Study of Images on Migrant Venezuelan Children in Brazil

Lucimara Fabiana Fornari◉ and Emiko Yoshikawa Egry$^{(\boxtimes)}$◉

Nursing School - São Paulo University, São Paulo, Brazil
{lucimarafornari,emiyegry}@usp.br

Abstract. This study sought to understand the reality of Venezuelan migrant children in Brazil, as shown on social media, in parallel with the images published by the official refugee protection agency. It is a documentary, exploratory and descriptive study, with a qualitative approach; made from 31 images published on the subject in electronic pages from January 2018 to February 2019. The Documentary Method of Interpretation, supported by the software webQDA, was used to analyze the images. The results, although partial, show children in refugee status in Brazil under the idealized perspective of the "role of the woman and the girl", leaving untouched gender and generational subalternities. It is concluded that the qualitative methodology through the method of image analysis is quite revealing of ideologies, prejudices and stereotypes, being able to give new meanings for the understanding of this social phenomenon.

Keywords: Gender · Generation · Migration · Qualitative research · Image analysis

1 Introduction

The Universal Declaration of Human Rights states that all children need social protection. To make this possible, children's right to life must be ensured, along with health and well-being, guaranteed through food, clothing, shelter, medical care, social services and security [1].

The economic and political crisis faced by certain countries has forced the emergence of adults and children in search of favorable living environments for the promotion of health and well-being. According to a report by the United Nations High Commissioner for Refugees (UNHCR), 65.5 million people in the world were pressured to leave their country of origin, out of which 22.5 million were refugees and more than half were children [2].

The current crisis in Venezuela is stimulating the exit of women and children in search of shelter in regions bordering Brazil and Colombia. According to a report by the International Organization for Migration conducted in the first semester of 2018, 63.5% of Venezuelan children and adolescents in Brazil were not attending school. In addition, out of the total number of Venezuelans interviewed, only 12% were aware of their rights and 6% were aware of the rights of children and adolescents as migrants or refugees [3].

© Springer Nature Switzerland AG 2020
A. P. Costa et al. (Eds.): WCQR 2019, AISC 1068, pp. 156–165, 2020.
https://doi.org/10.1007/978-3-030-31787-4_13

Regulation and implementation of migration policies aimed at guaranteeing human rights to children and adolescents is undoubtedly important. In Brazil, the law that approaches migrants' rights and duties is recent and presents as a principle the "integral protection and attention to the superior interest of the child and the migrant adolescent" [4].

The guiding questions outlined in this study were: (a) How does social media show the reality of Venezuelan migrant children? How are they seen by the official refugee protection agency? (b) What is the advantage of using qualitative methodology to study this phenomenon?

This research's goals were: (a) understanding how the Venezuelan children in Brazil are portrayed by the Brazilian social media and by the official body of refugee protection; (b) explaining the advantages of using qualitative methodology to study this phenomenon.

2 Methodology

This is an exploratory and descriptive documentary study employing a qualitative approach based on the Theory of Praxical Intervention of Nursing in Collective Health (TPINCH) [5]. This theory proposes that it is necessary to capture objective reality in three dimensions: structural, particular and singular, searching the historicity of the phenomenon's dialectical contradictions. In this research, the focus was the particular dimension, comprised of the social group of Venezuelan migrant or refugee children in Brazil. The analysis for the understanding of the object was based on the materialist-historical and dialectical world view [5], highlighting the analytical categories of gender, generation and ethnicity within the context of the studied Venezuelan population's social life [6].

The Documentary Method of Interpretation was used to analyze the images. This method makes an approximation between the theoretical fields of History of Arts and Sociology and is based on Karl Manheim's Theory, which seeks the analysis of world views underlying images. The method is divided into formulated and reflected interpretation, in which the plans, the iconographic elements, the formal composition and the iconic-iconological interpretation of images are evaluated [7].

This research was based on images published electronically on the Brazilian social media and by the official agency for the protection of refugees. Two national news portals, two national newspapers in their online versions, two non-governmental organizations and one official refugee protection agency were consulted.

The selected images were published in the period from January 1st, 2018 to February 23th, 2019, when Venezuela's political crisis worsened and its borders with Brazil and Colombia were temporarily closed. Initially, an internet search was conducted to see how Venezuelan women and children were portrayed by Brazilian news agencies. Subsequently, the criteria for image eligibility were listed in order to assemble the corpus.

The data collection was performed in February 2019. Seven electronic pages were consulted using as search descriptor "Venezuelan women". We selected images depicting Venezuelan children – especially girls – in a migration situation or in refuge in Brazil.

Taking into account their titles, 49 articles addressing the situation of Venezuelan women and children in Brazil were selected. Of these, two duplicate reports and a report referring to the refugee population of Central America were excluded, which led to a total of 45 reports as a final sample.

The image selection was carried out from the 45 reports. The presence of at least one child or pregnant women in the photograph was used as an inclusion criterion. The total was 31 final images: 19 photos from news portals, seven from the international official organization for protection of refugees, four from newspapers and one from a non-governmental organization.

As a first moment in data treatment, the images were coded by two independent researchers. In the second moment, divergent codings were discussed and validated by the research authors. In addition, the image analysis was supported by the webQDA qualitative analysis software [8]. Using this digital tool allowed importing images through the Internal Sources System, the emergence of empirical categories through the Codification System and the extension of the data analysis from the Questioning System. In addition, the software enhanced data organization and collaborative work.

We adopted "report" to identify all the texts published in the electronic pages. The images reproduced in this article were identified with the letters: N (newspapers), NP (news portal), NGO (non-governmental organization) and OF (official body for refugee protection), representing the type of electronic page from which they were obtained, followed by Arabic numerals to indicate their sequence.

This research has not been appreciated by the Research Ethics Committee, since images of publicly available electronic pages with free access to information were used.

3   Results and Discussion

Table 1 shows the number of images of Venezuelan children or pregnant women according to their dissemination vehicles. The news portals posted most of the pictures, followed by the official refugee's committee. Only one non-governmental organization presented data analyzed in the research.

Table 1. Number of images according to the type of electronic page

Electronic page	Images
News portal	19
Newspaper	04
International official organization	07
Non-governmental organization	01
Total	**31**

The study is still underway and therefore this article will address only one of the empirical categories that have emerged from image analysis. As can be seen, the views on the social media, the press and the official protection organization present idealized perspectives of gender and generation subalternity when it comes to migrant Venezuelan children.

3.1   Portraits of Venezuelan Childhood in Brazil: Gender and Generation Stereotypes

The photographic records selected for research composition revealed stereotypes of gender and generation. From image analysis, it was possible to perceive the different views on the Venezuelan children who sought shelter in the Brazilian territory as a guarantee for the maintenance of human rights.

Regarding the gender category, the idealized conception of colors associated with biological sex was verified. From the perspective of sexist patterns, it is understood that the blue color corresponds to the male sex and the color pink to the female sex. The maintenance of this stereotype occurs among Venezuelan women from gestation, being extended to childhood.

Gender stereotypes confer different social roles for men and women, mainly from the biological basis of the genotype and phenotype attributed to males and females. Culturally, the biological question based on the binary conception of the sexes determines the male gender and the female gender [9].

Image N1 (Fig. 1) has the bed as its bottom horizontal line. It is possible to identify children's clothes of blue color on it. The color shade of the baby's clothes indicates pregnancy with a boy. The clothes on the bed also suggest that the mother is prepared and awaiting the child's birth in Brazilian territory.

Fig. 1. Image N1.

The color of the clothes used by the child according to its biological sex is also verified after birth; this aspect is depicted in the NP18 image (Fig. 2). In addition, the photograph shows that the group of women, placed in the diagonal line of the image, is in the corridor of a hospital, performing their children's first care without the presence of a paternal figure or another companion.

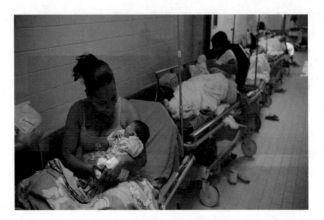

Fig. 2. Image NP18.

The absence of the paternal figure in child care is also illustrated in Fig. N2. In the photographs related to the journey undertaken by families presenting Brazil as their final destination, women were observed to be protagonists in the care, feeding and transportation of children. Regardless of whether women are alone or accompanied by other adults, the mother figure always carries her child in her arms (Fig. 3).

Representations of motherhood are often associated with the idea of maternal love. In this perspective, the role of women is being "good mothers" in order to ensure appropriate child development. The dissemination of this message is also observed through the media, which relegate to the female this same way of living and thinking about motherhood [10].

Fig. 3. Image N2.

The relationship of maternal care with children was identified in the public and private space. In the public space, it was possible to verify the presence of women with their children on the sidewalks and road shoulders. In the private space, it was verified that Venezuelan women and children lived in shelters or improvised housing. Even in pictures taken during educational activities, the presence of the female in the educator condition was noticed.

Migration experiences are mostly motivated by the need for fathers and mothers to engage in paid productive activity in order to ensure better living conditions for their children [11]. Hence, Venezuelan women usually work triple shifts, during which they are responsible for taking care of children and the household while performing paid productive activities.

Regarding gender stereotypes, the insertion of women into the means of social production is highlighted as not exempting them from responsibility for domestic activities and childcare [12]; in this way, they began to accept working double or triple shifts. In addition, it is emphasized that most of women's work is done free of charge, in the name of love and motherhood [13].

The OF7 image (Fig. 4) records the scene of an adult woman dressing a female child with a school uniform. The photograph shows the child in the center of the image, while the adult woman is marginalized. They are both on the same horizontal line. The woman shows responsibility and attention to her daughter, assuring her access to education and the necessary monitoring of her participation in educational activities.

Fig. 4. Image OF7.

A study carried out with migrant children in Chile found that entry into school is important for social interaction, since boys and girls have the possibility of experiencing moments of cultural inclusion routinely [14].

In addition, it is emphasized that free education consists of a principle of the Universal Declaration of the Rights of the Child [15]. Therefore, access to school is not the sole responsibility of the family, since the State also needs to offer this guarantee. In addition, it is the responsibility of the State to provide special protection and provision of services that guarantee health, social welfare and safety to children [15].

The photographs selected in this research revealed the establishment of intragender (woman-girl) and intergender (woman-boy) relations, as well as the prevalence of intergenerational relationships (adult-child).

With regard to intergenerational relations, some of the images reinforce the historical construction of the social role of women in society. The OF2 image (Fig. 5) shows an adult woman holding a girl in her arms, reinforcing the bonding relationship. They are centered in the picture. They both wear typical Venezuelan clothing. These clothes reaffirm the stereotype of femininity, since they are dresses. The scene emphasizes the generational and intra-gender transmission of customs even when incorporated into a new social context.

Fig. 5. Image OF2.

In addition to cultural issues, the transmission of gender stereotypes that are linked to the patterns of femininity constructed and reproduced socially was also observed. Picture NP8 (Fig. 6) centers on four female children. The foreground child is out of focus, wears pink and is moving by riding a bicycle. The three children in the background are presented with better clarity. The tallest child holds the youngest child in her arms while the other rests on her shoulder.

Image NP8 evidences a moment connected to the categories of childhood and gender. The childhood category is reinforced by the bicycle, an element of playfulness and recreation. The gender category is highlighted by means of female stereotypes that manifest in the color of their clothes and intragenerational care among the female children.

In this perspective, the determination of social phenomena such as migration are considered to lead to the articulation among the social categories of gender, generation, social class, and race or ethnicity. These categories present dynamical predominance, since they are related to social subjectivity [6]. In the analyzed images, it is possible to observe the combination between the categories of gender and generation.

Fig. 6. Image NP8.

The child-related generation category views children as subjects of rights and social actors and enables the understanding of the historical, social, economic, political and ideological influences that surround their universe. Children, as constituent agents of society, experience and suffer the consequences of social and historical changes. Generally, they are influenced by determinations of a hegemonic and dominant generational category, the adults [6, 16].

The gender category is based on the difference between the male and female sex. It is an inherent component of social relations and is primordial for the construction of meanings on the relations of power. As components of social relations, the difference between the sexes can be associated with four elements: the first is based on the culturally available symbols; the second on normative concepts; the third on social institutions and organizations; and the fourth on the construction of subjective identities, which comprises the complex relationships established in the process of human interaction [17].

The stereotypes of femininity can trigger violations, as the body of the woman and girl becomes objectified. A systematic review has found that despite the efforts of official organizations for refugees and NGOs, the support given to girls and young women for reproductive health is non-existent or incipient [18]. Considering the high rate of sexual violence pointed out in refugee studies, this is particularly worrying in the case of Venezuelan girls, since the issue of violence or sexual exploitation is not problematized in the images.

A multi-center and cross-sectional study on the prevalence and risk factors associated with violence against adolescent girls affected by conflict concludes that the prevalence of sexual violence is high and that considerable efforts should be made to address forms of violence, recognizing perpetrators as partners and caregivers in conflict situations [19]. In the case of Venezuelans, there is no data on caregivers, except for institutional neglect, since the state does not adequately protect migrant or refugee girls, what can be understood as a form of violation.

Given this, the vulnerability of Venezuelan children to different types of violations can be perceived. The study points out that refugee children are subject to exclusion both in political space and in decision-making, as they are seen as policy objects due to immaturity, as well as silent victims awaiting the benefit of governments or humanitarian agencies [20].

4 Conclusions

As a preliminary study, the conclusions to date point to some contradictions in the relationship between social media images and child protection agencies in general and of migrants in particular. Although child protection agencies portray more often achievements and overcomes, which reveals a disconnection from children's reality, the images do not portray the true suffering and needs of Venezuelan migrant children in terms of human rights and the needs of the child itself, including health.

In addition, the Documentary Method of Interpretation proved to be rich enough to explain the cultural diversity of the portrayed population as "read" by social media and what is shown by the official protection agency. Although the research consists of the analysis of images recorded at a certain time and scenario, it is possible to affirm that it represented an important way of giving voice to Venezuelan children who are daily vulnerable to the violation of their human rights.

Contrary to what has been done in qualitative studies in the health and social sciences, the photographs occupied the centrality of the empirical sources and the analysis was based on the appropriate qualitative methodology for its unveiling, especially the world views underlying them.

A limitation in this study is that the images were not produced for the research of living conditions, but rather portrayed by the media and protection organisms, each one under their purposes of exposing the facts.

Even so, this study was able to show the difficult living conditions of Venezuelan children who are vulnerable to the violation of their human rights on a daily basis, similar to Brazilian subaltern social classes, aggravated by the conditions of subordination of gender and generation.

Finally, it is important to state that these living conditions of migrant children living in Brazil have an impact on public health care services. They amplify the demand and make the projects of intervention in the health-disease processes of the population and the nursing care processes more complex.

References

1. United Nations Human Rights, Universal Declaration of Human Rights. https://www.ohchr.org/EN/UDHR/Pages/Language.aspx?LangID=por. Accessed 07 Mar 2019
2. ACNUR: 2017 Relatório de Impacto. Agência da ONU para Refugiados (2018)
3. Organização Internacional para as Migrações: Monitoramento do fluxo migratório venezuelano com ênfase em crianças e adolescentes. Agência das Nações Unidas para as Migrações, Brasília (2018)

4. Brasil, Lei No 13.445, de 24 de maio de (2017). https://www2.camara.leg.br/legin/fed/lei/ 2017/lei-13445-24-maio-2017-784925-publicacaooriginal-152812-pl.html. Accessed 07 Mar 2019
5. Egry, E.Y.: Saúde Coletiva: Construindo um novo método em enfermagem. Cone Editora, São Paulo (1996)
6. Egry, E.Y., Fonseca, R.M.G.S., Oliveira, M.A.C.: Science, Public Health and Nursing: highlighting the gender and generation categories in the episteme of praxis. Rev. Bras. Enferm. **66**(esp), 119–133 (2013)
7. Weller, W., Bassalo, M.B.: Imagens: documentos de visões de mundo. Sociologias **13**(28), 284–314 (2011)
8. Costa, A.P., Moreira, A., Souza, F.N.: webQDA - Qualitative Data Analysis. MicroIO and University of Aveiro, Aveiro (2019)
9. Couto, M.T., Schraiber, L.B.: Machismo hoje no Brasil: uma análise de gênero das percepções dos homens e das mulheres. In: Venturi, G., Godinho, T. (eds.) Mulheres Brasileiras e gênero nos espaços público e privado, pp. 47–61. Editora Fundação Perseu Abramo, São Paulo (2013)
10. Klein, C., Meyer, D.E., Borges, Z.N.: Social inclusion policies in contemporary Brazil and the education of motherhood. Cad. Pesqui. **43**(150), 906–923 (2013)
11. Conde, S.F., Alcubierre, K.S.L.: Sentidos e percepções de crianças migrantes em Florianópolis. Rev. Katálysis **21**(2), 358–368 (2018)
12. Vieira, A., Amaral, G.A.: The art of being a Hummingbird in women's triple-shift workday. Saúde Soc. **22**(2), 403–414 (2013)
13. Hirata, H., Kergoat, D.: Novas configurações da divisão sexual do trabalho. Cad. Pesqui. **37** (132), 595–609 (2007)
14. Soto, I.P., Valderrama, C.G.: Hijas e hijos de migrantes en Chile: derechos desde una perspectiva de inclusión social. Diálogo Andin **57**, 73–86 (2018)
15. Assembléia Nacional das Nações Unidas, Declaração dos Direitos da Criança. http://www. direitoshumanos.usp.br/index.php/Criança/declaracao-dos-direitos-da-crianca.html. Accessed 07 Mar 2019
16. Pretto, Z., Lago, M.C.S.: Reflexões sobre infância e gênero a partir de publicação em revistas feministas brasileiras. Rev. Ártemis **XV**(1), 56–71 (2013)
17. Scott, J.: Gender: A Useful Category of Historical Analyses. Gender and the Politics of History. Columbia University Press, New York (1989)
18. Ivanova, O., Rai, M., Kemigisha, E.: A systematic review of sexual and reproductive health knowledge, experiences and access to services among refugee, migrant and displaced girls and young women in Africa. Int. J. Environ. Res. Public Health **15**(8), 1583 (2018)
19. Stark, L., Asghar, K., Bora, G.Y.C., Baysa, A.A., Falb, K.: Prevalence and associated risk factors of violence against conflict–affected female adolescents: a multi–country, cross–sectional study. J. Glob. Health **7**(1), 1–11 (2017)
20. Martuscelli, P.N.: A proteção brasileira para crianças refugiadas e suas consequências. REMHU **XXI**(42), 281–285 (2014)

How the "Help" Feature Can Boost the Self-learning Process of CAQDAS: The webQDA Case Study

Fábio Freitas[1(✉)] ⓘ, Carla V. Leite[2] ⓘ, Francislê Neri de Souza[3] ⓘ,
and António Pedro Costa[1] ⓘ

[1] Research Centre on Didactics and Technology in the Education of Trainers
(CIDTFF), University of Aveiro, Aveiro, Portugal
`fabiomauro@ua.pt`
[2] Digital Media and Interaction Research Centre (DigiMedia),
University of Aveiro, Aveiro, Portugal
[3] Centro Universitário Adventista de São Paulo (UNASP), São Paulo, Brazil

Abstract. The inherent demand for scientific production commonly leads us to
rely on digital tools for the data analysis process, such as Computer Assisted
Qualitative Data Analysis Software. However, learning those tools can be
challenging for researchers, especially since it is commonly an autonomous task.
Taking that into account, webQDA software displays a "Help" feature which
intends to support its users to boost the self-learning process. Aiming to understand the potential of this feature, an evaluation workshop was carried out with 22
participants. It was possible to verify a significant user satisfaction, being the
"clarity of the instructions" and the "Ease of use" the highest valued aspects of
the "Help" feature in webQDA. It was possible to infer that this feature is a viable
option to support the self-learning process of this particular software.

Keywords: Self-learning · CAQDAS · webQDA · Qualitative analysis

1 Introduction

Currently, autonomy seems to be impelled to carry a great variety of tasks, namely, to
shop at the grocery store and to fill the tank of a car. We are living in the DIY (Do It
Yourself) Era being constantly challenged to perform tasks that used to belong to third
parties. Therefore, the process of autonomous learning is increasingly described as an
essential requisite of today's society. Moreover, it is important to highlight that
learning how to learn requires purpose, effort, discipline and responsibility, debunking
the idea of being a simple, easy and superficial process, and on the contrary, it is proved
to be fundamental for personal and for social welfare and progress [12].

This recent paradigm has been explored in the Education field, with higher education students being requested to develop self-learning skills, since they have to
identify their personal learning needs, and to search and use the resources they consider
the most effective ones, in a systematically and flexible way, through their cognitive,
social and creative individual capabilities [12]. This situation is already a reality in
some Higher Education institutions, where the Professor is dismissed and collaborative

© Springer Nature Switzerland AG 2020
A. P. Costa et al. (Eds.): WCQR 2019, AISC 1068, pp. 166–176, 2020.
https://doi.org/10.1007/978-3-030-31787-4_14

learning is promoted, thus enabling students to create autonomous learning environments based on more creative methods [15]. Therefore, self-learning is perceived as a key competence for proactive learning.

It should be emphasized that the process of self-learning it is supposed to be supported by a great sense of responsibility and autonomy, with the learners managing their own learning process and self-regulate their learning path, choosing the contents they want to acquire [6, 10, 11, 13]. Moreover, self-learning assumes and reveals the responsibility adoption since it is, *"marked by the individual construction of knowledge, in order to guarantee personal development and better adaptation to a constantly changing environment"* [2].

In this line of thought, considering CAQDAS (Computer Assisted Qualitative Data Analysis Software) learning, self-learning can be understood as an effective strategy for knowledge acquisition while using these qualitative analysis software packages. Nowadays, there are clear challenges to be addressed and gaps to be filled in the guidance for qualitative methodology, analysis techniques and technology usage [18], thence the adoption of CAQDAS can be demanding for researchers and supervisors, particularly during the data analysis tasks of a dissertation and thesis [5, 19]. This demand may be due to the inherent requirements of digital tools usage, since technical and methodological expertise might be needed. This circumstance is overtly crucial due to the short period of time while the majority of researchers can spend on learning software, namely who uses CAQDAS in Masters and Doctoral projects [7, 17].

The inclusion of tools that support self-learning into software for qualitative research, taking in consideration inherent factors of qualitative data, namely its variety and dimension, and a wide range of research contexts, makes this combination a viable alternative, turning the learning curve less steep [14], and consequently the research more efficient.

Taking into consideration the context, this study intends to explore how self-learning tools can truly support users for autonomous CAQDAS usage, by analysing the qualitative software webQDA, and focusing on its "Help" feature. For this purpose, a workshop for assessment was held with 22 participants, who had no previous experience with CAQDAS. The collected data was analysed and along with the discussion will be presented later on. This document is divided into four parts, with the first one being a brief introduction to the "Help" feature of webQDA; the second of them explains the methodology used to carry out this study; the third part explores the results and discussion; and lastly, the final considerations are presented that include a rumination about the study conducted.

Taking into consideration the context, this study intends to explore how self-learning tools can truly support users for autonomous CAQDAS usage, by analysing the qualitative software webQDA, and focusing on its "Help" feature. For this purpose, a workshop for assessment was held with 22 participants, who had no previous experience with CAQDAS. The collected data was analysed and along with the discussion will be presented later on. This document is divided into four parts, with the first one being a brief introduction to the "Help" feature of webQDA; the second of them explains the methodology used to carry out this study; the third part explores the results and discussion; and lastly, the final considerations are presented that include a rumination about the study conducted.

2 "Help" Feature of webQDA

In order to support the researchers during the process of learning and self-learning the software, webQDA provides a set of tools to guide the user, such as: methodological books; tutorial videos; frequently asked questions (FAQ); a forum; a blog; training courses; workshops; webinars; and consultancy opportunities [20]. Moreover, since the released version 3.1, this software includes a "Help" feature (Fig. 1), that allows the user to check descriptions and step-by-step instructions dynamically, regarding contextual possible actions. This information is complemented by hyperlinks to tutorial videos for further help.

Fig. 1. webQDA "Help" feature view

3 Methodology

In order to do a webQDA assessment, focusing on the "help" feature, and to understand the user needs, during the process of learning this software, a workshop was planned. The online form to apply was disseminated among researchers and academics from a university in Portugal. Only the applications from people who had no previous experience with CAQDAS were considered (exclusion factor).

The workshop lasted two hours, and each participant had to start by enabling the "Share Project" feature on webQDA, so it would be possible to collect all data they created. Each participant had to perform five tasks, and if help was needed, they could only use the "help" feature of webQDA, and no other means.

For the first task: "Importing Sources", the participant had to upload documents to be analysed on webQDA; During the second task: "Classifying Sources", the participant had to create descriptors and to classify each document; For the third task, "Creating Tree Codes", they had to define categories and subcategories of analysis); The fourth task was related to "Coding Sources", so the participant had to engage in the process of coding the previously imported documents; And finally, the fifth task "Word Searching and Cloud" was related to search the most frequent words of the imported documents, and the creation of a Word Cloud as the visual result. After concluding each task, each participant wrote on the webQDA log book, their positive and negative feedback, and pin pointed the aspects that need future improvement. This intended to make clear how the "Help" feature supported their learning path and to perform each task.

The workshop was conducted with 22 participants, plus 7 attendees that decided not to take an active role and not allowing the data collection by not sharing their project. This could be interpreted as lack of interest or experiencing some issues, however what was possible to understand in that moment and considering their sociodemographic factors, it seems they were interested in learning to use and have access to webQDA, but were not too keen on truly volunteering for this research. Therefore, only the data collected related to 22 active participants will be considered and presented. Regarding their characteristics, the majority (n = 17) identified themselves as women, and the remaining as men (n = 5). Almost half of them (45%, n = 10) were aged 20 to 30 years, around one-third were aged 31 to 40 years (32%, n = 7), close to 18% were aged 41 to 50 years (n = 4); and 5% were aged 51-60 years (n = 1). The most common occupation among the participants was being a Student (n = 16), followed by the low representation of Researchers (n = 2), Professors (n = 1) and having Other occupations (n = 3). Regarding their academic path, 10% were Doctorated (n = 2), slightly above one-third of the overall were doing a Doctoral degree (n = 8), and the exactly same number had their Masters ongoing, only 5% were doing a Bachelor (n = 1), and 13% had finished the Bachelor degree (n = 3).

As part of the adopted methods for this research, and emerging from the results obtained during the workshop, the following (Table 1) categories and sub-categories were established. This process was supported by webQDA use.

The data collected of the "help" feature usage during the workshop, was organised according to the categories presented in the previous table, the positive and negative feedback, plus the suggestions for future improvements. The results obtained, which resulted in the shown sub-categories, will be explored in the next section, along with some examples of the participants' notes.

Table 1. Internal consistency of the research analysis

Dimension	Category	Sub-category	Notes
Task 1	Positive aspects	Clear instructions	Regarding the "help" feature instructions' clarity
Task 2		Non-usage	When "help" feature was skipped while completing the task
Task 3		High ease of use	Regarding the good global efficiency and efficacy of the "help" feature to support the tasks completion
Task 4		Good experience	Regarding the positive emotional responses while using the "help" feature
Task 5		Easy access	Regarding the findability and discoverability of the "help" feature
		Don't know/No response (NR)	Regarding the findability and discoverability of the "help" feature the topic
		Not Applicable (NA)	When the comments are out of the "help" feature scope
Task 1	Negative aspects	Unavailability to read instructions	Regarding the avoidance to check the information provided in "help" feature while performing the tasks
		Lack of instructions	Regarding any limitation of the "help" feature instructions
Task 2		Hard to access	Regarding the poor findability and discoverability of the "help" feature
Task 3		Low ease of use	Regarding the poor global efficiency and efficacy of the "help" feature to support the completion of the tasks
Task 4		Bad experience	Regarding the negative emotional responses while using the "help" feature
Task 5		Don't know/No response (NR)	When no comment is provided by lack of knowledge or by skipping the topic
		Not Applicable (NA)	When the comments are out of the "help" feature scope
Task 1	Future improvements	Printed instruction manual	Regarding the "help" instructions being available in a printed user manual
Task 2		Tutorial videos display	Regarding the inclusion of tutorial videos on the "help" feature
Task 3		Visibility improvement	Regarding the visual display improvements of the "help" feature
Task 4		Ease of use improvement	Regarding the efficiency and efficacy improvements of the "help" feature to support the tasks completion
Task 5		Instructions improvement	Regarding the "help" feature instructions' improvements
		Search functionality	Regarding the suggestion of enabling to search inside the "help" feature
		Don't know/No response (NR)	When no comment is provided by lack of knowledge or by skipping the topic
		Not Applicable (NA)	When the comments are out of the "help" feature scope

4   Results and Discussion

For the results analysis, which will be presented in this chapter, two different approaches were used. Firstly, an overall review of all opinions regarding each task (Table 2) was conducted, followed by the detailed assay of the positive, negative and future implementations reported during each single task.

Table 2. Overall number of positive, negative and future improvements of the "Help" feature mentioned, by task

Tasks	Positive aspects	Negative aspects	Future improvements
Task 1 – Importing Sources	33	18	16
Task 2 – Classifying Sources	18	17	8
Task 3 – Creating Tree Codes	11	12	4
Task 4 – Coding Sources	10	8	3
Task 5 – Word Search and Cloud	16	4	2
Total	**88**	**59**	**33**

It is possible to observe that the workshop participants expressed a general satisfaction regarding the "help" feature during the learning process. From a total of 180 references, almost half were positive ones (n = 88), leaving 33% for negative (n = 59) and 18% for suggestions of improvements (n = 33). Notwithstanding the favourable results, it is important to highlight that the amount of negative aspects pointed out, along with the enhancement purposed ideas, makes it clear that are several aspects to be corrected and possible future implementations. Moreover, Table 2 does not include any reference to "Non applicable" or "Don't know/No response" numbers, since no comment could be coded at this point as such. Some examples of the gathered notes can be found along this section.

4.1   Positive Aspects

Regarding the favourable aspects, Table 3 shows that in all tasks the workshop participants pointed out the "Clarity of the instructions", turning it into the most common topic. Only in Task 4 this topic was slightly overpassed by the number of references to the fact that they waived the "Help" feature while performing it, mentioning that the task was obvious, so they did not need to use this feature to complete it.

The "help" feature made it easier to move forward the process. I completed the task.
                                                                                – Participant E

The (self-learning) tool is very handy for first time users. Easily accessible, it details all the steps to facilitate the performance and learning process. Very positive.
                                                                                – Participant N

The task was very simple to conclude, I did not feel the need to use the "help" feature.

– Participant A

Table 3. Number of mentions to positive aspects of the "help" feature mentioned, by sub-category, by task

Tasks	Clear instructions	Ease of use	No need to use "Help" feature	Experience	Access to "Help" feature	N/R	N/A
Task 1	17	7	0	7	2	0	1
Task 2	9	5	3	0	1	0	0
Task 3	5	1	3	2	0	1	0
Task 4	3	1	4	2	0	1	0
Task 5	6	5	4	1	0	0	0
Total	**40**	**19**	**14**	**12**	**3**	**2**	**1**

The importance of "Ease of Use" of the "help" feature is clear, with a total of 19 mentions to it, what can be interpreted as highly beneficial factor for the learning process and to complete the tasks on the webQDA software. The gathered data seems to comply with Preece, Rogers and Sharp [16], statements, who suggest that features carefully developed using usability standards would stand out for being effective, efficient and safe to use, useful, easy to remember and learn.

Throughout the analysis, considering the sequence in what the notes were made, it is possible to observe over the tasks, the gradual diminish of mentions to "Clear Instructions" and the increase of references to waving the "Help" feature to perform the specific task. This can mean that while progressing, the participants learn and understand how to use the software and start doing it empirically. This fact can be seen as a strong trait to boost the learning curve of webQDA, with basic concepts being easily available in the "help" feature.

(...) it facilitates the work being done, the size and font, along with showing restrictions in a obvious way, turns it beneficial.

– Participant T

4.2   Negative Aspects

According to the participants notes, the most common negative aspect was the lack of instructions of the "help" feature (Table 4). This is contradictory to what was found previously (Table 3) and can be explain by different expectations and needs. Nevertheless, the number of negative mentions was considerably lower than the positive ones about this aspect, showing the prevalence of favourable opinions. Notwithstanding, considering the general results, adding or detail more the instructions may be recommended, to facilitate the learning curve for those who expect more detailed instructions.

It is also noticeable that the number of positive notes regarding "clear instructions" diminishes along the tasks, but this is not true for the negative notes. Moreover, the first task (Importing Sources) seems to be the most positive one, contrasting with the second task (Classifying Sources) that generated the higher number of negative comments. The tasks did not have the same complexity; indeed, they were planned to increase gradually in difficulty. Therefore, with the second task being so negatively noted, it seems to represent a limitation worth to mention and be address in future versions of webQDA.

> The "help" feature should be explicit in how we can create sub-categories within the tree code and should allude to the fact that it is necessary to create one by one.
>
> – Participant H

> I found the "help" feature unintuitive, as it was not quite explicit the structure that needed to be created, namely the hierarchy.
>
> – Participant S

Table 4. Number of mentions to negative aspects of the "help" feature, by sub-category, by task

Tasks	Lack of instructions	Ease of use	N/A	N/R	Experience	Access to "Help" feature	Information density	Unavailability for reading instructions
Task 1	2	6	3	4	1	6	2	1
Task 2	11	3	1	1	3	0	0	0
Task 3	7	2	3	0	2	0	1	0
Task 4	6	1	2	0	1	0	0	0
Task 5	2	1	2	3	0	0	0	1
Total	28	13	11	8	7	6	3	2

Once again, "Ease of use" was also object of completely different opinions, being pointed out as one of the most positive aspects with 19 mentions (Table 3), but also got 13 negative comments (Table 4) considering all the notes took over the five tasks. A possible explanation for these results is found in the literature. It seems hard for the users to feel their needs fulfil when they lack basic concepts of qualitative research. If they don't have a common ground with the terminology and common process used in qualitative research, their perception of how easy it is to use CAQDAS features, will be affected. Gilbert, Jackson and Gregorio [9], state that the usage of CAQDAS implies, not only above basic digital literacy, but also wide knowledge of the methods used in qualitative research. Gilbert [8], in a previous work classifies the users of CAQDAS according to their levels of competence, as researchers and as computer users. Those authors go further, warning to effective risk, which the CAQDAS users with low competence in qualitative methods face, by limiting their work methodology to the available software features, instead of opting to use only the available features, thence, compromising their research goals.

4.3   Future Improvements

Not only corrections are needed to improve the learning process and usage of CAQ-DAS, its considering future implementations to keep up with the current information and communications technology systems demands and consequently with the user's expectations. Ergo, suggestions for possible developments of the "help" feature of webQDA were also collected and analysed during the workshop. However, as Table 5 shows, the majority (n = 25) of the suggestions did not concern directly the "help" feature, but rather other functionalities available on webQDA that are considered Not Applicable (NA) for this study.

A new feature to let you import Excel [Microsoft®] documents, and organise all the categories according to what was previously listed [on the spreadsheet], would definitely facilitate the data input.

– Participant G

[I would like] To have an agile and easier way to create subcategories.

– Participant L

Nonetheless, by analysing only the pertinent data regarding the "help" feature, "Instructions" and "Ease of use" were targeted once again, with 17 suggestions how to improve the guidance provided to the user better, and 8 to enhance the efficiency, effectiveness and usefulness. These two categories seem to prevail as the most relevant for the self-learning of CAQDAS, according to the workshop participants.

(...) to be more diversified and show [on "help" feature] what it possible to do after the classification and defining attributes, an extra help.

– Participant T

Further description of the Attributes creation process, and how they really work.

– Participant L

The only suggestion I have is to create an specific area in the software interface to turn this tool fixed, to avoid disturbing the work being done.

– Participant O

Table 5. Number of mentions suggesting future improvements of the "help" feature, by sub-category, by task

Tasks	NA	Instructions improvement	NR	Ease of use improvement	Visibility improvement	Tutorial videos display	Printed instruction manual	Search functionality
Task 1	2	5	3	5	4	2	0	0
Task 2	8	5	1	1	0	0	1	1
Task 3	6	3	0	1	0	0	1	0
Task 4	5	2	2	1	0	0	0	0
Task 5	4	2	2	0	0	0	0	0
Total	**25**	**17**	**8**	**8**	**4**	**2**	**2**	**1**

By comparing Tables 3 and 4, it is possible to notice that the "Visibility" of the "help" feature was negatively commented by some participants (n = 6), getting also suggestions for improvements (n = 4). Considering the participants notes, it is possible to understand that the symbol "?" on the top part of the interface, a commonly modern pictogram used to identify the "help" system in all types of software, seems to cause confusion, being reported as ambiguous. This means that this graphic communication is being missed, not only due to low digital literacy, but information and communication literacy in general. This symbol has a specific concept associated, used in several environments, not only digital, such as Public buildings, Tourism, Transportation. It means support in case of doubt and extra information, as acknowledged in AIGA/DOT [1] pictograms, globally used since their creation in 1979. The use and interpretation of the modern pictograms has been studied also according to the perceptions of various cultural and ethnic groups, as shown by Clawson, Leafman, Nehrenz and Kimmer [4], and by analysing the participants sociodemographic data, this was not a factor. Clara and Sawsty [3] suggest the use of short words along with the symbol, to reinforce of the message.

5   Final Thoughts

Based on the presented and discussed data on the previous sections, the "help" feature of WebQDA can be considered valuable and a useful tool in order to boost the self-learning process. It is possible to observe that all participants were able to perform almost all tasks, using only this option as support. The significant amount of positive references seems to bring to light the fact that the participants appreciated how the instructions were clear and objective, being satisfied how easy it is to use, and the fact that it can streamline and simplify the learning process. Notwithstanding, some improvements to the instructions are needed to leverage some functionalities.

It is important to mention that one limitation felt during the conducted study was the duration planned for the workshop to be conducted, since the participants did not have enough time to complete a sixth task, which was focused on the creation of matrices. However, this task was not considered, and no data collected, and it is believed that this fact did not condition the study.

Acknowledgments. This work was supported by Portuguese National Funding Agency for Science, Research and Technology (FCT - Fundação para a Ciência e a Tecnologia), under the project grant UID/CED/00194/2019. The presented data was originally collected in Portuguese, and this is a free proposed translation from the authors.

References

1. AIGA/DOT: Symbol Signs of American Institute of Graphic Arts for Department of Transportation
2. Albuquerque Costa, F., et al.: Recursos Educativos para uma Aprendizagem Autónoma e Significativa. Algumas Características Essenciais. In: Lozano, A.B., et al. (eds.) Libro de Actas do XI Congreso Internacional Galego-Portugués de Psicopedagoxía, pp. 1609–1615. Universidade da Coruña, A Coruña (2011)

3. Clara, S., Swasty, W.: Pictogram on signage as an effective communication. J. Sosiote-knologi. **16**(2), 166–175 (2017). https://doi.org/10.5614/sostek.itbj.2017.16.2.2
4. Clawson, T.H., et al.: Using pictograms for communication. Mil. Med. **177**(3), 291–295 (2012). https://doi.org/10.7205/milmed-d-11-00279
5. Davidson, J., Jacobs, C.: The implications of qualitative research software for doctoral work: considering the individual and institutional context. Qual. Res. J. **8**(2), 73–80 (2008). https://doi.org/10.3316/QRJ0802072
6. Eranki, K.L.N., Moudgalya, K.M.: Comparing the effectiveness of self-learning java workshops with traditional classrooms. Educ. Technol. Soc. **19**(4), 59–74 (2016)
7. Freitas, F., et al.: O Manual de Utilizador de um Software de Análise Qualitativa: as perceções dos utilizadores do webQDA. Rev. Ibérica Sist. e Tecnol. Informação. **19**(09), 107–117 (2016)
8. Gilbert, L.: Tools and trustworthiness: a historical perspective. In: Seventh Conference on Strategies in Qualitative Research. Methodological Issues and Practices in Using QSR NVivo and NUD*IST, University of Durham (2006)
9. Gilbert, L.S., et al.: Tools for analyzing qualitative data: the history and relevance of qualitative data analysis software. In: Spector, J.M., et al. (eds.) Handbook of Research on Educational Communications and Technology, 4th edn., pp. 347–248. Springer, New York (2014). https://doi.org/10.1007/978-1-4614-3185-5
10. Lima Santos, N., Faria, L.: Desafios da avaliação da auto-aprendizagem em contexto sócio-laboral. Rev. Psicol. Mil. **14**, 163–184 (2003)
11. Lima Santos, N., Faria, L.: Escala de auto-aprendizagem. In: Simões, M.R., et al. (eds.) Avaliação psicológica: instrumentos validados para a população portuguesa, pp. 137–148. Editora Quarteto, Coimbra (2007)
12. Lima Santos, N., Gomes, I.: Transformações e Tendências do Ensino-Aprendizagem na Era Digital: Alguns Passos para uma Arqueologia de um novo Saber-Poder. Rev. Antropológi-cas. **11**, 143–159 (2009)
13. Magalhães, M.S.N.: Auto-conceito de competência e auto-aprendizagem em alunos do ensino secundário: Comparação de cursos científico-humanísticos com cursos profissionais. Universidade Fernando Pessoa (2011)
14. Martin, B., et al.: Evaluating and improving adaptive educational systems with learning curves. User Model. User-adapt. Interact. **21**(3), 249–283 (2011). https://doi.org/10.1007/s11257-010-9084-2
15. Pickles, M.: Como funciona a universidade sem professores inaugurada nos EUA - BBC Brasil. http://www.bbc.com/portuguese/internacional-37797400
16. Preece, J., et al.: Interaction Design: Beyond Human - Computer Interaction. Wiley, New York (2002)
17. Silver, C., Rivers, C.: The CAQDAS postgraduate learning model: an interplay between methodological awareness, analytic adeptness and technological proficiency. Int. J. Soc. Res. Methodol. J. 1364–5579 (2015). https://doi.org/10.1080/13645579.2015.1061816
18. Silver, C., Woolf, N.H.: Five-Leves QDA. https://digitaltoolsforqualitativeresearch.org/2016/11/15/five-level-qda/
19. Silver, C., Woolf, N.H.: From guided-instruction to facilitation of learning: the development of five-level QDA as a CAQDAS pedagogy that explicates the practices of expert users. Int. J. Soc. Res. Methodol. **18**(5), 527–543 (2015). https://doi.org/10.1080/13645579.2015.1062626
20. webQDA: About webQDA. https://www.webqda.net/o-webqda/?lang=en

Refining the Conceptual Model with Software Support: Defining Categories for Collaboration Networks Analysis

Denise Leite[1] ⓘ, Isabel Pinho[2(✉)] ⓘ, Célia Caregnato[1] ⓘ,
Bernardo Miorando[1] ⓘ, Elizeth Lima[3] ⓘ, and Cláudia Pinho[2] ⓘ

[1] Universidade Federal do Rio Grande do Sul, Porto Alegre, Brazil
`denise.leite@hotmail.com.br`,
`celia.caregnato@gmail.com`, `bernardo.sfredo@gmail.com`
[2] University of Aveiro, Aveiro, Portugal
`{isabelpinho,claudiapinho}@ua.pt`
[3] Universidade Estadual de Mato Grosso (Unemat), Cáceres, Brazil
`elizeth@unemat.br`

Abstract. The need to construct and refine a conceptual model was identified in order to achieve two objectives: (1) to have a clear and functional communication tool for the collaborative work of a dispersed research team; (2) to have a data analysis tool, clearly structured for use in several projects but with a transversal theme (Collaboration Networks). A software (webQDA) was selected that provides a collaborative and ubiquitously distributed (web-based) environment, which is simple and intuitive to use, and suitable to integrate the various projects that the team is developing. This article focuses on the task of refining the conceptual model that structures the team's various projects under the umbrella theme called Collaboration Networks. After the introduction, we present the methodological trajectory of the research group and the methodology for the refinement of the conceptual tool.

Keywords: Qualitative research · Conceptual model · Education · Indicators · Networking

1 Introduction

From a reflection on the closure of two research projects on the theme "Evaluation and Collaboration Networks" and other associated projects, we tried to integrate the knowledge acquired by refining the conceptual model that transversally supported it.

Starting from the literature review, empirical work was developed by carrying out interviews with researchers. These subjects were selected from a universe of research excellence and leaders of research groups. Prior to the interviews, we built the co-authoring network of each of these researchers. This network – represented in a graph - was presented at the beginning of the interview and served as a starting point and a framework in exploring the subject. The interviews were analysed according to categories that integrated the interview questions. From the initial analysis, other categories emerged, and the conceptual model was reconstructed through the interaction of the

© Springer Nature Switzerland AG 2020
A. P. Costa et al. (Eds.): WCQR 2019, AISC 1068, pp. 177–188, 2020.
https://doi.org/10.1007/978-3-030-31787-4_15

various researchers associated with the research team, the results that had been obtained and the preparation of several publications.

In the present phase of the collaborative work, there is a need to refine the conceptual model in order to have a communication tool to be used among the various team members, as well as to have a reference to read and interpret the interviews conducted. This endeavour is collaborative work that will make it possible to bring together the main categories and subcategories, their definitions and clear examples in order to ensure the internal coherence of a solid collaborative and distance research work. The use of support software (webQDA) to organize information and work collaboratively proved to be an effective way of addressing this challenge [1, 2].

2   Theoretical Framework

Collaboration in research is still poorly understood and is undervalued by research evaluation although it is a reality of workspaces. Resources and products are measured but processes that turn resources into research products are often ignored. In this article, we have gathered elements of research accomplished by team members and their partners [3]. The conceptual framework that marked the research processes was leveraged by a deep review of the literature [4] on Research Networks. The results of the different studies carried out by the team were published and the methodology has been presented in specific texts [5]. In the next section, we gathered part of the methodological trajectory carried out in the network for the study of research and collaboration networks.

3   Methodology

3.1   Methodological Trajectory of the Research Group

Having defined the research problem, its objectives and the constructed the theoretical basis, the empirical works were developed in subgroups of researchers and over a period between 2010 and 2018. The main funding of the research came from the Brazilian National Council for Scientific and Technological Development (CNPq). In 2019, a decision was made to integrate the various projects associated with the theme "Collaboration Networks". An important step to address this challenge is the refinement of the conceptual model that structures this theme. Because perfecting a model that has been co-constructed and developed from the empirical work and its publications is the aim of this paper, it is deemed useful to explain the methodology that has brought us to this point of integration.

A Status report of the methodology used by the research group is demonstrated in the article titled "Evaluation of research and collaboration networks" [6]. In Fig. 1, we present a synthesis of the ground that has been covered, the definition or the concepts of the object of the research, aspects of the methodology, sources, instruments and materials and the results obtained as markers of research and collaboration networks evaluation. Briefly, the methodological trajectory began by identifying the object of study, interconnecting the themes "Network evaluation" and "Research and collaboration" with the literature review as structural support to all subsequent empirical work.

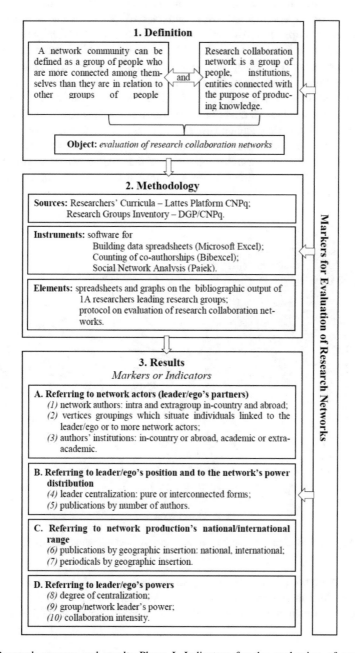

Fig. 1. Research process and results Phase I. Indicators for the evaluation of research and collaboration networks. Source: [5, 8]

In Phase I of the research, based on the assumption that this type of network is established when a group collaborates with the intention of producing knowledge and that this production is visible and explained in the co-authored publications, we set out to analyse the curricula of researchers, using software for the construction of spreadsheets, and counted co-authored work and carried out research network analysis (RNA). Several groups of Brazilian universities (such as the University of the State of Mato Grosso and the Federal University of Rio Grande do Sul) collaborated, as well as international partners (University of Aveiro, Portugal; National University of Colombia, Colombia; and University of the Republic, Uruguay) [7].

We are concerned with the number of publications and also with the intensity of collaboration as a qualitative-quantitative marker. The *intensity of collaboration* in a network refers to the variety and consistency of relationships that are woven in the web established between and among researchers within the research group under the leadership of one or more researchers. Aspects of this marker can be seen in Fig. 2, an example of the graphs constructed for each researcher who would be interviewed during the research.

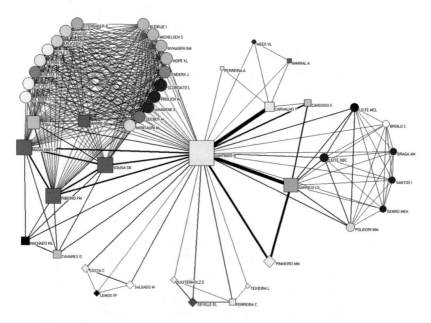

Fig. 2. Researcher collaboration network-1A Social Science (2001–2010)

RNA techniques allowed us to organise worksheets of the researchers' bibliographical production and graphs representing their networks and the construction of a protocol for the evaluation of collaboration and research networks (Table 1).

Table 1. Protocol of analysis. Source [9]

1. Project ID		
2. Researcher identification data		
Name of researcher under analysis: Institution Name: Researcher's area of knowledge: Group member: Analysis Date:		
3. Network structure, captured from articles (graph analysis)		
Network Authorship	*Network total*	
	– in Brazil/inside group	
	– in Brazil extra group	
	– outside of Brazil	
Groups Identification (clusters) of nodes (vertices)	– Isolated (connection only to the leader)	
	– With two components (one node in addition to the leader)	
	– With three or more components (two or more nodes in addition to the of the leader	
Institutions of authors	– Universities and others higher education institutions	– within the country – outside the country *Subtotal*
	– Extra-academic entities	– within the country – outside the country *Subtotal*
	Total	
Egocentric network	– Pure (relationship only between the leader and the other co-authors)	
	– Interconnected (relationships between secondary authors)	
4. Characterization of production based on the articles (spreadsheet analysis)		
Total of articles		
	Authors	
	1 (only ego), 2, 3, 4, 5, 6–10	
	> 10	
Articles by geographic location	– Within the country	
	– Outside the country	
Total of journals		
Journals by geographic location	– Within the country	
	– Outside the country	

The results contributed to identifying qualitative and quantitative markers/indicators for evaluating collaborative network research processes that were tested and validated in the application context. Such markers are described in other publications [5–7, 9].

3.2   Methodology for the Refinement of the Conceptual Tool

Although qualitative analysis is a flexible process, it is convenient to follow procedures in a structured way in order to perform a rigorous and replicable work of interpreting unstructured data. Data organization is fundamental in any research project, but when the research work is carried out in a team and this team works on multiple projects, it becomes central.

In Phase II of the research, the analysis of the interviews began with the reading in full of each interview, and the notes. Reviewing this material shows that it is possible to identify recurring themes or ideas that should form the basis for a conceptual construction, or "indices" that should be associated with the themes underlying the interview questions, which become a topic guide [10]. The research networks of the researchers were mapped by analysing their curricula and publications (Phase I) to deepen the questions raised. We interviewed 15 researchers of excellence CNPq Brazil,

Fig. 3. Tree codes

Table 2. Chart with conceptual adjustment of categories and subcategories with descriptions and definitions

Category	Sub category	Description (what to look for)	Definitions/means/references
01 Research collaboration network	01.1 Research group	Name and origin of the institution-based research group From what is defined by CNPq, institution, others Name, dates	RESEARCH GROUP consists of the knowledge production unit made up of leading researchers, senior researchers and assistant researchers, doctoral and master's students, as well as fellows of scientific initiation and technical personnel who share research on research lines [13]
	01.2 Network	Virtual collaboration Group and network development Forms of collaboration; as it develops and maintains Network Virtual network hierarchies Hierarchies in the Institutional group. There are, which are visible…	COLLABORATORS NETWORKS: A group of people who come together to produce teaching, research or extension by forming collaborative networks, co-authors, citation networks, training networks and guidance networks and institutionalized national and international partnerships RESEARCH COLLABORATION: "the working together of researchers to achieve the common goal of producing new scientific knowledge" [14]
02 Publications	02.1 Coauthorship	Positive Factors Worth it; positive; compensates; expands possibilities Negative factors. Not worth it; difficult; does not compensate; it's too much work	PUBLICATION is some "información registrada en formatos permanentes y disponibles para el uso común" [15]. PUBLICATIONS are products, visible results, the research work carried in magazines, journals, books, annals, events, interviews and others.
03 Knowledge	03.1 Knowledge production	*National* Inside the country of origin; only university itself; different universities in the some country *International*	ACADEMIC SCIENCE produced in different disciplinary areas and groups of collaboration in network KNOWLEDGE PRODUCTION, worldwide, can be captured from multidisciplinary databases as Web of Science/Web of

(*continued*)

Table 2. (*continued*)

Category	Sub category	Description (what to look for)	Definitions/means/references
		Various countries; international partners	Knowledge (http://apps. webofknowledge.com), Scopus, Scielo, Redalyc and others
	03.2 Vehiculation	Journals Articles Open Access; articles in journals Qualis Printers (Capes, Br); Articles in the country in magazines with national classification; Books and chapters preferred publication Other Annals Conferences Lectures	Digital platforms: Knowledge production can be shared by Google academic; Web of Science/Web of Knowledge (http://apps.webofknowledge. com), Scopus, Scielo, Redalyc and others National, regional, university, and international publishers
04 Influences	04.1 Endogenous	Network Interaction From the principal researcher; from the interaction within the group	KNOWLEDGE PRODUCTION NETWORKS COLLABORATION INFLUENCES the external evaluations of academics and balance universities of competition
	04.2 Exogenous	Pressures: Financing, political context and research evaluations; criteria of the calls for research and criteria resource allocation	KNOWLEDGE PRODUCTION in collaborative networks is impacted by external evaluations, by academics and universities, by competition for research funding. In the area of Human Sciences there is an individual hyperprodutivism and the partnerships and networks of collaboration are local, stronger at the national level than at the international level and may lack prestige. In the areas of Exact Sciences, productivism can be disseminated in the network of collaborators; the partnerships are usually long lasting, national and international, tend to confer prestige to academics and their universities

(*continued*)

Table 2. (*continued*)

Category	Sub category	Description (what to look for)	Definitions/means/references
	04.3 Others	Externals International literature flow; international financing	Criteria of calls from research support agencies influence the formation of networks LITERATURE highlights the formation of networks as part of 21st century science Co-authorship publications, including several institutions and countries, are more likely to be cited
05 Networks evaluation	05.1 Indicators	Indicators such as SINAES, Capes and CNPq direct or affect the work of research and production of knowledge in the network; interviewees make suggestions for network evaluation. Level of international excellence. Indicators of international rankings; of external funding agencies; requirements of companies and foundations; impact of newspapers and articles; indicators affecting networks and collaboration; make suggestions for improvement	EVALUATION gives answers, it tries to explain why things occur, the way they occur. Evaluation is a component of scientific policy [15] Scientific activity must be seen and interpreted within the social context in which it is framed. Therefore, the SCIENTIFIC PERFORMANCE ASSESSMENTS must be sensitive to the conceptual, social, economic and historical context of the society in which it operates. This means that science cannot be measured on an absolute and unique scale, if not in relation to the expectations of the society in which it develops and in them [15]
06 Institutional evaluation	06.1 Universities	Local and national level of the country SINAES and Capes indicators are considered; interviewers make criticisms, suggestions	BALANCE OF the institution's directions to qualify its actions (mission and functions of research, teaching, extension). Indicators are defined by government agencies in Brazil (Capes and INEP)

(*continued*)

Table 2. (*continued*)

Category	Sub category	Description (what to look for)	Definitions/means/references
	06.2 Rankings	*National Level* National Rankings and indicators like FOLHA (BR) are considered; interviewers make critiques, suggestions *Internacional Level* International Rankings and Indicators such as QS, THE are considered; make criticisms, suggestions	CRITERIA AND INDICATORS are defined by national and international market agencies. The rankings have a higher or lower reputation and their impact in Latin America may be relative Need to understand what each ranking means and measures
07 Changes in knowledge production	07.1 Innovation	Ways to do research Paradigm Shift Financing and markets Mode 1 Mode 2 applied research, multidisciplinary research, heterogeneity of institutions and networking	INNOVATION is a discontinuous process, breaking with traditional paradigms in teaching and research, or a paradigmatic transition with reconfiguration of knowledge and powers that occurs in different academic spaces and different universities [16]
	07.2 Futures	Forecast changes and innovations in research and for the researcher; science markets Researchers make criticisms, suggestions; Inferences Value Networks for Science of the Future Tribes and territories of the future	Trends based on Corsani [17]; Etzkowitz [18]; Feldman [19]

all leaders of research groups, 10 in Education, 3 in Physics, 2 in Production Engineering; 9 researchers of Graduate Programs, with or without links to the CNPq, 2 from Linguistics, 3 from Environmental Sciences, 2 from Education and 2 from Genetics. In addition, three leading researchers from research centres evaluated as excellent by the Portuguese Foundation for Science and Technology (FCT) were interviewed. The interviews were then grouped after treatment with specialized software.

The interviews began with the presentation of the graph with the map of the researcher's networks and the presentation of the objectives of the study. The interviewee signed the Informed Consent Form (ICF), in consonance with the principles of research ethics. The basic questions and queries were sustained by the theoretical categories and the results of the network mapping. These elements offered us a wide range of questions and allowed for the construction of new categories that emerged from the respondents' responses, as they spontaneously expressed their perceptions and experiences on the topic.

Please note that the procedures we usually use and take for granted are not always pacific and understood by all members of a network in the same way. Often, explaining the meaning of terminology can be a challenge, and a certain degree of agreement is required in the use of terms such as "codes", "categories" and "themes" to perform a coherent qualitative analysis [11, 12]. Thus, themes are classified and grouped into a smaller number of categories or "main themes" of a broader nature and inserted into a general structure. These categories can have subcategories. From the initial categories, a tree of categories was established using the software chosen for analysis and can be seen in Fig. 3.

Given the large volume of data from the interviews, a clarification of the categories was established, for the production of adequate analysis, by constructing a framework with the categories, their definitions and/or meanings according to the original design and the additions from the interviews themselves and the literature. In the Table 2 below we have the category identifier label, the subcategory on which we focused, its description, what we seek out to understand with the category and some definitions provided by the literature. The grouping shows the conceptual refinement that allows the analyser to organize, aided by the software, the "reading" of the interviews in order to meet the research objectives.

4   Conclusions

In this article, we have gathered the experience of researchers from different universities and countries working in research networks, to reach an understanding of how collaboration networks operate and to define evaluation procedures - qualitative and quantitative markers/indicators. The work shows that the methodological path was not finalised when the expected results were obtained. It was important to take the research corpus and to re-examine the conceptual scheme, in order to work the studies that had been carried out in different places by different researchers, coherently. Furthermore, we analyse the fine-tuning between different applications of the protocols and the guidelines. The need to draw up the interpretation scheme of data and information was in line with the use of the software we used (webQDA). We understood that it was not enough to insert the material we collected in the software and that the analysis of the material that the research team had in hand required the use of resources from different sources - methodological, technical, conceptual and semantic - to produce analytical results. It was this 'methodological conversion' that we show, step by step, by means of images in this text, emphasizing the importance of data systematisation, organisation of documentation and use of software as a research aid, especially when it comes to research that takes place in networks, that is, between different nuclei of researchers and with the support of a great diversity of collected data.

References

1. Neri de Souza, F., Costa, A.P., Moreira, A.: webQDA: manual do Utilizador. Universidade de Aveiro, Portugal (2016)
2. Costa, A.P., de Souza, F.N., Moreira, A., de Souza, D.N.: webQDA 2.0 versus webQDA 3.0: a comparative study about usability of qualitative data analysis software. In: Rocha, Á., Reis, L.P. (eds.) Developments and Advances in Intelligent Systems and Applications, pp. 229–240. Springer, Cham (2018)
3. Leite, D., Lima, E.G.S.: Conhecimento, avaliação e redes de colaboração. Produção e produtividade na universidade. Editora Sulina, Porto Alegre, Brasil (2012)
4. Pinho, I., Leite, D.: Doing a literature review using content analysis-research networks review. In: CIAIQ 2014, pp. 377–378, vol. 3, pp. 686–691. Oliveira de Azeméis - Aveiro - Portugal (2014)
5. Leite, D., Miorando, B.S., Pinho, I., Caregnato, C.E., Lima, E.: Research networks evaluation: indicators of interactive and formative dynamics. Comunicação & Informação **17**, 23–37 (2014)
6. Leite, D., Caregnato, C.E., Lima, E.G.d.S., Pinho, I., Miorando, B.S., Bier da Silveira, P.: Avaliação de redes de pesquisa e colaboração. Avaliação: Revista da Avaliação da Educação Superior **19**, 291–312 (2014)
7. Leite, D., Caregnato, C. (org.), Cruz, O., Pinho, I., Dominguez, E., Lima, E., Genro, M.E.: Redes de Pesquisa e colaboração-Conhecimento, avaliação e o controle internacional da ciência Editora Sulina, Porto Alegre, Brazil (2018)
8. Leite, D., Miorando, B.S., Caregnato, C.E., Lima, E., Pinho, I.: Constructing markers to evaluate research collaboration networks. In: CIAIQ 2014, pp. 111–115, Ludomedia (2014)
9. Leite, D., Pinho, I.: Evaluating Collaboration Networks in Higher Education Research: Drivers of Excellence. Springer, Palgrave Macmillan, New York (2017)
10. Ritchie, J., Spencer, L., O'Connor, W.: Qualitative Research Practice: A Guide for Social Science Students and Researchers. SAGE, London (2003)
11. Costa, A.P., Amado, J.: Content Analysis Supported by Software. Ludomedia (2018)
12. Bardin, L.: Análise de Conteúdo Edições 70, Lisboa (2004)
13. Mocelin, D.G.: Concorrência e alianças entre pesquisadores: reflexões acerca da expansão de grupos de pesquisa dos anos 1990 aos 2000 no Brasil. RBPG - Revista Brasileira de Pós-Graduação **6**, 35–64 (2009)
14. Katz, J.S., Martin, B.R.: What is research collaboration? Res. Policy **26**, 1–18 (1997)
15. Spinak, E.: Indicadores cienciométricos. Comunicação e Informação **27**, 141–148 (1998)
16. Leite, D., Cunha, M.I., Lucarelli, E., Veiga, I., Fernandes, C., Braga, A.M., Genro, M.E., Ferla, A., Campani, A., Campos, M., Alves, E., Nolasco, L.: Inovação na Universidade: a pesquisa em parceria. Interface - Comunicação, Saúde, Educação **3**, 41–52 (1999)
17. Corsani, A.: Elementos de uma ruptura: a hipótese do capitalismo cognitivo. In: Galvão, A., Silva, A.G., Cocco, G. (eds.) Capitalismo cognitivo: Trabalho, redes e inovação, pp. 15–32. DP&A, Rio de Janeiro (2003)
18. Etzkowitz, H.: Research groups as 'quasi-firms': the invention of the entrepreneurial university. Res. Policy **32**, 109–121 (2003)
19. Feldman, J.M.: Towards the post-university: centres of higher learning and creative spaces as economic development and social change agents. Econ. Ind. Democracy **22**, 99–142 (2001)

Handcrafted and Software-Assisted Procedures for Discursive Textual Analysis: Analytical Convergences or Divergences?

Isabel Cristina dos Santos Martins[1] ,
Valderez Marina do Rosário Lima[2] ,
Marcelo Prado Amaral-Rosa[2(✉)] , Luciano Moreira[3] ,
and Maurivan Güntzel Ramos[2]

[1] PUCRS – Pontifícia Universidade Católica do Rio Grande do Sul,
Escola de Humanidades, Porto Alegre, RS 90619-900, Brazil
[2] PUCRS – Pontifícia Universidade Católica do Rio Grande do Sul,
Escola Politécnica, Porto Alegre, RS 90619-900, Brazil
marcelo.pradorosa@gmail.com
[3] CIQUP, DEI, Faculdade de Engenharia da Universidade do Porto,
Porto, Portugal

Abstract. The goal of this research was to evaluate the procedures of the Discursive Textual Analysis implemented in a handcrafted manner or assisted by the IRAMUTEQ software in order to identify analytical convergences and/or divergences. The theme was the expectations of young students from military colleges in Rio Grande do Sul (Brazil). Participants were High School students ($n = 7$). The data collection instrument was the semi-structured interview. The data analysis method was the Discursive Textual Analysis, both in a handcrafted manner and assisted by the IRAMUTEQ software. The main result was the convergence of 62% between both ways to generate categories. The convergences and divergences between the procedures of the Discursive Textual Analysis method presented satisfactory parameters for the context of the research and the interpretative assumptions of qualitative research.

Keywords: Discursive Textual Analysis · Qualitative analysis · Handcrafted procedures · IRAMUTEQ

1 Introduction

The Discursive Textual Analysis is a qualitative method that aims to produce new perspectives on textual data [1]. It is a movement with a robust interpretative nature, because, in the process, it takes into consideration the context of the speaker. It is organized around three procedures: (i) disassembling of the texts into unities of meaning; (ii) forming categories; (iii) producing new perspectives through meta-texts [1–3]. This way, it enables to deepen interpretations about the themes of research and new perspectives about the object begin to appear [2].

© Springer Nature Switzerland AG 2020
A. P. Costa et al. (Eds.): WCQR 2019, AISC 1068, pp. 189–205, 2020.
https://doi.org/10.1007/978-3-030-31787-4_16

The IRAMUTEQ software is the acronym for the Franch *Interface de R pour les Analyses Multidimensionnalles de Textes et de Questionnaires* [4, 5]. The software was developed under a free license and, thanks to his algorithm, enables to perform analyses that take into consideration the context in which the words are inserted [5]. One of the functions of the software gives the possibility of performing textual analysis in different levels [4, 5, 7]. To our knowledge, the analytical resources of IRAMUTEQ [4–6] are used, since 2018, to assist the demands of the Discursive Textual Analysis method [1]. Nonetheless, in Brazil, the use of this software for qualitative analysis, in general, started in 2013 [5]. As such, one might say that the software is still is in an embryonic phase of implementation in qualitative research, in the area of education/teaching. In this research, we will focus on the Descending Hierarchical Classification (DCH) and the Correspondence Factorial Analysis (CFA) [4, 5, 7].

The significance of this work is on the identification of analytical convergences and/or divergences between the handcrafted procedures [8] and the procedures assisted by the software IRAMUTEQ in the implementation of the Discursive Textual Analysis method [1]. To our knowledge, we suggest an inedited analytical, methodological perspective [1–3, 6]. Moreover, we claim that the interpretations of the researchers will impact on the convergence and/or divergence between the analytical procedures [9].

Representations about the life experiences of students in military colleges in Rio Grande do Sul, Brazil, are the theme of this research. The Military Police is the armed wing of the State that aims to maintain the public order and protect citizens [12]. Nonetheless, its functions in society have become more complex as problems have also become more severe, e.g., the labyrinthic phenomenon of violence [8].

As they look for solutions that meet up the expectations of the population, the Military Policy of Rio Grande do Sul[1] implements educational projects, as PROERD - *Programa Educacional de Resistência as Drogas e a Violência* (Educational Program of Resistance to Drugs and Violence)[2], developed for private and public schools. The contemporary military police aim to promote shared actions with responsible structures, social and educational institutions regarding public security in order to form individuals, with a particular focus on preventive actions on citizenship [13].

As the process of adopting a citizen conception by the military police unfolds, research in military institutions identified some essential themes to discuss, such as the importance of challenging conventional views on reality [13]. Among the particularities that go along with the construction, development and consolidation of the profile policemen-citizen, educational training might promote reflective movements that might actualize military pedagogical practices [8]. Given the social context of Brazil, such discussions are timely.

Public school is the state institution that requires individuals to live long periods according to social rules developed over time [14]. Therefore, to investigative the youth multiplicity [15] under training in Military colleges in Rio Grande do Sul through a methodological strategy is relevant, timely, and inedited. To understand youth [16, 17]

---

[1] The *Polícia Militar* (Military Police), in Rio Grande do Sul/Brasil, is known by the nickname *Brigada Militar* (Military Brigade).

[2] For more information about PROERD: https://www.brigadamilitar.rs.gov.br/Servicos/Proerd.

views about the context of their life experiences, which includes growing violence [18, 19] and political conjuncture [20], contributes to reflect on the training of future citizens in consonance with the demands of the formative context.

As such, we resorted on the Discursive Textual Analysis to obtain the desired understanding of the theme above exposed [1]. Our attention lays on two ways to perform the method: handcrafted [1–3] and assisted by the qualitative software IRA-MUTEQ [6], because specific programs are more and more available for qualitative data analysis [9, 21–23].

This way, the guiding question of this research was: *Do the analyses of qualitative data, based on the Discursive Textual Analysis method, performed by the handcrafted procedure and assisted by the software IRAMUTEQ, present analytical convergences and/or divergences?* In this context, the goal was to evaluate the method of the Discursive Textual Analysis, performed by the handcrafted procedure and assisted by IRAMUTEQ, in order to identify convergences and/or divergences.

This paper is organized in three other sections: (i) *methods,* with a particular focus on the procedures; (ii) *results and discussion,* in which results of both analytical procedures are reported and discussed in the light of the literature; and (iii) *conclusions,* in which we resume the central question and identify the convergences and/or divergences between the analytical procedures, based on the methodological assumptions of the Discursive Textual Analysis.

2   Methods

2.1   Theme and Context of the Study

The goal of the study is to compare the results of handcrafted and software-assisted (IRAMUTEQ) procedures, based on Discursive Textual Analysis [1]. The analysis is centered on students' representations [10, 11] about the training in military colleges in the state of Rio Grande do Sul, Brazil. The functioning of the colleges is ruled by a series of rules, such as *Lei de Ensino da Brigada Militar*[3] (The Law of the Teaching of the Military Brigade) [24–26].

It is worth noting that the colleges of the Military Police of Rio Grande do Sul have an administrative connection with the State Secretariat for Public Security and the State Secretariat of Education. They are educational institutions with and hybrid pedagogical identity, with two pedagogical fields in the same physical location: civil and military [24–26]. Teachers come from both spheres, and educational strategies include civil and military pedagogies [27]. Here, we focus on the military pedagogy that aims to mediate moral and ethic values through themes, such as history, symbols, and military rituals and disciplinary aspects belonging to military training [8].

Students are admitted to the colleges after applying to selective open annual calls, that include intellectual and physical exams: (i) exams on mathematics and Portuguese; (ii) medical exams; and (iii) physical tests (for both male and female candidates). Approved candidates, after completing registration, are called to come one week earlier

---

[3] For more information: https://www.brigadamilitar.rs.gov.br/.

to take part in the adaption period to the military school life. At the end of each school year, each student is evaluated. Students that failed to pass for the second time are expelled [25, 26].

In Rio Grande do Sul, there are seven military colleges, with around 1.500 students in every year of High School, in day and night modalities [8]. The goal was to collect school experiences, interrelated with social aspects, in a specific school context [14], that differs from civil public schools in many regards, e.g., severe discipline, after-school obligations and the process of application. The understanding of the military youth universe, in the scope of education, is relevant because individual representations are images that give actual meaning to the world, being archetypes that generate patterns, actions and social behaviors [10, 11].

2.2 Participants and *Corpus* of Analysis

Participants were seven students enrolled in military colleges in Rio Grande do Sul, aged between 15 and 16 years old, 43% (03) males and 57% (04) females, attending the different years of High School: 28.5% (02) attended the first year, 43% (03) the second year, and 28,5% (02) the third year. They participated voluntarily and gave their informed consent for the usage of data. Data were anonymized by assigning the letter P and sequential numeration (P1 to P7) to each participant.

We must highlight that we analyzed the same *corpus* with two different analytical procedures, based on the Discursive Textual Analysis method [1]: (i) handcrafted procedure [8], and (ii) IRAMUTEQ software-assisted procedure [6]. The first was performed during the second term of 2016 [8], whereas the second was performed in the second term of 2018.

The total lexical *corpus* consisted of 29.483 words, with an average of 4.211 per text. In the handcrafted procedure, the *corpus* included transcripts of seven interviews (n = 07), in a total of 36 pages and an average of five pages per interview (Δ 02 to 11). The process of *unitarization* is the first step of a Discursive Textual Analysis [1–3], aiming at dissembling the texts into units of meaning. In the process of *unitarization*, it is important to codify all material [1–3]. This procedure facilitates the localization of fragments in the original text, allowing for reanalyzes if necessary. To assign meanings, it is sometimes necessary to rewrite the units of meaning while keeping in mind the statement of the participant [28]. Rewriting is a genuine and spontaneous process, on the part of the researcher, given that the ideas in the statement need to be interpreted for the sequence of analyses. Besides this, it allows the researcher to *impregnate* in the *corpus* of analysis, which favors the future analytical interpretations [28], including what was not explicitly said [1–3].

In the handcrafted *unitarization*, we used a single archive (36 pages) containing all interviews (n = 07) processed with Microsoft *Word* software. The process generated 589 units of meaning, with an average of ~ 84 per interview. To illustrate the construction, we present two excerpts of interviews from two participants and respective unitization (Tables 1 and 2).

Table 1. Excerpt of the *corpus* and respective handcrafted *unitarization* - Participant 2.

Excerpt of the *corpus* - Participant 2		
[...]. *I studied in a public school in a municipality school until the eight. I did not know the Military Police College. I have never heard of. I knew the Federal Instituto of Santa Rosa[a] that offers the technical program on architecture and engineering, but I not very close to that field of engineering, as such, if I am not wrong, it was in 2013, there the Soletrando[b] and it was a girl[c] from Porto Alegre who won de Porto Alegre and I cannot remember well if she studied in the Military Police College[d] or the Military College of Porto Alegre[e], but she studied in one of those* [...] (P2)*		
N.	Unities of meaning	Rewriting	Keywords
1	*I studied in a public school in a municipality school until the eight*	The participant studied in a public school of the municipality until the eight grade of Basic Education	Training Basic Education
2	*I did not know the Military Police College. I have never heard of*	The participant did not know the Military Police College.	Choice of the Military Police College (pelo Colégio da Polícia Militar (Ignorance)
3	*I knew the Federal Instituto of Santa Rosa[a] that offers the technical program on architecture and engineering,*	The participant only knew the Federal Institute that offers technical programs of Architecture and Engineering	Choice of the Federal Institute
4	*but I not very close to that field of engineering,*	The participant has no affinity with the field of Engineering, which did not make the program attractive	Lack of affinity with the programs of the Federal Institute
5	*as such, if I am not wrong, it was in 2013, there the Soletrando[b] and it was a girl[c] from Porto Alegre who won and I cannot remember well if she studied in the Military Police College[d] or the Military College of Porto Alegre[e], but she studied in one of those*	In 2013, one girl of one of the military colleges in Porto Alegre won the *Soletrando* contest	Moving closer to military colleges

Notes:
[a]The Federal Institute of Santa Rosa is a federal educational institution that covers secondary, technical, and higher education levels. The municipality of Santa Rosa is located in the northwest of Rio Grande do Sul at 490 km from Porto Alegre, the state capital.
[b]*Soletrando* was a national spelling contest, held every year by an open television program in Brazil. Students from the public school system from all over the country could participate. The prize was 100 thousand reais that should be invested in the educational formation of the winner. It had seven editions, the last being in 2015.
[c]The girl's name is Yasmin Esswein. At the time of the contest, she was a student at the Military College of Porto Alegre.
[d]To avoid confusion, we decided to adjust to the original discourse differentiation between the military colleges of Porto Alegre, since P2 called both military colleges.
[e]The Military College of Porto Alegre is under the hedge of the Brazilian army. The Military Police Colleges are under the hedge of the state government of Rio Grande do Sul.
Source: Adapted from Martins [8].

Table 2. Excerpt of the *corpus* and respective handcrafted *unitarization* - Participant 7.

Excerpt of the *corpus* - Participant 7

[…]. *Private school from first to eighth grade. It was a series of events, there was a dilemma I was thinking of continuing to study in my private college or going to a public school. I did not want to continue in my private school because it was always very weak. It would not grant me a place in the federal university, for example. The decision of going to another college even when I saw the list of people approved to the university in my college* (P7)

N.	Unities of meaning	Rewriting	Keywords
1	*Private school from first to eighth grade*	The participant studied in a private school from first to eighth grade	Training Basic Education
2	*It was a series of events, there was a dilemma I was thinking of continuing to study in my private college or going to a public school*	The participant was questioning the quality of the private school, thinking to change to a public school	Training Basic Education
3	*I did not want to continue in my private school because it was always very weak*	The private school was not taking care of the participant's training	Beginning of the process of choosing another school
4	*It would not grant me a place in the federal university for example*	The school was not going to grant the participant access to Higher Education	Beginning of the process of choosing another school
5	*The decision of going to another college even when I saw the list of people approved to the university in my college*	The participant decided to change school	Beginning of the process of choosing another school

Source: Adapted from Martins [8].

With relation to the entries of data in the IRAMUTEQ software [4, 5], it is worth declaring that we adopted the following options: (i) language: Portuguese language; (ii) method of constructing Text Segments (TS): occurrences; (iii) length of TS: *40*. Furthermore, both in the preparation of the material as well as in the entry options of the software, we adopted the default options [4, 5, 7]. In the classic lexicographical description, the glocal *corpus* of analysis consisted of 826 TS, with the software considering 83,05% (686 TSs). The TSs have, on average, 3,25 lines [4, 5], with ~ 35 words, 810 being single occurrences (Hapax coefficient = 2.75% of the occurrences).

2.3 Data Collection and Data Analysis

Data were collected via semi-structured interviews [9, 21, 29–31]. The interview script consisted of 26 open questions, comprehending dimensions about the life experiences, expectations, and youth challenges after the admittance to military colleges. In the discourse of the participants there were representations about the decision process to apply for the military educational system; adaptation difficulties; relationship with colleagues, teachers, and superiors; feelings towards the family, friends, and society; engagement in internal activities; prejudices; social participation and plans for the future [8].

In the handcrafted data analysis procedure, we used inductive analysis, in which the categories emerge from the *corpus* and were regrouped according to the units of meaning [1]. All the interviews took place in the facilities of the military colleges, with an average duration of ~ 45 min. One student from each school year in each college was randomly invited to participate, forming a total sample of 21 students.

The final number of participants (n = 07) represented in the *corpus* of analysis considered, in principle, to the criterium of data saturation about the youth representations [32, 33]. The final intention is not the comparison of discourses between colleges or groups of students, but, instead, the global mapping of their representations as High School students of the military system.

In the analytical procedure of Discursive Textual Analysis assisted by the IRAMU-TEQ software [6], we used the Descending Hierarchical Classification (DHC) and the Correspondence Factorial Analysis (CFA) [4, 5]. Despite the quantitative analytical base of the software, the analysis stresses the interpretation of the results, pointing to relevant explorations associated with the method of Discursive Textual Analysis [1–3, 6, 28].

Finally, considering the three procedural phases of Discursive Textual Analysis (*unitarization, categorization*; and production of *meta-texts*), we highlight that this paper is focused on the phases of *unitarization* and *categorization*. The reasons are the following: (i) the length of the paper, since the presentation for meta-texts would require more space; (ii) the relations between the handcrafted and the software-assisted procedures are explored in the phases of unitarization and categorization; and (iii) the focus is the convergences and/or divergences between the ways of implementation of the method. Therefore, it is explicitly outside the scope of this paper the communication of the new emergent approach to analyze the representations of the participants about their life experiences in the school context of the military colleges in Rio Grande do Sul.

3   Results and Discussion

To obtain the results with the handcrafted procedure of the Discursive Textual Analysis [1], we used the interviews (n = 07) of the participants, which formed the *corpus* of analysis. In Tables 1 and 2, the difference of representations about the *Ensino Fundamental* (elementary School) of participants P2 and P7 is self-evident. While P2 *"[...] have never heard [...]"* about military colleges, obtaining the first news through a contest in public television, P7 was interested in changing of school to get the necessary training to go to University, considering that his old school *"[...] would not grant him a place in the federal university [...]"*. Rewriting is an important component of this phase of Discursive Textual Analysis [1, 2] because it enables the interpretative process f the enunciates, which will be very important in the phase of producing new readings of the representations of the participants, i.e., the meta-texts [1–3]. For the reasons previously reported, the meta-texts will not be presented in this paper.

During the handcrafted procedure of the Discursive Textual Analysis, *unitization* is first concerned with identifying each idea in the context of the discourse(s), forming the units of meaning. In the examples, we used the results of P2 (Table 1) and P7 (Table 2).

The selected excerpt of P2 has 99 words, and an average of ~ 20 words/unit of meaning, and the excerpt of P7 has 77 words and an average of ~ 15 words/unit of meaning. In the software-assisted procedure, the identification is more standardized since each TS is identified according to the relevance of the words assigned by the researcher (Fig. 1).

**** *P2
because so-and-so said such a thing about you a few years ago there is no future it's hard to say because society thinks many *young* people have no future but I think the youth should start to become sensitized

**** *P2
she should even help herself to change this concept that all *young* people are rebellious and should be detained by society and further continue to evolve increasingly in the participation of society at large

**** *P7
refreshing ideas but in those ideas they are representing the *young* is what I stopped to think lately about this they are representing the *young* in some way because the *young* person in political propaganda is rare

**** *P7
is changing before it should be worse but it is rare that one so another so is defeated because adults still think *young* is immature and the *young* has no ability

---

Fig. 1. Excerpts of P2 and P7 in the IRAMUTEQ - word "young". Source: adapted from IRAMUTEQ data.

The main difference between the units of meaning (handcrafted procedure) and TS (IRAMUTEQ), in the method of Discursive Textual Analysis [1], is *unitarization* procedure the *corpus* of analysis. The handcrafted procedure, by definition, requires more attention from the researcher. The tendency is that it will show differences according to the style of the researcher: (i) analytical researcher, who is leaned to fragmentize and present units of meaning center in sentences with nuclear ideas; and (ii) the contextual researcher, who is leaned to identify more comprehensive units of meaning, preferring context over nuclear ideas. The *unitarization* with different "intensities" is not a decisive aspect for the ulterior process, the categorization, because, in the handcrafted procedure, the interpretative approaches of the researcher occur intuitively.

Instead, the production of the TS in the software IRAMUTEQ is an automatized procedure, based on statistical operations, the interference of the researcher in the process being impossible. Each TS may have more than one unit of meaning [1] in the same TS. The *unitarization* and *categorization* phases of the Discursive Textual Analysis happen at the same time, with a focus on the formation of classes, which in the handcrafted procedure are named categories.

However, before moving on to the results of the procedure of categorization, we call the attention to data saturation. In the handcrafted phase, after 21 interviews in different colleges, municipalities, and regions, $\sim 66\%$ of the information contained in the interviews were considered to be similar. The homogeneity of the discourse is one aspect emphasized by the military pedagogy, even covering the internalization of automatic answers to situations that comprehend the context in which one lives [27].

In IRAMUTEQ [4, 5, 7], the Hapax coefficient might give clues about data saturation. Only 2.75% of the words of the *corpus* had a single occurrence, suggesting that even the selected discourses of the participants (n = 07/21) have similar terms. If they did not, the Hapax coefficient data would be higher. Nonetheless, we claim that to increase the analytical precision regarding the Hapax coefficient; it is necessary to perform comparative tests with other *corpora*.

The next step of the Discursive Textual Analysis is categorization [1]. In developing the handcrafted procedure (Tables 3 and 4), we grouped the similar units of meaning, forming the initial categories and, subsequently, successive regroupings occur to form intermediary and final categories [1–3, 28]. In the inductive categorization, we name each category after the common meanings of the units of meanings that compose each (re)grouping [1–3, 28].

Table 3. Rewriting of the units of meaning and handcrafted categorization - P2.

N.	Rewriting	Categories		
		Initial	Intermediate	Final
1	The participant studied in a public school of the municipality until the wight grade of Basic Education	Training Basic Education	Training Basic Education	Before entering in the Military Police College
2	The participant did not know the Military Police College	Choice of the Military Police College (unique choice)	Choice of the Military Police College	
3	The participant only knew the Federal Institute that offers technical programs of Architecture and Engineering	Choice of the Military Police College (unique choice)	Choice of the Federal Institute[a]	
4	The participant has no affinity with the field of Engineering, which did not make the program attractive	Choice of the Federal Institute (lack of affinity)		
5	In 2013, one girl of one of the military colleges in Porto Alegre won the *Soletrando* contest	Moving closer to military colleges	Moving closer to military colleges[a]	
Total		05	04	01

[a]These intermediate categories were refined and fit the final intermediate categories (Table 5).
Source: Adapted from Martins [8].

Table 4. Rewriting of units of meaning and their handcrafted categorizations - P7.

N.	Rewriting	Categories		
		Initial	Intermediate	Final
1	The participant studied in a private school from first to eighth grade	Training Basic Education	Training Basic Education	Before entering in the Military Police College
2	The participant was questioning the quality of the private school, thinking to change to a public school	Training Basic Education		
3	The private school was not taking care of the participant's training	Beginning of the process of choosing another school	Beginning of the process of choosing another school[a]	
4	The school was not going to grant the participant access to Higher Education	Beginning of the process of choosing another school		
5	The participant decided to change school	Beginning of the process of choosing another school		
Total		05	02	01

[a]This intermediate category has been refined and fits the final intermediate categories (Table 5).
Source: Adapted from Martins [8].

Tables 3 and 4 present five unities of meaning which have been rewritten for each participant. Regarding P2, it is possible to assign the same initial category to the unities of meaning no. 3 and no. 4. Three unities of meaning keep their independence until the formation of the final categories. Regarding P7, on the contrary, all regroupings occur before the final regrouping. We should highlight that both tables are just examples of the complete handcrafted process, which consisted of 79 initial categories, 13 intermediary categories, and three final categories (Table 5).

For obtaining results in the IRAMUTEQ software [4, 5, 7], the same interviews of the handcrafted process (n = 07) were submitted to a DCH and CFA [5, 6, 34]. DCH resulted in a dendrogram with five classes (Fig. 2). One considers that the classes correspond to the intermediary categories in the handcrafted methods fo Discursive Textual Analysis [6].

The classification of the text segments is conducted according to the vocabulary and frequency of the reduced forms, allowing for bringing groupings together or pushing groupings apart in the more stable configuration [5, 34]. The names of the five intermediary categories resulted from the interpretations of the meaning of the 10 texts

Table 5. Intermediary and final categories of the handcrafted process.

Synthesis of the unities of meaning	Categories	
	Intermediary	Final
How did it go?	01. Training Basic Education	Before entering in the Military Police College
	02. Choice of the Military Police College	
	03. Expectations about the Military Police College	
	04. Preparation for entering the Military Police College	
How is it like now?	05. Challenges of the admittance and saying in the Military Police College	During the stay in the Military Police College
	06. Routines of the Military Police College	
	07. Interpersonal relationships	
	08. Sociability (leisure and friendships)	
What do the participants think about the themes?	09. Prejudice and discrimination	
	10. Violence	
	11. Being young	
	12. Youth participation	
How will it be?	13. Future projects	After graduating from entering in the Military Police College
Total	13	03

Source: Adapted from Martins [8].

segments with higher score[4] in each intermediary category: (i) *predisposition to learn in the college* (category 2 - Δ score 140 to 201.18); (ii) *demanding college in the studies* (category 3 - Δ score 215 to 299.33); (iii) *demands, fears and perspectives* (category 4 - Δ score 193 to 261.62); (iv) *activities and interpersonal relationships* (category 5 - Δ score 144 to 194.15); and (v) *views of the society on the youth* (category 1 - Δ score 290 to 357.57). The higher the punctuation, the higher the density and importance of the segment within the category [6].

By reading the dendrogram (Fig. 2) from left to right, one observes two branches: (i) branch 1 (B1), subdivided in B1(1) (categories 2 and 3) and B1(2) (categories 4 and 5); and (ii) branch 2 (B2), with category one alone. Regarding the formation of the final

[4] Adopted score in this paper: *Absolute* (sum of the χ^2 of the entries highlighted in the text segment). For the inclusion of the words in each category, the association of the word with each category is determined by a χ^2 score equal to or higher than 3.841 [7, 34].

categories in the perspective of the Discursive Textual Analysis, the similar the intermediary categories, the greater the contextual affinity and probability of (re)-groupings, when necessary [1, 6].

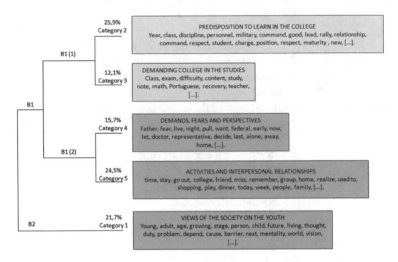

Fig. 2. DCH dendrogram about students' representations. Source: Adapted data from IRAMUTEQ.

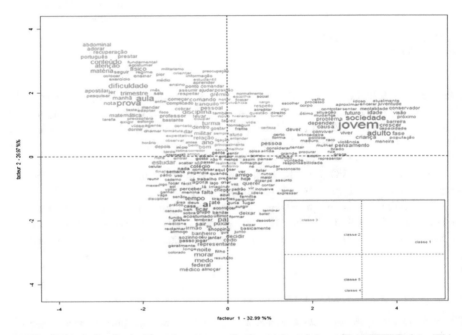

Fig. 3. CFA of the words of each category. Source: Adapted from IRAMUTEQ data (This image is not translatable).

In the FCA, one can verify the relations of dependency between each intermediary category (Fig. 3). It is an adequate analysis to visualize the affinities among categories according to their coordinates in the cartesian plane [34].

The proximity/distance, in the horizontal or vertical lines of the cartesian plane, demonstrate the level of relation between categories [34]. Therefore, variables present a relation of dependency or independency when: (i) projected categories in the same quadrant, (ii) projected in opposite quadrant; and (iii) projected categories near to the same horizontal or vertical line. Furthermore, regarding the relations of dependency, it is possible to identify four levels of intensity: (i) null; (ii) weak; (iii) moderate; and (iv) strong.

Based on the data of the cartesian plane (Fig. 3), the relations of dependency are the following: (i) between categories 2 (gray) and 3 (green): dependency of moderate intensity, because, although they are projected in the same quadrant, there is a visible distance between them; between categories 4 (blue) and 5 (violet): dependency of high intensity, because they are projected in the same quadrant, close between them and close to the vertical line; (iii) between categories 2 (gray) and categories four (blue) and 5 (violet): dependency of weak intensity, because they are close to the vertical line, but in opposite projected in opposite quadrants with considerable distance; and, lastly, (iv) between category 3 (green) and 1 (red): null relation of dependency, because the distance between them is too large to exist any dependency.

Table 6. Cohesion force between the handcrafted and IRAMUTEQ procedures.

Converging categorizations				
Handcrafted		IRAMUTEQ		Cohesion force
Intermediate	Final	Intermediate	Final	
05. Challenges of the admittance and saying in the Military Police College	During the stay in the Military Police College	Category 4 Demands, fears, and expectations	Expectations, fears and interpersonal relationships	Weak
06. Routines of the Military Police College				Moderate
07. Interpersonal relationships		Category 5 Activities and interpersonal relationships		Strong
08. Sociability (leisure and friendships)				
09. Prejudice and discrimination		Category 1 Views of the society on the youth	Views of the society on the youth	
10. Violence				
11. Being young				
12. Youth participation				

Based on the analyses and interpretations of the five intermediate categories that emerged from the IRAMUTEQ software [4, 5, 7], three final categories were formed: (i) *demands and predisposition to learn in the college,* which resulted from the merging of categories 2 and 3; (ii) *expectations, fears and interpersonal relationships,* which resulted from the merging of categories 4 and 5; and (iii) *views of the society on the youth,* which resulted from category 1 alone.

The convergences between the handcrafted and IRAMUTEQ Software-assisted occur both in the intermediary categories as in the final categories. From cross-comparing data from both processes, it is possible to determine, in an interpretative manner, the force of cohesion between the handcrafted and IRAMUTEQ software-assisted categorization procedures (Table 6).

In the handcrafted procedure, there are 13 intermediary categories [8], eight of which showed *cohesion force* with the intermediate categories that resulted from the IRAMUTEQ software-assisted procedure (five categories intermediaries), representing 62% of the *corpus* of analysis. The cohesion between procedures was concentrated in the intermediary categories 1, 4 and 5, formed with the IRAMUTEQ software and the intermediary categories under the final category *During the stay in the Military Police College* of the handcrafted procedure. The handcrafted intermediary categories that do not show cohesion force with the categories created with the software are related with other two final categories: *before admittance to the Military Police College* (01, 02, 03 and 04) and *after graduating in the Military Police College* (13).

On the other hand, considering the intermediate categories of the IRAMUTEQ software-assisted procedure, the intermediate categories of the branch B1(1), category 2 – *Predisposition to learn in the College* and category 3 – *Demanding college in the studies,* which formed the final category *Demanding and predisposition to learn in the College,* do not showed cohesion force with the handcrafted procedure, which corresponds to 38% of the *corpus* of analysis. One considers, for now, de difference found between categorization as acceptable, given the interpretative nature of both analytic procedures and their contexts of implementation.

Finally, we claim that the central point is not to create a divide between handcrafted qualitative data analysis procedures and software-assisted procedures. We highlight that the convergence between the Discursive Textual Analysis and the resources of the IRAMUTEQ [4, 5] is an endeavor to consider the affordances that a qualitative analysis software, among many others available in the market, can contribute to the processes of analysis historically consolidated, as the Discursive Textual Analysis, point out new perspectives.

4   Conclusions

The purpose of this study was to identify convergencies and/or divergencies between the handcrafted and IRAMUTEQ software-assisted procedures for the method of Discursive Textual Analysis. With this in mind, we tried to answer to the question: *Do the qualitative data analyses, based on the method of Discursive Textual Analysis, implemented through handcrafted procedures as well as through the IRAMUTEQ software-assisted procedure, present convergences and/or analytical divergences?* Considering the results from both procedures, we reached the following conclusions:

(i) *on analytical convergences*: convergences between handcrafted and IRAMU-TEQ software-assisted procedures, in the method of Discursive Textual Analysis, presented a satisfactory score in the context of this research. We think that the concomitant use of both procedures can present qualitative evidence, on the object under analysis, with high level of reliability and interpretative accuracy for the scientific community.

(ii) *on analytical divergences:* interpretative divergences are expected in an approach that involves multiple researchers. The score obtained is the result of factors, such as the temporal gap between the analyses, variable levels of expertize on the method of the researchers, and context of the implementation of the analyses. We think that if the context is designed from the beginning to include both procedures, the interpretative divergences will lessen.

(iii) *on the cross-comparison between handcrafted and software-assisted procedures*: procedures are complementary. The handcrafted procedures give room to impregnating in the data, personalized interpretations and the perception of what has not been said regarding the corpus of research data; the software-assisted procedure offers flexibility, hidden relationships between and analytical strength. Therefore, the body of information afforded by the software regarding the formation of intermediary categories seems to be a viable, reliable and free alternative for qualitative researchers, because it "illuminates the black box" [6], which are the data construction resulting from tools like questionnaires and interviews.

Lastly, in order to foster discussions in future work, we argue for the development of research project from the beginning to use together with the handcrafted and IRAMUTEQ software-assisted procedures in the field of Education and/or teaching of Sciences and Mathematics with the goal of clarifying and improve the accuracy of data interpretation. In particular, for the improvement of this research, it is necessary: (i) analyze in-depth the internal data of each intermediary categories; (ii) pay attention to the detailed analysis of the subcorpus of each branch of categories; and (iii) give examples of meta-texts regarding the representations of the participants.

Acknowledgments. Thanks are due to CAPES – Coordenação de Aperfeiçoamento de Pessoas de Nível Superior for the post-doctoral grant awarded to the third author and Fundação para a Ciência e a Tecnologia (FCT) for the Ph.D. grant awarded to the forth (PD/BD/114152/2015).

References

1. Moraes, R., Galiazzi, M.C.: Análise Textual Discursiva (Discursive Textual Analysis). Ed. Unijuí, Ijuí (2015)
2. Lima, V.M.R., Ramos, M.G.: Percepções de interdisciplinaridade de professores de Ciências e Matemática: um exercício de Análise Textual Discursiva (Perceptions of interdisciplinarity of Science and Mathematic's teachers: an exercise of Discursive Textual Analysis). Revista Lusófona de Educação **36**, 163–177 (2017)

3. Moraes, R., Galiazzi, M.C., Ramos, M.G.: Aprendentes do aprender: um exercício de Análise Textual Discursiva (Learning learners: an exercise in Discursive Textual Analysis). Indagatio Didactica 5(2), 868–883 (2013)
4. Ratinaud, P.: IRAMUTEQ: Interface de R pourlesAnalysesMultidimensionnelles de Textes et de Questionnaires - 0.7 alpha 2 (2014). http://www.iramuteq.org. Accessed 15 Apr 2019
5. Camargo, B.V., Justo, A.M.: IRAMUTEQ: um software gratuito para análise de dados textuais (IRAMUTEQ: a free software for analysis of textual data). Temas em Psicologia 21 (2), 513–518 (2013)
6. Ramos, M.G., Lima, V.M.R., Amaral-Rosa, M.P.: IRAMUTEQ software and discursive textual analysis: interpretive possibilities. In: Costa, A., Reis, L., Moreira, A. (eds.) Computer Supported Qualitative Research. WCQR 2018. Advances in Intelligent Systems and Computing, vol. 861, pp. 58–72. Springer, Cham (2019)
7. Camargo, B.V., Justo, A.M.: Tutorial para uso do software IRAMUTEQ (Tutorial for using IRAMUTEQ software) (s.d.). https://goo.gl/22jP4X. Accessed 10 Jan 2019
8. Martins, I.C.S.: O ingresso de jovens nos colégios Tiradentes da Brigada Militar/RS: um sonho dos jovens ou só um meio para se atingir a um fim? (The admission of youth in the Tiradentes colleges of the Military Brigade/RS: a youth dream or just a means to reach an end?) 2017. 237 p. Dissertação (Mestrado) - Programa de Pós-Graduação em Educação, PUCRS - Pontifícia Universidade Católica do Rio Grande do Sul (2017)
9. Stake, R.E.: Pesquisa qualitativa: estudando como as coisas funcionam (Qualitative research: studying how things work). Penso, Porto Alegre (2011)
10. Olsen, W.: Coleta de dados: debates e métodos fundamentais em pesquisa social (Data collection: fundamental debates and methods in social research). Penso, Porto Alegre (2015)
11. Pesavento, S.J.: História & História Cultural (History & Cultural History). Autêntica, Belo Horizonte (2008)
12. Azevedo, E.F.: A polícia e suas polícias: clientela, hierarquia, soldado e bandido (The Police and its Polices: Customers, Hierarchy, Soldier and Criminal). Psicologia: ciência e profissão 37(3), 553–564 (2017)
13. Bengochea, J.L.P., Guimarães, L.B., Gomes, M.L., Abreu, S.R.: A transição de uma polícia de controle para uma polícia cidadã (The transition from a control police to a citizen police). São Paulo Perspec. 18(1), 119–131 (2004)
14. Elias, N.: A sociedade dos indivíduos (The society of individuals). Jorge Zahar, Rio de Janeiro (1994)
15. Margulis, M.: La juventud es más que una palabra (Youth is more than a word). In: Margulis, M. (org.) La juventud es más que una palabra: ensaios sobre cultura y juventud. 3ª ed. Editorial Biblos, Buenos Aires (2008)
16. Dayrell, J.: O jovem como sujeito social (The young as social subject). Revista Brasileira Educação 24, 40–52 (2003)
17. Dayrell, J.: A escola "faz" as juventudes? Reflexões em torno da socialização juvenil (Does the school "make" youths? Reflections on youth socialization). Revista Educação e Sociedade, especial 1105–1128 (2007)
18. Almeida, M.G.B. (org.): A violência na sociedade contemporânea (Violence in contemporary society). EDIPUCS, Porto Alegre (2010)
19. Cerqueira, D., et al.: Atlas da violência (Atlas of Violence). Fórum Brasileiro de Segurança Pública, Rio de Janeiro (2017)
20. Carvalho, J.M.: Forças armadas e política no Brasil (Armed forces and politics in Brazil). Jorge Zahar, Rio de Janeiro (2005)
21. Gray, D.: Pesquisa no mundo real (Real World Search). Penso, Porto Alegre (2012)

22. Mayring, P.: Qualitative content analysis: theoretical foundation, basic procedures and software solution. Klagenfurt, Austria (2014). https://goo.gl/pNjubm. Accessed 10 May 2018

23. Costa, A.P., Amado, J.: Análise de conteúdo suportada por software (Software Supported Content Analysis). Ludomedia, Aveiro/POR (2018)

24. Rio Grande do Sul. Lei n° 12.349, de 26 de outubro de 2005. Institui o Ensino na Brigada Militar do Estado do Rio Grande do Sul e dá outras providências. https://goo.gl/nD4Nbb. Accessed 10 May 2015

25. Rio Grande do Sul. BRIGADA MILITAR. Portaria n° 535/EMBM/2012. Aprova o Manual do Aluno e Institui o Regulamento de Uniformes, Insígnias, Distintivo e Apresentação Pessoal dos Colégios Tiradentes da Brigada Militar (RUAP/CTBM) e o Regulamento Disciplinar dos Colégios Tiradentes da Brigada Militar (RD/CTBM) (2012)

26. Rio Grande do Sul. BRIGADA MILITAR. DEPARTAMENTO DE ENSINO (DE). Edital n° 035/DE/2015. Processo seletivo de admissão e classificação para ingresso nos Colégios Tiradentes da Brigada Militar de Porto Alegre, Passo Fundo, Santa Maria, Ijuí, Santo Ângelo, São Gabriel e Pelotas para o ano Letivo de 2016. DE, Porto Alegre (2015)

27. Neto, A.F.: A Pedagogia no Exército e na Escola: a Educação Física (1920–1945) (Army and School Pedagogy: Physical Education). Revista Motrivivência, Ano XI, n° 13 (1999)

28. Ramos, M.G., Ribeiro, M.E.M., Galiazzi, M.C.: Análise Textual Discursiva em processo: investigando a percepção de professores e licenciandos de Química sobre aprendizagem (Discursive Textual Analysis in process: investigating the perception of chemistry teachers and undergraduates about learning). Campo Aberto 34(2), 125–140 (2015)

29. Amado, J.: Manual de investigação qualitativa em educação (Qualitative research manual in education). Imprensa da Universidade de Coimbra, Coimbra (2013)

30. Minayo, M.C.: Pesquisa social: teoria, método e criatividade (Social research: theory, method and creativity). 15ª ed. Vozes, Petrópolis/RJ (2000)

31. Minayo, M.C., Costa, A.P.: Fundamentos teóricos das técnicas de investigação qualitativa (Theoretical foundations of qualitative research techniques). Revista Lusófona de Educação 40, 139–153 (2018)

32. Creswell, J., Miller, D.: Determining validity in qualitative inquiry. Theory Pract. 39(3), 124–130 (2000)

33. Barros, N.F., Rodrigues, B.S., Teixeira, R.A.G., Oliveira, E.S.F., Silva, L.F.: "Quantas entrevistas são suficientes?": reflexões sobre a técnica da saturação dos dados na pesquisa qualitativa ("How many interviews are enough?": Reflections on the technique of data saturation in qualitative research). In: Oliveira, E.S.F., Barros, N.F., Neri de Souza, D.C.D. B. (org.) Metodologias qualitativas em diferentes cenários: saúde e educação, pp. 19–33. Gráfica UFG, Goiânia (2017)

34. Veraszto, E.V., Camargo, E.P., Camargo, J.T.F., Simon, F.O., Miranda, N.A.: Evaluation of concepts regarding the construction of scientific knowledge by the congenitally blind: an approach using the correspondence analysis method. Ciênc. Educ. 24(4), 837–857 (2018)

Ideas About Plagiarism and Self-plagiarism with University Professors and Researchers: A Case Study with WebQDA

Lina Melo[(✉)] [iD], Luis Manuel Soto-Ardila [iD], Ricardo Luengo [iD], and José Luis Carvalho [iD]

Department of Sciences and Mathematics Education, Faculty of Education, University of Extremadura, Badajoz, Spain
{lvmelo, luismanuel, rluengo, jltc}@unex.es

Abstract. Plagiarism is a problem that has more presence in our society, and it is increasingly. It is an attack on the rights of authors, it does not respect for research ethics. For this reason, the objective of our study, with qualitative focus, was to know what teachers and researchers of the University of Extremadura do who participated in the Workshop Introduction to Qualitative Analysis Software webQDA, think about the plagiarism and self-plagiarism of scientific documents. Four categories emerged from the analysis carried out: (a) concept of plagiarism and self-plagiarism, (b) motives for plagiarism, (c) use of self-plagiarism software and (e) frequency of plagiarism. The most significant results suggest that sometimes self-plagiarism is not considered plagiarism and from the perspective of the investigation the intentions of those who commit plagiarism determine its definition and impact. In addition, participants do not express concern about evaluating the originality of a text using specialized computer media.

Keywords: Plagiarism · Self-plagiarism · WebQDA · Teaching staff

1 Introduction

Plagiarism and self-plagiarism are issues that are increasingly present daily. There are many people, who in a planned way, use these texts from other people to do their own scientific research. Currently, we do find some cases of plagiarism in Doctoral Theses, final works, scientific studies of personalities from Public offices, as may be the case of politicians. However, this is not a remote thing that only happens in Spain, but we can find it in influential personalities from all over the world. Some of these examples can be: the president of Hungary, Pál Schmitt, who plagiarized more than 20% of his Doctoral Thesis; or in the case of Karl-Theodor zu Guttenberg (the German defence minister from 2009 until 2011), who lost his Doctor of Law degree, due to having plagiarized his investigations. These are just some of the examples of the presence that plagiarism has in our society.

As one can think, plagiarism has been quite present throughout the past centuries, but it is now, in the society in which we find ourselves, where there are particular

© Springer Nature Switzerland AG 2020
A. P. Costa et al. (Eds.): WCQR 2019, AISC 1068, pp. 206–215, 2020.
https://doi.org/10.1007/978-3-030-31787-4_17

situations that make plagiarism much more present. Before addressing the different conditions that facilitate the appearance of plagiarism, it is important to define what plagiarism and self-plagiarism are.

Authors such as Girón [1], explain that plagiarism is the taking of words or ideas written by others without there being any recognition of the rights of the authors, in addition to those complete texts that a person is attributed, without being the true author of it. Imran [2] explains under what conditions self-plagiarism occurs, including those in which the author copies a particular work that he has done previously, modifying the words he used the first time, to achieve a different appearance.

After clarifying the definitions of these two concepts, we can move onto the conditions in which plagiarism is present. As it can be seen, in a society of information and communication, in which the Internet is a mass phenomenon, we can find practically all of the existing information, where plagiarism becomes a widely used practice, as we currently known as "copy and paste" [3]. Nevertheless, this is not the only existing reason that can "justify" the presence of plagiarism, there are other reasons that lead professional researchers to perform this practice. One of them is the increase in the demands of scientific work. Nowadays, researchers are required to publish a number of publications that, if it is a wordy text, they will achieve better results in the various evaluations. This means that the need to mass produce articles which leads them to the use of foreign texts that are not cited as they should be. Throughout this document, we will name several investigations that address this issue and detail the results as well as the conclusions that have been obtained with this study.

2   Theoretical Framework

As indicated in the Introduction, plagiarism is a problem that increasingly has a presence in our society. It can be considered as an attack on the rights of authors that shows no respect for research ethics. This means that there are more and more studies that address this issue, trying to create awareness among people and making them reflect on what may be the reasons for plagiarism and how it can be avoided.

Soto [4] shows in his study, an analysis of the concept of plagiarism and the impact it has at a global and national level. This study aims to demonstrate the importance of this problem and how it is getting worse each time, being a common practice on many occasions. Other authors, such as Balbuena [5], detail the damage that plagiarism causes to the author's rights of the works, explaining how an author may be economically affected, by the fact that they have their work plagiarized. On the other hand, Ramírez and Jiménez [6] give an explanation about plagiarism and self-plagiarism from a neutral point of view, taking into account the different investigations and the different cases of plagiarism that have arisen within the pass of time.

Other authors such as Rojas [7] collect information about the different types of plagiarism that exist, to get an approximate view of how much, a text can be considered plagiarism or not. The following typology stands out:

- Direct plagiarism: it is considered direct plagiarism the moment you take a complete text without citing it.

- Plagiarism for inappropriate use of paraphrasing: it occurs when you take a text and modify some words to obtain a small change.
- Complex plagiarism using a reference: it occurs the moment in which a text is referenced, but it is done in an erroneous or incomplete way (for example, not indicating the pages where the reference appears).
- Plagiarism with loose quotes: It occurs when we close a quote (using the quotation marks) but we still use phrases copied from the text outside the quotation.
- Paraphrasing: it happens when words of a copied text are modified and when the text is of a great extension and at no time material is added that enriches the text information.
- Self-plagiarism: it occurs when the appearance of a text that the author has written in order to appear to be another text has been modified or, if the text has been used from another study by the same author and it is not cited.

After seeing some of the research that has been performed in recent years, a clear vision of what will be studies can be obtained, the reason why it is studied and why it is done, allowing us to acquire a scientific basis on which to set our research on.

3 Research Problem

The objective of our study is to know what the professors and researchers of the University of Extremadura think about the plagiarism and self-plagiarism of scientific documents. This objective was broken down into the following research questions:

1. What perception do professors and researchers have about plagiarism and self-plagiarism? Is self-plagiarism really a plagiarism?
2. What perception do professors and researchers have about the reasons for plagiarism and self-plagiarism?
3. What knowledge do professors and researchers have about the use of anti-plagiarism software?
4. What perception do professors and researchers have about who and in what scientific field plagiarism more frequent?

4 Methodology

The present research, with a qualitative approach, was developed during a course included in the training plan for the teaching and research staff, whose objective was to deepen the knowledge of the participants about the handling of non-numeric and unstructured data through of the use of webQDA software for qualitative data analysis. Fifteen professors and researchers were part of our sample.

The participants have on average a teaching experience of 4 years and a research experience of 6 years. The research fields are mostly social sciences (53%) as well as science and technology (27%). About gender, 53% are women and 47% are men.

To give an account of our research problem, we use the contributions of the participants in a forum created in the Virtual Campus of the course (Introduction to the webQDA software for qualitative data analysis) as a data collection tool. The objective of the forum was to gather the opinions of professors and researchers about plagiarism as well as self-plagiarism through a driving question. According to Carvalho, Luengo, Casas y Cubero [8], the virtual forum is a space for the construction of meanings through interaction and collaboration.

The systematization of the data and its analysis was carried out following the techniques of content analysis [9, 10] which include the following steps (a) identification of the units information from each instrument or tool used to collect the data; (b) coding of the information units in the categories and subcategories; (c) categorization of the information units; (d) analysis of the information units; (e) incorporation of emerging categories to the description based on the analysis carried out [11].

In order to develop the categorization scheme (Table 1), subsequent in-depth readings were made on the contributions of the participants in the Forum. The validation of the category system was ensured by having a consensus among three researchers.

Table 1. Category systems

Dimensions	Categories
A. Concept	A1. Plagiarism
	A2. Self-plagiarism
B. Reason	B1. Lack of training
	B2. Belief of impossibility to detect plagiarism
	B3. Ease and the impunity to copy from the Internet
	B4. Have a good curriculum vitae/Increase academic production
	B5. Need for promotion
	B6. Error if it occurs involuntarily
	B7. Ethic/Values
	B8. Reserved rights/Self-plagiarism
C. Software	C1. Software knowledge for plagiarism detection
	C2. Examples/Software type for plagiarism detection
	C3. Need for use
	C4. Frequency of use
	C5. Availability
D. Frequency	D1. Research experience
	D2. Scientific field

In order to identify the information units, we resorted to the manifest content, in order to maintain the parameters of reliability and verifiability, we sequentially placed in each category the corresponding information units, through the webQDA software for qualitative data analysis. Agreements of 92% were achieved (disagreements were resolved by consensus). The webQDA software was used for the possibility of working collaboratively online as well as being in real time. In addition, it is on the cloud (computing cloud) and it allows the researcher access to his project anywhere [12].

5 Results

The results will be presented following our research questions. We will start by describing the ideas that participants have about plagiarism and self-plagiarism, followed by the reasons. Concluding, we will finish with the description indicating the knowledge or beliefs that participants have about the use of the anti-plagiarism software, their perceptions about who and in what scientific field plagiarism is more frequent.

5.1 Ideas on the Meaning of Plagiarism and Self-plagiarism

Most of the participating, professors and researchers consider that plagiarism entails appropriating the work of others and attributing merits with the intention to deceive, an ethically reprehensible behavior being even more when referring to the scientific field. Only one of the participants alludes to paraphrasing as an extended practice of academic plagiarism, another to the low percentage of originality in a work, and three of the participants typify it as a crime since it implies supplanting the identity of another person. However, they do not investigate the actions that must be taken to face plagiarism when it is detected.

In the case of self-plagiarism, only two of the professors-researchers do not explicitly make distinctions between their considerations about self-plagiarism and plagiarism. The rest of the participants make the distinction indicating the origin of the source and its relationship with the author. 73% of participants (eleven participants) suggest that the self-plagiarism involves the partial or total reproduction of previous work with the intention of passing it off as new. One of the participants considers that socially, self-plagiarism does not seem to have the same implications and severity as if it had plagiarism and another one points out that the partial reproduction in sections as materials and methods of an investigation, cannot be considered as self-plagiarism whenever the researches of the different sections are replicated. An example of it is:

[…] On the other hand, self-plagiarism is very frequent in certain scientific works since there are parts in an article (such as bibliographic reviews or the material and method section) that are repeated between works necessarily and I do not consider it to be a self-plagiarism: the work is different, even if it has a common part with the one(s) that is supposed to have been plagiarized (Participant 12).

5.2 Reasons for Plagiarism

Regarding the reasons for plagiarism, participants indicate more than one option as we summarized in Fig. 1 and their opinions vary depending on the years of research experience and the field of research. The most representative reasons are to have a good curriculum vitae or increase scientific production, since the one that has the most is rewarded in the university and scientific field, the lack of training as a researcher that leads to committing plagiarism naively and a lack of ethics or values fuelled by the competitive culture in which we find ourselves immersed. Considering this, one of the participants states:

[…] It would be necessary to examine in depth to what extent the academic system of competition by merits as it is now configured, is related to the practices of plagiarism and self-plagiarism. Explaining it metaphorically, it is insufficient to only hold cyclists responsible for the problems of doping that have flooded sport. In the same way, we could talk about the academic competition (Participant 4).

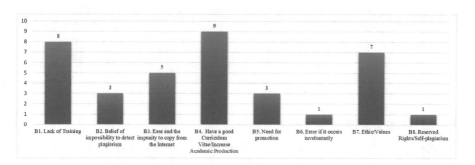

Fig. 1. Reasons for plagiarism and self-plagiarism

Another reason of relevance is the ease and impunity that one has to copy due to having easy access to information through the internet. The participants justify plagiarism for this reason due to ignorance of how the academic world works, sparing oneself the work for an author who wants to publish quickly and not wanting to try hard enough to achieve what they want.

Three of the professors and researchers consider that another reason for plagiarism is based on beliefs about the impossibility of detecting plagiarism, and three others point out that plagiarism is due to the need for promotion. In the case of self-plagiarism, the main reason stated is that in which "the author considers that his creation is a right that can be exploited and reused" (Participant 7) as he considers it appropriate.

As Table 2 shows, there are differences according to the years of research experience in terms of the stated motivation to carry out plagiarism and self-plagiarism. Professors and researchers, with more than 11 years of experience, point out that the lack of training is the main reason for plagiarism, followed by a good curriculum vitae and the belief that plagiarism is impossible to detect.

On the other hand, professors with less than five years of research and teaching declare as the main reasons as having a good curriculum vitae and lack of ethics or values of the person who commits the plagiarism, followed by the lack of training and the ease and impunity of the Internet to copy. In particular, one of the participants indicates that plagiarism can be considered an error if it occurs involuntarily. Finally, the professors and researchers, whose field of research are the social sciences and humanities, point out that the main reason is to have a good curriculum vitae (29%), while those who research in science, technology or health areas indicate the lack of training (27%).

Table 2. Reasons for Plagiarism and self-plagiarism based on years of research experience of the participants

Reasons for plagiarism and self-plagiarism	Frequency %(N) depending on years of research experience	
	<= 5	11>
B1. Lack of training	17% (4)	31% (4)
B2. Belief of impossibility to detect plagiarism	4% (1)	15% (2)
B3. Ease and the impunity to copy from the Internet	17% (4)	8% (1)
B4. Have a good curriculum vitae/Increase academic production	25% (6)	23% (3)
B5. Need for promotion	8% (2)	8% (1)
B6. Error if it occurs involuntarily	4% (1)	0%
B7. Ethic/Values	25% (6)	8% (1)
B8. Reserved rights	0%	8% (1)

5.3   Anti-plagiarism Software Use

In regard to the use of anti-plagiarism software, as shown in Table 3, only 20% (three participants) mention any knowledge in the handling of anti-plagiarism software, compared to 33% (five participants) who do not express concern to evaluate the originality of a text using specialized computer media. In some cases, participants indicate that they have detected plagiarism without the use of such means. Among the most recognized software are Turnitin and Plagiarisma, followed by the Google search engine, and there does not seem to be a general perception of a greater need to justify using specialized tools in the context of research. In particular, one of the participants doubts how this type of software really works:

[…] As the author of research papers that are published in scientific journals, I have to say that my opinion about this type of software is very negative. My experience tells me that they do not do a good screening and that they give as plagiarism percentages of works based primarily on the coincidence of words than to real plagiarism. Specifically, on occasions, editors of magazines have accused us of plagiarism when in no case, we intended to commit it (Participant 12).

On the other hand, the rest of the participants who allude to the need of having to use it, justify it from their teaching work, and require the University to use these types of tools, since the current conditions in which these types of programs operate are under license and hinders their accessibility and use.

Table 3. Knowledge about the use of anti-plagiarism software

	Frequency %(N)
C1. Software knowledge for plagiarism detection	20% (3)
C2. Examples/Software type for plagiarism detection	47% (7)
C3. Need for use	27% (4)
C4. Frequency of use	7% (1)
C5. Availability	20% (3)

5.4   Frequency of Plagiarism According to the Investigative Experience and Scientific Field of the Investigator

Regarding who and research field, plagiarism is more frequent; participants have similar opinions as shown in Fig. 2. Four of the participants indicate that the frequency of plagiarism is the same regardless of the scientific field, while four other participants (three of which research in the area of social sciences) point out that the most prone field is social sciences and humanities. The reasons to which they allude are the characteristics of this field of knowledge or the great existence of descriptive studies that are published in this area, which can be more easily manipulated. However, one of professors and researchers believes that in the scientific community "there is no plagiarism problem" because it would not make sense since "a researcher who wants to advance in knowledge if he plagiarizes, he does not advance" (Participant 12).

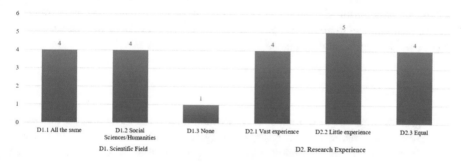

Fig. 2. Plagiarism frequency

Regarding the years of research experience, the participants have opinions divided as to the frequency of plagiarism. Five of the professors and researchers consider that plagiarism is more frequent in those researchers with little experience in research while four (three of whom have less than 5 years of research experience) point out that it is more frequent in those who have a lot of experience. Finally, four participants indicate that plagiarism does not depend on the years of investigative experience.

6 Discussion

In the investigations where we have reviewed Plagiarism and Self-plagiarism, they are treated in a joint and directly related way. As for the points of view, the damage that occurs to the author as well as the damage to the entity or person who owns the rights to commercialize a work [4], is generally considered. Another distinction that is usually made, is concentrating the focus on the subject who commits the plagiarism, so that it is not the same when plagiarizing another than when plagiarizing oneself. Tudela and Aznar [13] in this regard establish "a clear distinction between self-plagiarism and the use of the author's material, which is necessary to support research that is later developed" (p. 18).

Contrasting our results with those of other authors, we detected quite a few coincidences regarding the perception of professors and researchers about the reasons for plagiarism and self-plagiarism [13, 14, 15], the inefficient use of programs for their detection [4] and above all, the consideration of plagiarism as a "misconduct" in research [5, 7].

Post et al. [14] and Tudela and Aznar [13] in this regard, point out that the "pressure" to publish in the university environment facilitates practices that can be identified as plagiarism conscientiously. In addition, and in agreement with Soto [4], we consider that the use of programs for the detection of plagiarism only solves the problem in the short term, therefore, "it should be invested in other actions to decrease each time more the incidence of this problem in both students and professionals" (p. 10).

But we also find that our results differ to the perception that professors and researchers have about who and in what scientific field, the plagiarism is more frequent, and we estimate that there are not enough studies to reach decisive conclusions on the subject, so the argumentative discussion has only just begun.

7 Conclusions

The objective proposed in our study was to contextualize the ideas that the professor-researcher in the university field has about plagiarism and self-plagiarism in order to deepen reflections on its definition, causes and possible measures to prevent it. The most relevant conclusions are:

- Most professors make a clear distinction between plagiarism and self-plagiarism, defining it respectively: (i) plagiarism involves taking over the work of others and attributing their merits with the intention of cheating; (ii) Self-plagiarism implies the partial or total reproduction of previous work with the intention of making it as new.
- The most representative reasons for plagiarism for professors and researchers with more than 11 years of research experience are focused on the lack of training, while for those with less than five years of teaching and researching experience, they are due to the rush to get a good curriculum vitae.
- Regarding the use of anti-plagiarism software, they distinguish the field of research from that of teaching. Participants do not express concern about evaluating the

originality of a text using specialized computerized media, due to their image of the operation of anti-plagiarism software in the field of research, although they claim their use in the teaching context.

- While there are informants who consider that more experience is likely to lead to more plagiarism, there are others who do not consider that the investigative experience has a direct relationship with plagiarism, so principally we do not reach any conclusion on this.

References

1. Girón, S.J.: Anotaciones sobre el Plagio. Universidad Sergio Arboleda, Colombia (2008)
2. Imran, N.: Electronic media, creativity and plagiarism. SIGCAS Comput. Soc. **40**(4), 28–32 (2010)
3. Miranda, A.: Plagio y ética de la investigación científica. Rev. Chil. de Derecho **40**(2), 711–726 (2013)
4. Soto, A.: El plagio y su impacto a nivel académico y profesional. E-Cien. de la Información **2**(1), 1–13 (2012)
5. Balbuena, P.: El plagio como ilícito legal. Revista Ventana Legal. En Red. (http://www.ventanalegal.com/revista_ventanalegal/plagio_ilicito.htm)
6. Ramírez, R., Jiménez, H.D.: Plagio y "auto-plagio". Una reflexión. HistoReLo, Rev. de historia Reg. y Local **8**(16), 271–284 (2016)
7. Rojas, M.E.: Plagio en textos académicos. Rev. electrónica Educare **16**(2), 55–66 (2012)
8. Carvalho, J., Luengo, R., Casas, L., Cubero, J.: What is better to study: the printed book or the digital book?: An exploratory study of qualitative nature.: En: Costa, A., Reis, I., Moreira, A. (eds) Computer Supported qualitative Research. New trends on Qualitative Research, pp. 34–45. Springer, Switzerland (2019)
9. Bardin, L.: El análisis de contenido. Akal, Madrid (1986)
10. Fraenkel, J., Wallen, N.: How to Design Evaluate Research in Education. McGrawHill, New York (2009)
11. Solís, E., Porlán, R., Rivero, A.: ¿Cómo presentar el conocimiento curricular de los profesores en ciencias y su evolución? Enseñanza de las Cienc. **30**(3), 9–30 (2012)
12. Neri de Souza, F., Costa, A. P., Moreira, A.: WebQDA: Software de apoio à análise qualitativa. En: Rocha, A. (ed.) 5a Conferência Ibérica de Sistemas e Tecnologias de Informação, CISTI 2010. Universidade de Santiago de Compostela, Santiago de Compostela, pp. 293–298 (2010)
13. Tudela, J., Aznar, J.: ¿Publicar o morir? El fraude en la investigación y las publicaciones científicas. Persona y Bioética **17**(1), 12–27 (2013)
14. Post, D., Stambach, A., Ginsburg, M., Hannum, E., Beanavot, A., Bjork, C.: Rank scholarship. Comp. Educ. Rev. **56**(1), 1–17 (2012)
15. Sureda, J., Comas, R., Morey, M.: Las causas del plagio académico entre el alumnado universitario según el profesorado. RIE **50**(1), 197–220 (2009)

Integrated High School: A Possibility for Emancipating Education in Nursing?

Elaine Kelly Nery Carneiro-Zunino[1](✉) [iD],
Gilberto Tadeu Reis da Silva[1] [iD], Juliana Maciel Machado Paiva[1] [iD],
Juliana Costa Ribeiro-Barbosa[1] [iD], and Silvana Lima Vieira[2] [iD]

[1] School of Nursing, Federal University of Bahia,
R. Basílio da Gama, 241, Salvador, Bahia 40110-907, Brazil
lanenery@hotmail.com
[2] Department of Life Sciences, University of the State of Bahia,
R. Silveira Martins, 2555, Salvador, Bahia 41000-150, Brazil

Abstract. The aim is to analyze the existence of elements that enable emancipating education for nursing in the Integrated High School (IHS) documents. Documentary research with qualitative approach that used political data, decrees and other regulatory documents as a source, which were described according to type and year of publication. The theoretical reference was Paulo Freire, and webQDA® software was used for analysis. During the treatment of the data, the following categories emerged: aspects approaching IHS to and distancing it from emancipatory education. Elements that permeate emancipatory education have been evidenced, recognizing the individual as a historical-social subject, the individual's importance in the teaching-learning process, and also the teacher's change as a mediator of the knowledge construction process. But we also found aspects that stiffen this modality. Emancipatory education is necessary for nursing to build its ideals of autonomy and power through reflective practice.

Keywords: Integrated High School · Nursing · Emancipating education · Paulo Freire

1 Introduction

Technical professional education in nursing has faced challenges throughout history. These challenges include the overcoming of the structural duality in Brazilian education, with a clear demarcation of the educational trajectory of those who have performed intellectual or instrumental functions and those responsible for the implementation of technical procedures [1].

It has also been repeatedly affirmed that technical schools often adapt their pedagogy to the new market needs, transforming the workers' skills into practical activities, a knowledge addressed to the world of capitalist production. Ira Shor [3] calls this reality positivist logic, which inverts the theory-practice relationship, where the approaches and methodologies "imported" from economics and finance are applied to the educational policies and practices and to teacher training, disseminating the culture of meritocracy in schools, stimulating reproductivism and individualism among educators.

In the educational field, Freire has long called this banking education because it is based on the use of static and universal content; in the student's passivity; and in decontextualized practices of memorization and repetition, among others [4]. This banking education represents the counterpoint of emancipatory education.

For Freire, emancipatory education seeks, among other things, to enhance the human vocation of being more, a characteristic through which the human beings are in constant search, curiously venturing in the knowledge of themselves and the world, besides fighting for the affirmation/conquest of their freedom [5].

In July 2004, Decree No. 5.154 was issued, which regulates the Integrated High School (IHS), technical education integrated into secondary education. Education experts such as Saviani, Frigotto and Ramos recognize it as a modality capable of counteracting the structural duality of the education system that was historically implemented in Brazil [6].

This perspective represents a major challenge in view of the need to link general education and vocational training to a unique training itinerary, which requires an appropriate infrastructure and a differentiated and innovative curriculum structure [7].

In the area of nursing, Miranda and Barroso [8] list the ideas-strengths of **Freire's** pedagogy related to health education carried out by nursing, highlighting: the recognition of the historical, temporal, creative and cultural subject; the incorporation of dialogue and problematization and reflexive practice, as contributions to the construction of a liberating, critical nursing practice and in the valuation of the subject. Furthermore, according to the conclusions of Vasconcelos' study [9], training professionals in order to be able to overcome mechanical action requires overcoming the apparent restraint of thinking, reflecting, diverging and resisting.

Therefore, considering that this modality provides better conditions for citizenship, work and social inclusion to young people and adults in search of high-quality professional education and new horizons for their lives, in this study, we seek to investigate technical training in nursing in this modality. In this sense, the following question is raised: What elements are found in the documents about IHS that approach or distance it from an emancipatory education for nursing? The objective is, therefore, to analyze the existence of elements that permit an emancipatory education for nursing in the documents about the Integrated High School (IHS).

2 Method

In order to reach the proposed objectives, we opted for an eminently qualitative method, carried out through a documentary research technique that had used the official documents published on the integrated high school as its data source.

Electronic searches were carried out with the keywords: *"ensino médio integrado"* (integrated high school); *"legislação"* (legislation); *"documentos oficiais"* (official documents). In addition, the keywords were combined; boolean operators were used, as well as quotation marks in the search expressions; and empty words were suppressed. The results were only used in Portuguese because the interest of the study was focused on the local reality. We found publications between 1997 and 2014, accounting for just under 20 years. The sources of information were mainly found in the gray literature, which is the entire literature produced at all government, academic, business and industry levels, in print and electronic format, not controlled by commercial publishers [10].

In the organization of the material, webQDA® software was used, supporting the pre-analysis, exploration of the material and data processing. Documents not regulated by ordinances, regulations or laws/decrees were excluded. Two categories emerged after the exhaustive reading of the material: aspects that approached IHS to emancipatory education and aspects that distanced IHS from emancipatory education, from a nursing perspective.

For Bardin [11], the positive consequences of using the computer, specifically from the content analysis viewpoint, include speed, investigative rigor, flexibility to introduce new analysis instructions, storage that permits the reproduction and exchange of information with whom one is working on the project and the possibility to handle complex data. Creativity and reflection can be used as key elements to achieve the desired results.

In the data analysis, Paulo Freire's theoretical framework was used, aiming to approximate the theoretical-methodological aspects of IHS to the challenge of emancipating professional education in nursing.

3   Results

The documents selected for the study were described in Chart 1 according to type and year of publication. In the pre-analysis stage, we elaborated a chart with the most frequent words in the documents (Chart 2). In the exploration stage of the material, the conceptual map with the coding of the document contents was elaborated, which permitted, in the data analysis phase, the definition of the categories and subcategories: aspects that approach IHS to emancipatory education and aspects that distance IHS from emancipatory education, as shown in Table 1.

Name	Type	Year
Regulates § 2 of Art. 36 and arts. 39 to 41 of Law No. 9.394, from December 20th 1996	Decree No. 5.154, from July 23rd 2004 [12]	2004
Secondary Technical-Professional Education Integrated into High School	Basic Document [13]	2007
National Curricular Guidelines for Secondary Technical-Professional Education	CNE/CEB Resolution No. 6, from September 20th 2012 [14]	2012
National Education Plan 2014-2024	Law of the National Education Plan, from June 25th 2014 [15]	2014

Chart 1. Documents selected for analysis

Words	Repetition
Teaching	3112
Education	2657
Average	2041
Training	1578
Work	1526
Professional	1317
Integrated	1094
Brazil	852
History	707
School	588
Curriculum	577
Project	548
Practices	528
Technique	490
Social	487
Teachers	454
Curricular	448
Foundations	444
Knowledge	438

Source: *Software webQDA*

Chart 2. Frequent words

Table 1. Coding of Results

Emancipation for nursing	Categories	Subcategories
Overcoming the structural duality	Aspects approaching IHS to emancipating education	Critical Education
Development of different dimensions of the Human Being (work, science, culture and technology)		Integral Education
Emancipation of the Individual		Dialogical and Dialectical Education
Intentionality of professional education	Aspects distancing IHS from emancipating education	Politicality of Education

3.1   Aspects Approaching IHS to Emancipating Education

Critical Education. Elaborating an education plan in Brazil today implies making commitments to the continuous effort to eliminate the historical inequalities in the country. Therefore, the goals are intended to confront the access and permanency barriers;

the educational inequalities in each territory with a focus on the particularities of its population; training for work, identifying the potentials of local dynamics; and the exercise of citizenship [15]. **F1**

The identity of High School is defined as overcoming the dualism between propaedeutic and professional. It is important to configure a model that gains a unitary identity for this stage and that assumes diverse and contextualized forms of the Brazilian reality [14]. **F2**

Integral Education. It expresses a conception of human education, based on the integration of all dimensions of life in the educational process, aiming at the omni-lateral education of the subjects [13]. **F3**

From the organizational point of view, this relationship needs to integrate in the same curriculum the comprehensive education of the learner, enabling high-level intellectual constructions; the appropriation and concepts necessary for conscious intervention in reality and the understanding of the historical process of knowledge construction [13]. **F4**

The idea of integrated education suggests overcoming the human being historically divided by the social division of labor between the action of performing and the action of thinking, directing, or planning [13]. **F5**

Increasing the supply of education for workers is an urgent action but, in order to ensure its quality, this supply should be based on the principles and understanding of universal and unitary education, aimed at overcoming the duality between general cultures and technique [15]. **F6**

Dialogical and Dialectic Education. In pedagogical work, the lecture method should re-establish the dynamic and dialectical relations between concepts, reconstituting the relations that shape the concrete totality they originated from, so that the object to be known gradually reveals itself in its own peculiarities [13]. **F7**

These new demands require a new behavior of the teachers, who change from being knowledge transmitters to being mediators, facilitators of knowledge acquisition; they should stimulate research development, knowledge production and group work. This necessary transformation can be translated by the adoption of research as a pedagogical principle [14]. **F8**

The first principle is to understand that men and women are historical-social beings who act in the concrete world to satisfy their subjective and social needs and, in this action, produce knowledge [13]. **F9**

3.2 Aspects Distancing IHS from Emancipating Education

§ 4. The Ministry of Education, through the National Network of Professional Certi-fication and Initial and Continuing Training (CERTIFIC Network), shall elaborate national standards of professional certification to be compulsorily used by the Pro-fessional and Technological Education institutions of the federal education system and of the state public networks, when in certification processes [14]. **F10**

According to data from the National Household Sample Survey [16], Brazil had a population of 45.8 million people aged 18 years or over who did not attend school and did not complete elementary school [17]. **F11**

The School Census of Basic Education of that year also shows that the students who attended elementary school years of the youth and adult education (EJA) had a much higher age than those who attended the final years of elementary school and high school in this modality. This fact suggests that the early years are not producing demand for the final years of elementary education in EJA. It also represents strong evidence that this modality is receiving younger students from regular education. Another factor to be considered in this modality is the high dropout rate, caused, among other reasons, by the unfitness of the curricular proposals to the specifics of this age group [17]. **F12**

It incorporates ethical-political values and historical and scientific contents that characterize human praxis. Therefore, providing professional training does not mean providing exclusive preparation for work practice, but offering an understanding of the socio-productive dynamics of modern societies, with their conquests and setbacks, and also enabling people to exercise professions autonomously and critically, without ever reaching their limits [13]. **F13**

4 Discussion

4.1 Aspects Approaching IHS to Emancipating Education

In this analysis, we seek to approximate the documentary aspects of IHS to Paulo Freire's critical-reflexive view of educational practice, understanding that there is no neutrality in education, which can serve as a form of preserving the dominant ideology that strengthens oppression, as well as of proposing ways of resisting the oppression and liberation from the structural duality.

We begin our analysis by arguing [18] that education can be both a way of making knowledge common and of reinforcing inequality between men. This is because educational processes both influence and are influenced by social, economic, political and cultural aspects.

In the analyzed documents about professional technical education at the secondary level, we evidence concepts of education that are consistent with an educational practice concerned with overcoming the structural duality the traditional model of education maintains, as in the first F1 fragment extracted from the National Education Plan 2014-2024. In fragment F2, extracted from the National Curricular Guidelines for Technical Professional Education at the Secondary Level, we also observed the importance of the integration between the propaedeutic and professional for this overcoming.

But Freire alerts to the fact that education should represent a practice that liberates from this reality, although it is not enough for the oppressed to recognize themselves as such and to overcome their situation by becoming oppressive. A liberating educational practice implies the critical recognition of the individual, the "reason" for this situation and the search for liberation and no longer oppression [4].

In professional education, this dichotomy is even more blatant. It reflects, throughout our history, the typical power relations of a society divided into social classes, in which the exercise of intellectual and leadership functions is attributed to the elites and vocational education to the workers. Often, technical schools adapt their pedagogy to new market needs, transforming workers' practice into competencies for practical activities, a knowledge addressed to the world of capitalist production [2].

Authors like Bevs and Watson [19] reflected that professional education in nursing as an assembly industry produces nurses who at most follow the "status quo". These nurses can provoke waves, but remain within rules, while living lives that are circumscribed by inflexibility, and carry the inevitable label of banality and mediocrity in their thoughts and actions.

Finally, in the contemporary world, it also intends to offer new educational practices to workers in order to attend to the complexity of social life, which requires greater appropriation of scientific, technological and socio-historical knowledge. Thus, public policies need to commit to integral education, an intention evidenced in fragments F3, 5 and 6.

Within the foundations of integrated secondary education, omnilaterality is established as one of the pillars. This concept, coming from Marxism, aims at integrating the contents to approach the student to the whole.

For Freire [20], the Real, being Real, is a transdisciplinary totality. The analytical process of splitting the Real through disciplinary biases should be followed by transdisciplinary re-totalization through an interdisciplinary epistemological process. This path explained by Freire can be evidenced in the pedagogical work described in the basic document highlighted in fragment F7.

In addition to Freire, authors such as Ramos [21] and Saviani [22] point to curricular integration as the most appropriate path for the possibility of overcoming the duality of Brazilian education, educating for work and for the continuity of studies, in a dialogical and dialectical way at the same time. The fragment F4 is in line with what these authors have discussed, ratifying the need for comprehensive student education and the possibility of intervention in reality, which according to Freire represents the liberation of the individual.

Although the dialogue is interwoven in several objectives and methodological paths in the documents studied, we did not find the spaces for its occurrence. In Freire's work, the proposal of dialogic education, as opposed to banking education, seeks to oppose educational models that, by fragmenting the analysis of reality, make it impossible to understand the totality and, therefore, the projection of a different reality [18].

Another important foundation in the legislation studied was the conception of the individual as a historical-social subject (F9), which permits the establishment of a new flow of the teaching-learning process, where the learner is also an educator, and the construction of knowledge departs from the learner's historical perspective.

Thus, the intention is to put in practice the change of focus from the teacher who transmits banking education to a facilitator of knowledge acquisition. In fragment F8, extracted from the national curricular guidelines for secondary education, we can see the expansion of this movement, when it is argued that teachers should stimulate research, knowledge production and group work. This demonstrates a direct relationship with Freire's educational-critical practice, where teaching is not to transfer knowledge, but to create the possibilities for its production or its construction [23].

Still in the context of this transformation, Freire describes that the path from transmission to mediation is based on:

proposing to the people, through certain basic contradictions, their existential, concrete, present situation as a problem which, in turn, challenges it and thus demands a response, not only at the intellectual level, but also at the level of action. Never just discuss it and never give the people content that has little or nothing to do with its yearnings, doubts, hopes, fears. Contents that sometimes raise these fears. Fears of oppressed conscience [20].

According to Freire [4], for an educational activity to have emancipatory characteristics, it should necessarily be preceded by a reflection on man and an analysis of the concrete means of life of the concrete man whom we want to educate (or rather, who we want to help to educate himself).

The IHS also contains this aspect through the elaborations on the praxis as in fragment F13, recognizing that it is in the action that man produces the transformation of the world. For Freire, this is always a dynamic relationship of action-reflection-action.

In view of the above, we also highlight the compliance of the pedagogical proposal to changes in the world of work, which seeks a worker capable of responding to the specific characteristics imposed by the great transformations in social work practice, with greater schooling, flexibility, greater access to information, decision-making ability in the face of complex problems and critical thinking [24].

Thus, we find convergence between the aspects of emancipatory education in Freire and which are important for the transformation of nursing practice in the IHS documents.

4.2 Aspects Distancing IHS from Emancipating Education

For many years, Secondary-Level Professional Technical Education in nursing has been impregnated with the ideology of a dominant class, influences from hegemonic groups, with a strong content-based and compartmentalized approach, keeping these professionals intentionally in a sphere of execution of procedures and not of thought and reflection on practice and care [25].

Thus, the challenge and proposal of IHS is to overcome this excluding and fractional logic of education towards emancipatory education. The documents highlight positive but still incipient initiatives, as detected by fragments F11 and F12. In the documents, we found the difficulty to reduce the high school dropout rate and the demand of an increasingly younger population in professional education, seeking early insertion in the field of work. This is a reality of the young Brazilians and the integrated high school seeks to contribute to the crossing towards the polytechnic high school.

Also, within the analyzed documents, the proposal to establish evaluative criteria that prioritize qualitative aspects in student evaluation was identified. Next, however, they seek to submit the schools to a framework of national indicators, aiming to obtain metrics for the evaluation of secondary-level professional technical education as shown in fragment F10.

5 Conclusions

In the findings of this study, we analyze the official IHS documents in a critical approach based on emancipatory education as discussed by Paulo Freire and nursing practice. This approach made it possible to identify aspects of this education modality that approach emancipatory education, as well as to highlight the aspects that distance it.

Among the aspects that approximate the integrated modality to emancipatory education, we highlight the need to overcome the traditional model of education that perpetuates the structural duality of society in class divisions and contributes to high school dropout rates among young people.

Next, we show elements of the proposed model, such as the recognition of the individual as a historical-social subject, and its importance in the teaching-learning process, which converges to a dialogical and dialectical education proposed by Freire. Also related to this aspect was the role of the teacher as a mediator/facilitator of the knowledge construction process.

In addition, of these characteristics, the integrated modality is based on foundations such as polytechnics, omnilaterality and unitarian school, which commune with Freire's principle of totality, in that totality causes a better understanding of reality, reading of the world, reading of the word and (re)reading of the world.

Although the intended model approaches Paulo Freire's framework, we highlight aspects impregnated in the organization form of the model in force, such as the evaluation and competition mechanisms stimulated by national indicator standards, without respecting the subjectivities of the country's different regions.

In the reality of nursing, the search to overcome a nursing that is not only technically qualified, but also critical and politically prepared has been built for many years. Thus, we approach the need for nursing professionals to construct liberating practices, through the key ideas of Paulo Freire's emancipatory education, based on the principles of dialogue, problem, comprehensiveness and totality, which will make it possible to understand the inferences of the social, historical and political context about the way of doing and transforming reality.

References

1. Brasil.: Lei das Diretrizes e Bases da Educação Nacional. MEC, Brasília (1996)
2. Dias, FAC.: Por uma educação emancipatória: a concepção da Escola de Formação em Saúde da Família Visconde de Sabóia – Sobral/CE na formação técnica em Enfermagem. [Dissertação de Mestrado] Fundação Oswaldo Cruz (2016)
3. Shor, I., Saul, A., Saul, A.M.: O poder que ainda não está no poder: Paulo Freire, pedagogia crítica e a guerra na educação pública - uma entrevista com Ira Shor. Educar em Rev. **61**, 293–308 (2016)
4. Freire, P.: Pedagogia do Oprimido, 17th edn. Editora Paz e Terra, Rio de Janeiro (1987)
5. Streck, D., Redin, E., Zitkoski, J.J.: Dicionário Paulo Freire, 3rd edn. Autêntica editora, Belo Horizonte (2016)
6. Viamonte, P.F.V.S.: Ensino profissionalizante e ensino médio novas análises a partir da LDB 9394/96. Educação em Perspectiva **2**(1), 28–57 (2011)

7. Ferreira, EB., Garcia, S.R.O.: O Ensino Médio Integrado à Educação Profissional: um projeto em construção nos estados do Espírito Santo e Paraná. In: Ensino Médio Integrado: concepções e contradições. 3rd ed. Cortez, São Paulo (2012)
8. Miranda, K.C.L., Barroso, M.G.T.: A contribuição de Paulo Freire à prática e educação crítica em enfermagem. Rev. Latino-Americana de Enfermagem **12**(4), 631–635 (2005)
9. Vasconcelos, C.M.D.C.B.: Prática pedagógica na educação profissional de enfermagem – expressões autocríticas de um grupo de enfermeiras (os) educadoras (es). [Dissertação de Mestrado] Universidade Federal de Santa Catarina (2002)
10. Botelho, R.G., Oliveira, C.C.: Literaturas branca e cinzenta: uma revisão conceitual. Ciência da Informação **44**(3), 501–513 (2017)
11. Bardin, L.: Análise de Conteúdo. Edições 70. São Paulo, Portugal (2011)
12. Brasil Homepage. http://www.planalto.gov.br/ccivil_03/_Ato2004-2006/2004/Decreto/D51-54.htm. Accessed 04 Mar 2019
13. Brasil.: Educaçao Profissional Técnica de Nível Médio Integrada ao Ensino Médio - Documento Base. MEC, Brasília (2007). In: Recuperado de. portal.mec.gov. br/setec/arquivos/pdf/documento_base.pdf
14. Brasil.: Diretrizes Curriculares Nacionais para a Educação Profissional Técnica de Nível Médio. MEC, Brasilia . In: (2012). http://portal.mec.gov.br/index.php?option=com_docman&view=download&alias=11663-rceb006-12-pdf&category_slug=setembro-2012-pdf&Itemid=30192
15. Brasil.: PNE - Planejando a Próxima Década. Conhecendo as 20 metas do Plano Nacional de Educaçã. MEC, Brasília. In: (2014). http://pne.mec.gov.br/images/pdf/pne_conhecendo_20_metas.pdf
16. PNAD/IBGE Homepage. https://ww2.ibge.gov.br/home/estatistica/populacao/trabalhoerendimento/pnad2015/default.shtm. Accessed 04 Mar 2019
17. Brasil.: Plano Nacional de Educação PNE 2014-2024 : Linha de Base. 1st ed. INEP, Brasília (2015)
18. Rocha, SFM.: A Interdisciplinaridade em Freire como práxis de Leitura de Mundo: uma proposta de educação concebida como formação humana. Universidade Federal de Pelotas (2017)
19. Bevis, E., Watson, J.: A caring curriculum: a new pedagogy for nursing, 1st edn. National League for Nursing, New York (1989)
20. Freire, P.: Pedagogia do Oprimido, 59th edn. Editora Paz e Terra, Rio de Janeiro (2015)
21. Ramos, M.: Concepção do Ensino Médio Integrado. Seminário promovido pela Secretaria de Educação do Estado do Pará nos dias 08 e 09 de maio de 2008, p. 30 (2008)
22. Saviani, D.: Trabalho e educação: fundamentos ontológicos e históricos. Rev. Brasileira de Educação **12**(34), 152–165 (2007)
23. Freire, P.: Pedagogia da Autonomia, 57th edn. Editora Paz e Terra, Rio de Janeiro (2018)
24. Garay, ABS.: Reestruturação Produtiva e Desafios de Qualificação : Algumas Considerações Críticas. . In: pp. 1–18 (2014). http://www.cefetsp.br/edu/eso/globalizacao/desafioqualifi cacao.htmlhttp://read.adm.ufrgs.br/read05/artigo/garay.htm
25. Vieira, SL.: Movimento ensino-aprendizagem no curso técnico de enfermagem: educando(a)s em contexto de vulnerabilidade social. [Tese de Doutorado]. Universidade Federal da Bahia (2017)

Government Communication - The Dubai and United Arab Emirates Ministry of Happiness

The Objective of the Creation of the Ministry of Happiness – Content Analysis

Diamantino Ribeiro[1]([:envelope:]) [iD], António Pedro Costa[2] [iD],
and Jorge Remondes[3] [iD]

[1] CEFAGE – Centro de Estudos e Formação Avançada em Gestão e Economia
da Universidade de Évora, Evora, Portugal
diamantinojtribeiro@gmail.com
[2] University of Aveiro, Aveiro, Portugal
apcosta@ua.pt
[3] Lusophone University of Porto, Porto, Portugal
jorge@jorgeremondes.eu

Abstract. This article presents a study based on government communication regarding the creation of the Dubai and United Arab Emirates Ministry of Happiness. Under the scope of this paper, from the news analysed, we have chosen that which reflects the main objectives of the creation of the Dubai and United Arab Emirates Ministry of Happiness. Using the technique of content analysis, through the use of webQDA software, the aim was to understand how the Government communicated those objectives. Consequently, the aim was to understand whether there was discursive consistency between the content of the text analysed and the message that the Government wished to convey. The results enabled us to identify emphasis on the Government's objective of creating genuine and authentic happiness for the whole of society through the creation of this Ministry.

Keywords: Happiness · Qualitative analysis · Government communication

1 Introduction

The study of government communication, in the logic of communication for development, is an academically challenging area due to the constant updating of research content and the thematic scope, diversity, typology, implications (positive or negative) of government policies, among other aspects. The beginning of the study that underlies this work coincided with the initiative of the governing authorities of the United Arab Emirates to create the Ministry of Happiness in the year 2016.

It was understood that there could be academic and social interest in deepening the study of communication carried out by the Ministry of Happiness and in understanding

A. P. Costa et al. (Eds.): WCQR 2019, AISC 1068, pp. 226–238, 2020.
https://doi.org/10.1007/978-3-030-31787-4_19

its contribution to Development. At the same time, it was considered interesting to understand the objectives of the creation of this Ministry, the communication model used by the UAE Government and how an abstract concept such as Happiness is translated into concrete actions that can effectively contribute to happiness and the well-being of citizens and to Development.

The investigation begins by analysing government communication within the general communication framework, communication theories and the mainstream media today, followed by a case study through content analysis of the news released by the Governments of Dubai and the United Arab Emirates in one year (February 2016 to February 2017), specifically about the Ministry of Happiness.

Soon after the creation of the Ministry of Happiness (February 2016), it was possible to gather news in international media related to the objectives that were the basis of the creation of this government body. The option to select a time period of 1 year, in addition to the analysis of the communication, allowed for the understanding of the evolution of communication and the use of human and material resources. Following a qualitative approach, the content of this news was analysed through the qualitative analysis software webQDA [1]. In this article, the content of one of the most emblematic texts was analysed in the scope of the research carried out, which concerns the launching of the Ministry for Happiness.

In the following sections we will undertake a brief overview of Government Communication, the Ministry of Happiness, Methodology, Analysis and Discussion of Results.

2 Theoretical Framework

2.1 Government Communication

When one thinks of government communication, taking into account where its nomenclature directs us, one thinks of the exchange and sharing of information between the State and the people, the citizens.

This particular form of communication, in fact, is a legitimate way for a Government to render accounts and bring the projects, actions, activities and policies that it performs and which are of public interest, to the knowledge of the public [2]. Brandão [2] argues that government communication can be understood as a form of public communication. These two forms of communication share some resemblances, because government communication aims to be an instrument for building the public agenda, as well as a mechanism of accountability and a stimulus for social participation.

The duty of government communication is to inform citizens of what is happening within the Government and, for this reason, it is an instrument that allows them to learn about government actions and, simultaneously, that they can convey their expectations to the Government [3].

It is necessary that government communication be guided by some standards and rules, so as to fulfil its duty - to inform society - because the communication system is essential for the processing of the internal administrative functions and for the relationship with the external environment [4].

Since government communication is not an easy process, the Government must make an effort to improve its communication, making it understandable and accessible to all. In this sense, organisations have to be convinced that communication needs to be worked on and managed by specialised professionals. Otherwise, they will always be improvising, thinking that they are communicating when they are merely reporting. You cannot plan organisational communication without basic foundations [4].

From a strategic point of view, government communication should be planned for its target audience, studies should be made to assess the image of the administration with the public and events should be planned to disseminate the information.

Assuming that there is political will and a decision by senior managers to develop a communication plan, the planning process should be organised into three phases: strategic diagnosis, strategic planning and strategic management [4].

In short, government communication should be transparent, be well acquainted with its interlocutors and be carried out on the basis of defined theoretical principles, supported by good planning.

2.2   The Dubai and United Arab Emirates Ministry of Happiness

In 2014, Sheikh Mohammed bin Rashid Al Maktoum launched the Happiness Index to measure how satisfied citizens are with government services.[1]

In the start of 2016, he surprised the media, via Twitter, with the indication that he would appoint a Minister of Happiness [5]. Days later, he appointed Ms Ohood Al Roumi to take the position of Minister of State for Happiness as an integral part of the Governor's office and whose main mission would be to oversee "plans, projects, programmes and indices" that would improve the overall climate of the country (www. happy.ae/en).

At the inauguration, the Minister stated that the purpose of her work was to create authentic and genuine Happiness in public services. A little more than a month after taking office, the Minister presented a package of positive initiatives and institutional happiness in the Federal Government. The National Plan for Happiness and Positivity (PNF) was approved on International Day of Happiness on 21 March. The PNF comprises 3 main areas:

1. Inclusion of happiness in the policies, programmes and services of all government agencies as well as in the work environment;
2. Consolidation of values of positivity and happiness as a way of life in the UAE community;
3. Development of tools and indices to measure happiness levels.

The programme is based on a *"scorecard"* of happiness and positivity, and all national government agencies will have to work according to this instrument. The Government approved programme also includes:

(a)   The appointment of a CEO for happiness and positivity in all government agencies;
(b)   The establishment of happiness and positivity boards in federal entities;

---

[1]   https://www.khaleejtimes.com/nation/government/shaikh-mohammed-launches-happiness-index.

(c) Certain hours allocated to programmes and activities related to happiness in the Federal Government;
(d) Creation of offices of happiness and positivity;
(e) Customer service centres will be transformed into customer happiness centres;
(f) Special programmes are tailored to change the culture of Government employees, to serve the clients and make them happy;
(g) The programme also includes annual indices, surveys and reports to measure happiness in all sectors of the community.

Since the approval of the Programme, the Government, and in particular the Minister, were involved in initiatives ranging from the scientific training of managers specialised in Happiness to the integration of women and children into actions aimed at promoting Happiness and Positivity. Meanwhile, the programme has been extended to the private sector and has attracted the support of the country's large economic groups.

One of the concerns of the Government and the Minister is the measurement of results, with the aim of bringing policies closer to the real desire of citizens. This measurement includes, among other tools, a scientific study that is being carried out by the University of Abu Dhabi, based on citizens' tweets in the year 2015.

In concrete terms, the Government has endeavoured to make the most of its adherence to the concept created in the 1970s in the Kingdom of Bhutan. To do this, it has changed the name of public services (the citizen's bureau was named the Happiness Centre), amusement parks and sports and, for example, the new area of the city that is being built next to the future airport of Dubai (e.g. Dubai World) will be named the City of Happiness. In addition, it has encouraged several organisations to launch initiatives for the happiness of their workers, as in the case of "Dubai Culture", which implemented the "Make It Happen" programme. The Government believes that happy workers are contagious to customers and also wants private companies to work to make customers happy.

Overall, the Government believes that the model can be replicated internationally. It promotes the constant collection of opinions from the public and from experts. In this context, an international event was held in 2017 called "Global Dialogue for Happiness", on the eve of one of the biggest political events of the year in the region, the World Government Summit.[2] The "Global Dialogue for Happiness" was attended by more than 30 experts (scientists, economists, governors, psychologists, etc.) with the responsibility of exchanging ideas and encouraging discussion about trends and Happiness for the people of the world. The two events had a new edition, in February 2018. During the 2018 "World Happiness Forum", the first Global Happiness Policy Report was launched.

3 Methodology

In this study, the work was structured as recorded in the following Diagram (Fig. 1) as a way to obtain the data to be analysed:

---

[2] https://www.worldgovernmentsummit.org/.

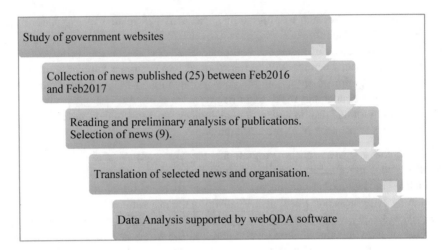

Fig. 1. Evolution and structure

The news was collected for the analysis of content published in the media during the period of analysis (February 2016 to February 2017) in order to understand the evolution of the concept, the form and direction of communication, and actions on the ground of the UAE Government.

Table 1 summarises the dates, sources and titles related to the texts selected for analysis.

Table 1. Listing of texts in chronological order

#	Date	Title	Source
1	8 February 2016	"Genuine happiness is the ambition of the United Arab Emirates Minister of Happiness"	Gulf News
2	7 March 2016	"Mohammed analyses the UAE Happiness and Positivity Programme"	Emirates 24/7 News
3	21 March 2016	"United Arab Emirates Happiness Programme approved on International Day of Happiness"	Emirates 24/7 News
4	15 May 2016	"Happiness, positivity through the eyes of children of the UAE - Children will inspire the logo of the Happiness and Positivity programme"	Emirates 24/7 News
5	14 June 2016	"UAE reveal the formula for Happiness"	Emirates 24/7 News
6	30 June 2016	"Joyful task for 60 Emirati named pioneers of positive thinking in the government"	The National UAE
7	1 September 2016	"Abu Dhabi teachers study Tweets to assess Happiness in the UAE"	The National UAE

(continued)

Table 1. (*continued*)

#	Date	Title	Source
8	26 December 2016	"The Crown Prince of Dubai welcomes the Year of Donation"	The National UAE
9	2 February 2017	"Dialogue on Happiness welcomes 300 experts"	The Gulf Today

After a free translation into Portuguese, an analytical reading of the texts was carried out without any kind of formalities. Then the texts were imported to the webQDA platform. In a second reading, the contextualization was carried out and the categories adapted for each of the texts were created. Using the selection, collection and integration tool available in the software, the most representative contents of the communication were selected and integrated within the categories. At the same time, a cloud of the keywords of each of the texts was constructed. From the selections made, conclusions were drawn for each text. At the point of discussion of results, the link between all the texts is made and the conclusions drawn.

The analysed data were collected from the Internet or *corpus latente* - the *corpus latente* refers to the existence of large databases with which everyone can work, namely the Internet, which accumulates more and more information day after day in the form of texts, images and videos, among others Pina, Souza and Leão [6]. The *corpus latente* is a set of content, available on the Internet for those who wish, and have the necessary skills and qualifications, to extract them [6].

In order to analyse the data, the analysis of content was the recommended technique. Content analysis consists of a set of methodological tools that aims to analyse different sources of content, both verbal and nonverbal. With regard to its practical implementation, it covers several stages, particularly to confirm the full significance of the data collected. According to Costa and Amado [7], these stages are, in turn, organised into seven distinct phases:

1. Definition of the problem, work objectives and theoretical bases
2. Organisation of the information gathered
3. Reading of the information
4. Categorisation and codification
5. Formulation of questions
6. Analysis of matrices
7. Presentation of findings

For Krippendorf [8] content analysis is a research technique that allows one to make valid and replicable inferences of the data to its context. Hence, inferences are made about what can affect the type of interpretation of analysis, on the basis of establishing a relationship between the data obtained [9].

As support for the achievement of this establishment of relationship between data, technological tools can and should be used to obtain results that go beyond traditional observation/interpretation.

Bardin [10] proceeds to a definition of codification, assigning it the meaning of transformation, especially through extraction, aggregation and enumeration, and based on certain precise rules on all the textual information, that end up representing all the characteristics of the content.

Table 2 presents the proposal that was followed in this study.

Table 2. Proposal of organisational model by Bardin [10]

Category	Subcategory	Registration unit	Context unit
Here the major themes of the analysed data are joined together, (in this case the titles of the texts)	Most important subtopics within a certain major theme	Fragments of text taken by an indication of a characteristic (category and subcategory)	There are fragments of the text that encompass the registration unit, contextualising the registration unit

The use of the webQDA tool to analyse the data of this study allowed the generation, as outputs, of tables (matrices) with the encoded data. On the other hand, according to Costa, Linhares and Neri de Souza [11], Computer Assisted Qualitative Data Analysis Software (CAQDAS) extend the possibilities of communication among researchers in the definition of analysis (categorization, coding, and recoding, etc.), that are fundamental for the construction of synthesis and analysis.

In the case of the research carried out, the option to use specific software of qualitative analysis was important in the deepening of the analyses carried out.

4   Results and Discussion

Since it is not possible to present all the results of the enlarged study in this article, it was decided, as already mentioned, to describe the analysis performed to the so-called "Text 1"; this text, published two days after the appointment of the Minister of Happiness, makes known the basic objectives of the Ministry of Happiness.

An analysis of the content of the news found two categories directly related to the objectives themselves: Genuine happiness and authentic happiness, as may be observed in the following excerpts of the text:

- *"The new Ministry of Happiness seeks to create genuine happiness..." [Ref. 1]*
- *"(...) create authentic happiness of the whole of society" [Ref. 2]*

Simultaneously, it can be seen that there is a clear objective of change with respect to behaviour and attitudes:

- *"(...) what we are trying to achieve here is to create a real change..." [Ref. 1]*

Moreover, it can be seen that there is are parallel objectives of measuring results:

- *"The new Ministry of Happiness seeks .... is measurable and tangible for the whole of society in the United Arab Emirates" [Ref. 1]*

From this, it may be inferred that, from the outset, the Government of the UAE and Dubai effectively intends to introduce happiness policies, but without neglecting to measure their results. A strategy that, in itself, represents an objective and focused attitude on the part of those executing it. However, the word that receives the most emphasis in the entire text is "happiness", as can be seen in Fig. 2:

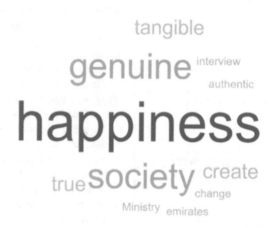

Fig. 2. Result of the study of words most frequently found in Text 1

The study of this text is in line with the one recommended by Torquato [12] regarding the role of government communication as a means of informing citizens about what happens within the Government, and that is why it is an instrument that allows them to learn about government actions [12].

Thus, with regard to government communication of the Ministry of Happiness, this paper enables us to show two main elements:

(1) that it is based on theoretical principles of state communication;
(2) that the strategies have been defined with the specialised advice and support of prominent international bodies;

To prove and support our opinion, we drew up a table (Table 3) in which we systematised the theoretical concepts with the lines of communication and action obtained from the present research.

The examples evidenced, then, a systematisation of the theoretical principles and demonstrate that the communication strategies are based on very well-structured plans. Thus, in short, it appears that:

- The message is focused and has well-defined recipients. The content is created and conveyed in such a manner as to foster involvement and commitment on the part of the Government itself, public and private institutions and the citizenry.

Table 3. Systematisation of the conclusions

Theoretical concepts	Lines of communication
• Adjust, internally, the environment	• 'Sheikh Mohammed stressed that, "The final objective of all our initiatives is to make people happy and make their lives simpler"'
• Express the identity of the Government, via its image and credibility	• 'His Highness Sheikh Mohammed bin Rashid AI Maktoum, vice-president and prime minister of the United Arab Emirates and leader of Dubai, said that a commitment to happiness and a positive lifestyle are the commitment of the government that unites the communities of the United Arab Emirates'
• Express the values and culture of the Government	• 'The commitment of the Cabinet (...) to establish positivity as a value to guarantee the happiness and well-being of individual, families and society'
• Clarify, to the citizenry, government initiatives	• 'The cabinet also approved a series of initiatives and projects presented by Ohood Khalfan AI Roumi, Minister of State for Happiness, as part of the national happiness programme'
• Formulate a strategy that assists those executing the policy to communicate	• '"His Highness Sheikh Mohammed bin Rashid AI Maktoum, (...), reiterates that it is the task of the government to drive positivity as a key facet of society", affirms AI Roumi, Minister of Happiness'
• Understand the social interests behind opinion polls	• 'Justin Thomas and Ian Gray, associate professors in psychology at the Zayed University in Abu Dhabi, are measuring happiness around the emirate based on data collected from social media'
• Take account, in its strategic plan, of the fact that information should be educational for society	• 'The initiative invites the team to send their wishes to the 'Happiness Team', which is then tasked with ranking, filtering and presenting the wishes of all employees who would like to offer and support a colleague's wish, harnessing their skills, experience, knowledge and connections'
• Takes on a political function by sharing information with the entire administrative structure	• 'The training programme is intended to be another step towards implementing the vision of the governments of the UAE and Dubai that the government should play a part in people's lives, and ultimately, function to make them happier and build a promising future for their children'

(continued)

Table 3. (*continued*)

Theoretical concepts	Lines of communication
• Assumes a social function, as it should form part of groups through its information links	• "'With this message, the UAE wishes to reiterate its humanitarian approach, adopted decades ago to help peoples in need, spread hope and ensure the happiness of people everywhere'"
• Assumes an ethical function, as truth should shine through as the primary value for society	• 'We in the UAE believe that people's happiness is a renewable and sustainable asset and an indicator of a positive and coherent society' • 'The country is focused on the concept of happiness and intends to achieve 95% happiness in 2021'
• Adjust, internally, the environment	• 'Sheikh Mohammed stressed that, "The final objective of all our initiatives is to make people happy and make their lives simpler"'
Theoretical concepts	Lines of communication
• Adjust, internally, the environment	• 'Sheikh Mohammed stressed that, "The final objective of all our initiatives is to make people happy and make their lives simpler"'
• Express the identity of the Government, via its image and credibility	• 'His Highness Sheikh Mohammed bin Rashid AI Maktoum, vice-president and prime minister of the United Arab Emirates and leader of Dubai, said that a commitment to happiness and a positive lifestyle are the commitment of the government that unites the communities of the United Arab Emirates'
• Express the values and culture of the Government	• 'The commitment of the Cabinet (…) to establish positivity as a value to guarantee the happiness and well-being of individual, families and society'
• Clarify, to the citizenry, government initiatives	• 'The cabinet also approved a series of initiatives and projects presented by Ohood Khalfan AI Roumi, Minister of State for Happiness, as part of the national happiness programme'
• Formulate a strategy that assists those executing the policy to communicate	• "'His Highness Sheikh Mohammed bin Rashid AI Maktoum, (…), reiterates that it is the task of the government to drive positivity as a key facet of society", affirms AI Roumi, Minister of Happiness'

(*continued*)

Table 3. (*continued*)

Theoretical concepts	Lines of communication
• Understand the social interests behind opinion polls	• 'Justin Thomas and lan Gray, associate professors in psychology at the Zayed University in Abu Dhabi, are measuring happiness around the emirate based on data collected from social media'
• Take account, in its strategic plan, of the fact that information should be educational for society	• 'The initiative invites the team to send their wishes to the 'Happiness Team', which is then tasked with ranking, filtering and presenting the wishes of all employees who would like to offer and support a colleague's wish, harnessing their skills, experience, knowledge and connections'
• Takes on a political function by sharing information with the entire administrative structure	• 'The training programme is intended to be another step towards implementing the vision of the governments of the UAE and Dubai that the government should play a part in people's lives, and ultimately, function to make them happier and build a promising future for their children'
• Assumes a social function, as it should form part of groups through its information links	• "'With this message, the UAE wishes to reiterate its humanitarian approach, adopted decades ago to help peoples in need, spread hope and ensure the happiness of people everywhere'"
• Assumes an ethical function, as truth should shine through as the primary value for society	• 'We in the UAE believe that people's happiness is a renewable and sustainable asset and an indicator of a positive and coherent society' • 'The country is focused on the concept of happiness and intends to achieve 95% happiness in 2021'

- In terms of the Communication Strategies, it may be inferred that the same canons are followed. With regard to communication media, it became clear that traditional media were employed, in particular the press, in this case, online.

So, it can be inferred that government communication is carefully elaborated and follows the organisation presented by some experts. As indicated by Kunsch [4], government communication is not an easy process, and the Government must make an effort to improve its communication, making it understandable and accessible to all. The research carried out allows us to conclude, in our view, that the principles of theoretical application suggested by Torquato [12] are fulfilled in their entirety, almost in compliance with a previously established script.

5 Closing Remarks

The analysis carried out allowed us to show the discursive consistency between the text analysed and the message that the Government wished to convey. By creating the Ministry of Happiness, the Government intends to create genuine and authentic happiness for the whole of society, and this is clearly shown in the message.

It was found that, by creating this Ministry, the Government also has the objective of 'creating real change'; this enables us to infer that the Government wishes to transform or improve the state of mind of its citizenry, making them happy, or happier, and beginning this change in the governmental bodies through the creation of this Ministry. Transposing the results gathered for a broader view on the communication of the Ministry of Happiness, it is possible to highlight three main elements: (1) that it is based on the theoretical principles of government communication; (2) that the strategies are defined with specialised advice with the support of international reference organisations; (3) that it systematically refers to social media and social networks.

This study, as well as the studies carried out in the scope of the broader research that has been developed on this subject, allows for the conclusion that, effectively, government communication has a strong influence on Happiness, it also being possible to conclude that the communication can contribute greatly to the success of the policies defined by the Government.

With regard to happiness, based on the concept originating from the Kingdom of Bhutan, the Government of Dubai and the UAE has managed to place (both internally and externally) great emphasis on the concept, encouraging debate and involving renowned entities and personalities through, for example, the inclusion of a day dedicated to happiness on the agenda of the World Government Summit, which is held annually. In conceptual terms, the very scope of the Ministry of Happiness has been developing. This has been proven by a content analysis performed on articles published in the *corpus latente*, based on technological tools of qualitative research.

The results allow the conclusion to be reached that the concept has been worked on exceptionally well, with implications, but above all, practical applications, capable of creating a positive impact on the lives of the citizenry, actively and assertively communicating "Happiness and Positivity". In terms of theoretical implications, the results allow it to be shown that it is possible to begin with a concept that is very difficult to define, given the individual nuances that each of us ("individual" to each one of us) may attribute to it, and transform it into a programme with a political base, the primary objective of which is for current and future generations of citizens to be happy.

References

1. Costa, A.P., Moreira, A., de Souza, F.N.: webQDA - Qualitative Data Analysis (2019). www.webqda.net
2. Brandão, E.P.: Conceito De Comunicação Pública. Comun. Estado, mercado, Soc. e Interess. Público, 1–21 (2009)
3. Torquato, G.: Tratado de comunicação Organizacional e Politica. Pioneira Thomson Learning, São Paulo (2002)

4. Kunsch, M.: Planejamento de relações públicas na comunicação integrada. Summus, São Paulo (2003)

5. Ribeiro, D., Remondes, J., Costa, A.P.: Comunicação governamental: o exemplo do ministério da felicidade dos Emirados Árabes Unidos. Ámbitos. Rev. Int. Comun. 54–72 (2019). https://doi.org/10.12795/Ambitos.2019.i44.04

6. Pina, A.R.B., Neri de Souza, F., Leão, M.C.: Investigación educativa a partir de la informacion latente en internet. Rev. Eletrônica. Educ. **7**, 301–316 (2013)

7. Costa, A.P., Amado, J.: Content Analysis Supported by Software. Ludomédia, Aveiro (2018)

8. Krippendorff, K.: Metodología de análisis de contenido: teoría y práctica. Ediciones Paidós (1990)

9. Amado, J., Costa, A.P., Crusoé, N.: A técnica da análise de conteúdo. In: Amado, J. (ed.) Manual de Investigação Qualitativa em Educação, pp. 303–352. Imprensa de Coimbra, Coimbra (2017)

10. Bardin, L.: Análise de conteúdo. Edições 70, Lisboa (2009)

11. Costa, A.P., Linhares, R., de Souza, F.N.: Possibilidades de Análise Qualitativa no webQDA e colaboração entre pesquisadores em educação em comunicação. In: Linhares, R., de Ferreira, S. L., Borges, F.T. (eds.) Infoinclusão e as possibilidades de ensinar e aprender, pp. 205–215. Editora da Universidade Federal da Bahia, Universidade Tiradentes, Aracaju, Brasil (2014)

12. Torquato, G.: Marketing Político E Governamental um roteiro para campanhas políticas e estratégias de comunicação. Summus, São Paulo (1985)

Destination Image(s) Formed of the Alentejo Litoral/Southwest Region from the Experience of Participating in the MEOSUDOESTE Music Festival

Sandra Saúde[1]([⊠]) [iD] and Ana Isabel Rodrigues[2] [iD]

[1] Polytechnic Institute of Beja and CICS.NOVA (Interdisciplinary Centre of Social Scienes), Beja, Portugal
ssaude@ipbeja.pt
[2] Polytechnic Institute of Beja, Beja, Portugal
ana.rodrigues@ipbeja.pt

Abstract. The central objective of the study that is based on this article is to characterize the destination image(s) formed by festivalgoers about the Alentejo Litoral/Southwest region based on their experience of participating in the (MEO) Sudoeste festival. We analysed the answers, of a qualitative and quantitative nature, given by a sample of 122 festivalgoers on June 2018, 11 months after their participation in the 2017 edition of the festival. The dimensions under study were the object of combined analysis with T-LAB and webQDA software packages. The results seem to show that the images conceived are very conditioned by the sporadic relationship that the festivalgoers maintain with the territory surrounding the festival grounds. There are distinctive elements of the territory that do not emerge in the shared images, bordering on the predominant idea of beautiful beaches.

Keywords: Destination image · Impacts · Region · Music festival

1 Introduction

The (MEO)Sudoeste festival is considered one of the oldest and most popular music festivals held in Portugal in the last 20 years. It takes place at the Herdade da Costa Branca, located in Zambujeira do Mar, in the parish of São Teotónio, Odemira municipality, in Alentejo Litoral[1]. As a festival site, Herdade da Casa Branca offers "unique and distinctive" features, according to the different stakeholders and festivalgoers. The event is considered innovative because in the same venue it brings together a relevant line-up of musicians, a campsite with a variety of offers, as well as an extensive canal and the emblematic nearby beaches. In addition, these conditions, the

---

[1] The municipality of Odemira is located in southern Portugal and is part of the NUTS II Alentejo and NUTS III Alentejo Litoral. It is the largest Portuguese municipality, in terms of territory, with 1720.60 km², and has a resident population of 24 917 inhabitants, according to the latest estimates of INE (2016).

© Springer Nature Switzerland AG 2020
A. P. Costa et al. (Eds.): WCQR 2019, AISC 1068, pp. 239–252, 2020.
https://doi.org/10.1007/978-3-030-31787-4_20

festival benefits from the umbilical connection with Zambujeira do Mar, of which it has become a brand – (MEO)Sudoeste is associated with Zambujeira do Mar and the town is immediately identified when the festival is mentioned. The festival was first held in 1997 and has taken place every year since then.

The starting point of the study, presented below, was the economic and socio-cultural impact study carried out on the 2017 edition of the (MEO)Sudoeste festival, when it celebrated its 20th anniversary. Given its history and the fact that until then there had been no study that explored the festival's relationship with the territory, the Municipality of Odemira decided to contract the Polytechnic Institute of Beja to carry it out. Following the results obtained and focused principally on the transversal evaluation of the impacts, we proceeded to a complementary follow-up phase, of which the main objective is to characterize the impact that participation in the festival in 2017 had/has on the image(s) formed by festivalgoers about the region where it takes place: Alentejo Litoral/Southwest region. This includes:

- knowing if participation in the (MEO)Sudoeste festival influenced the image that the festivalgoer has come to have of the region and how he/she characterizes the image that he/she has now;
- three characteristics which, according to the festivalgoer, make the Alentejo Litoral/Southwest region unique and different.

For this purpose, we chose to use a combination of two content analysis software packages – T-LAB and webQDA, which on the one hand allowed us to enhance the definition of dimensions, categories and subcategories of analysis, which are usually very laborious processes [1] and, on the other hand, allowed an improvement of the holistic interpretation of the qualitative material collected.

2   The Images Formed About a Region

The present study is based on destination image theory (DI), produced since the early 1970 s. The prominent work of Hunt [2] was decisive for the materialization of DI studies, stating that "through travellers' perceptions we can learn more about how land qualities become tourism resources" (p. 1). It was a milestone in the history of DI literature, highlighting the importance of studying the perceptions of tourists when they visit a given destination for the development of the territory. Ideas associated with the concept of DI as "organized representations", "the sum of beliefs and ideas", "a complex combination", "a general impression or attitude" and "a visual or mental impression" have been used to define DI. As a consequence of its holistic nature, Gallarza, Saura and Garcia [3] outlined four characteristics of the DI concept: complexity (not unambiguous), multiplicity (in elements and processes), relativistic (subjective and generally comparative) and dynamic (varying in time and space).

Along with this more conceptual discussion, a debate also arose in the literature regarding how to measure and evaluate the DI concept. The approach that was advanced in the 80 s/90 s was based on a more "quantitative" assessment (as mentioned by Gartner [4]), which is done item-by-item. It is a more "closed" and deterministic approach in which a set of statements/items or attributes are listed and where the respondent is asked to evaluate the image of the destination (re)taken from a list. In the 2000s, a more holistic assessment approach to DI emerged and was consolidated, due to the nature of the concept, based on more "naturalistic" and therefore qualitative forms of measurement such as open-ended questions using "free elicitation of words" techniques [5], among others. A measurement model that starts from the combination of both approaches (quantitative and qualitative) was proposed by Echtner and Ritchie (1991) [6] and corroborated by Jenkins [7]. Later, Beerli and Martin [8] proposed a 9-dimensional image measurement scale that allowed the organization and classification of attributes by categories (e.g. "Natural Resources", "General Infrastructures", "Tourist Infrastructures", "Tourism, Leisure and Recreation"," Culture, History and Art", "Political and Economic Factors", "Natural Environment", "Social Environment" and "Atmosphere of the Place"). This is the theoretical frame adopted in this research.

3   The Experience of Participating in the (MEO)SUDOESTE Festival

More than two decades since its first edition in 1997, the (MEO)Sudoeste festival is not just a music festival and has broadened its ontological vocation; according to Guerra [9], "the Sudoeste [festival] has been, from the beginning, a typical "summer festival", associated with vacations, the beach, sun and camping, transforming these aspects into real visiting cards that notably differentiates it from other Portuguese festivals" (p. 11).

In 2017, with a total of 200 000 participants, Sudoeste was considered the third most popular festival in Portugal by the Portuguese Music Festivals Association. The typical (MEO)Sudoeste festivalgoer was:[2]

- Portuguese (80%) – of the 20% of foreigners, most were Spanish.
- not born or resident in the region (municipality of Odemira);
- male (57.1%), aged between 19 and 24 (53.4%) and 14 to 18 (28.9%).

Based on the answers of participants in the 2017 edition, the great majority (70.1%) were there for the first time and most had purchased a 5-day ticket, combining entrance and camping accommodation (70.6%) [10].

When asked about the dynamics experienced during the festival, with impact on the community of Odemira at large and of Zambujeira do Mar in particular, the non-resident participants highlighted:

(i) better knowledge of the region: Zambujeira do Mar and other places in the municipality;
(ii) the fact that the festival helped establish a closer relationship with the resident population of Zambujeira de Mar and the whole municipality (pp. 117–118) [10].

Of the festivalgoers who are Odemira residents, the majority considered the impact of the festival to be very positive. According to the respondents [10], "the festival

---

[2] Only about 10% are from the Odemira municipality.

contributes very positively to advertising the territory and to consolidating its image abroad, providing, at the same time, an opportunity for participation and having new experiences." (p. 125).

4 Methodology

The present work continues the study of economic and socio-cultural impact made on the (MEO)Sudoeste festival in 2017 [10]. It now proceeds with another dimension of research, characterizing the impact that participation in the 2017 festival had/has on the image(s) formed by festivalgoers about the region where it takes place.

This is an exploratory study in which the option for the research-action (RA) methodology [11] "as a means of approaching and gaining knowledge of a certain socio-organizational phenomenon" (p. 217) proved to be the most appropriate. The objectives are not only about the production of knowledge (from the results obtained in the study), but also for responding to a concrete situation; the results of the work will be presented to the Municipality of Odemira in order to contribute to reflection about the effects of the (MEO)Sudoeste festival on the image of the region and, accordingly, promotion strategy for the territory which can be adopted.

The study developed was based on the following research questions (RQ):

- *RQ1: What image do festivalgoers have of the region after participating in the (MEO)Sudoeste festival?*
- *RQ2: What are the three characteristics that, according to the festivalgoers, make the Alentejo Litoral/Sudoeste region unique and different in Portugal?*

4.1 Data Collection: Instrument and Procedure

The data collection technique used was an online questionnaire, structured in closed and open questions. In order to answer the research questions, data were collected on the basis of three questions, namely:

- Q7: Did your participation in the 2017 edition of MEOSudoeste influence the image that you came to have of the region (Alentejo Litoral/Southwest)? (select the option that matches your opinion):

  1. Yes, I came to have a different, more positive image;
  2. Yes, I came to have a different, more negative image;
  3. I did not change the image I had.

- Q7.1: If you answered Yes in the previous question (that is: I have now a different image more positive or negative), how would you characterize the image you have now? (open-answer question);
- Q.8: Without thinking too much, give three characteristics that make the Alentejo Litoral/Southwest unique and different from all other places you have visited so far in Portugal? (open-answer question).

The collection of data to respond to *RQ1* focused on the combined use of questions 7 and 7.1. If the respondent stated that he/she had come to have a different image of the region after his/her stay at the festival, he/she was asked in the form of an open question (Q7.1) to freely express his/her thoughts. The response to *RQ2* is supported by the data collected through Q8, based on the free elicitation technique [5]. The option for this technique is intrinsically related to the concept of DI, which, as was verified in the literature review section, is assumed to be a holistic, complex and comprehensive concept composed of sensations, emotions and perceptions [3].

In total, the questionnaire had nine questions, the last three being those referenced above; the other six questions are related to other dimensions of analysis which not explored in this study. The questionnaire applied had a selection question at the beginning, in which respondents were asked if they had already been/visited the region before their participation in (MEO)Sudoeste in 2017. Only those who stated that (MEO)Sudoeste 2017 edition was their first time in the territory advanced to the following questions; thus, it was assured that the impressions, opinions and images shared were only influenced by this experience. The questionnaire was sent by email, to 231 festivalgoers between June 15 and 30, 2018, that is, those who authorized contact to be made for further data collection at the time of the study carried out in 2017. Authorization for data collection was reinforced through detailed information shared with the respondents on the objectives of the study, the justification and the framework of the contact made by email, as well as the guarantee of confidentiality in the responses. After three requests encouraging responses to the questionnaire, the reception period was closed on July 20, 2018.

4.2   Participants

Responses were collected from 122 festivalgoers, which corresponds to a return rate of 53%. The ex-post sample, based on nationality, is representative of the sample of festivalgoers in 2017: of the 122 valid questionnaires, 100 are from Portuguese and 22 from foreign festivalgoers.

4.3   Data Analysis: Strategy and Techniques

The three key questions for the study carried out led to the collection of quantitative and qualitative data. The answers given to question Q7 were quantitatively analysed through descriptive statistics. The answers given to the two open questions Q7.1 and Q8 of the questionnaire constitute the corpus of the content analysis carried out. In total, and for each question, 122 answers or syntactic units were collected, each corresponding to the answer given to the question by each respondent. The answers given in question Q7.1 have a greater extension than those of Q8, involving one or two phrases of between 1 and 32 words through which sensations, experiences, representative aspects and justifications of the image(s) taken on the Alentejo Litoral/Southwest are characterized in the form of articulated discourse. The answers given in question Q8 are phrases ranging from 1 to 21 words, centred on aspects/characteristics that associate/distinguish the region visited.

Taking the defined objectives into account, the option for an analysis of thematic content is justified [12], as we needed to identify the structuring categories of the images of the region formed by the festival participants. The analytical approach was first exploratory and then descriptive, based on two phases that arose from the combined use (in a complementary relation) of two CAQDAS (computer-assisted qualitative data analysis software):

- 1st phase: the use of the CAQDAS T-LAB[3] in order to perform an exploratory analysis of the corpus, extracting the first ideas, in order to understand the relationship between words and keywords. A co-occurrence analysis was done, in particular an "association of words" and a "analysis of sequences". Through this analysis, it was possible to identify which words the respondents repeated and/or associated equally in their answers (e.g. words that appear equally before or after others in the collected syntactic units);
- 2nd phase: Use of the CAQDAS webQDA[4] with the objective of organizing and systematizing the thoughts resulting from the previous phase through a deductive analysis, based on the perception and image dimensions by means of the typology proposed by Beerli and Martin [8]. The coding and interpretation phase were done following the deductive method.

The combined use of two CAQDAS is justified by the fact that they have complementary analytical potentials, one more in the exploratory component of the data and the other in the dimension of the holistic interpretation of their meaning. T-LAB software consists of a set of linguistic, statistical and graphic tools that allow various types of text analysis. T-LAB uses an automatic approach that allows meaningful patterns of words and/or subsets of themes in the texts under review to be highlighted (https://tlab.it/en/presentation.php, accessed on February 25 of 2019).

webQDA is software that supports qualitative analysis for researchers who need to analyse qualitative data, either individually or collaboratively and in a synchronous or asynchronous way in several contexts [13]. The simplicity of its use and adaptation to various types of research is seen as one of its differentiating factors. In this particular case, the use of webQDA essentially enabled coding of the text units, using the "descriptive" coding method in which "the basic topic of a passage" is summarized "in a word or short phrase – most often as a noun" [14].

The coding was done based on the classification scale of destination attributes most used in the theory of DI developed by Beerli and Martin [8]. Descriptive coding was applied by matching each text unit found to the category of target attributes associated (e.g. "beautiful landscape" was encoded in the "Natural Resources" category); the grid defined by the abovementioned authors was fully followed. All categories were duly described in order to better clarify the categorisation/coding process. Prior to the corpus analysis, a filtering and data-cleansing process was performed which eliminated orthographic errors and put a full stop at the end of each sentence as a way to identify the end of each syntactic unit under analysis. These are fundamental procedures to guarantee the fulfilment of the criteria required by the software applied.

---

[3] More information available at: https://tlab.it.

[4] More information available at: https://www.webqda.net.

5 The Images Formed of the Alentejo Litoral/Southwest Region by MEOSUDOESTE Festivalgoers

5.1 Images that They Came to Have After Participation in the MEOSUDOESTE Festival

To answer RQ1: *What image do festivalgoers have of the region after participating in the (MEO)Sudoeste festival?*, participants in the study were asked to indicate whether or not their participation in the festival influenced their image of the Alentejo Litoral/Southwest region (Q7). From the results obtained it can be concluded that:

- 85.2% came to have a more positive image of the region resulting from the festival;
- 1.6% came to have a more negative image, and
- 13.1% did not change the image they had.

The 86.8% who said that they came to have a different image (more positive or negative) were then asked to characterize the image that they have now. In order to characterise the image they currently have of the Alentejo Litoral/Southwest region, the most frequent words used by respondents'/festivalgoers were: "beaches"; "like"; "great"; "beautiful/pretty"; "animation/entertainment"; "good" and "meosudoeste"; these are the lemmas that are most often repeated in the syntactic units analysed. Among the most frequent words, the "centrality" of "beaches", around which the lemmas[5] are more markedly associated, is evident, as can be seen in Fig. 1.

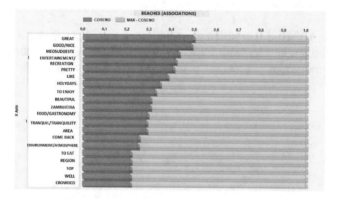

Fig. 1. Association of lemmas with the keyword: beaches.
Legend: COEF: Coefficient of Association between the lemma "beaches" and other lemmas. The association index ranges from 0 to 1.
Source: T-LAB 2019, questionnaire applied in July 2018.

Based on the analysis of Fig. 2, we can conclude from the analysis of how the words appear sequentially in the syntactical units analysed, that in the images taken on

---

[5] Labels assigned to groups of lexical units classified according to linguistic criteria (lemmatization) (Manual T-LAB PLUS, 2019, available at: https://tlab.it/en/allegati/help_en_online).

the Alentejo Litoral/Southwest, expressions such as "great beaches"; "like the beaches"; "beautiful beaches" and "good beaches" stand out.

Fig. 2. One-to-one relationship between the lemmas "EGO NETWORK".
Legend: blue line: precession relationship (preceding word); orange line: succession relationship (consecutive word). The weight of the line is greater or less depending on the intensity of the relationship.
Source: T-LAB 2019, questionnaire applied in July 2018.

It can also be noted that in the characterisation made, and before beaches being highlighted, the festivalgoers also identified items such as: "entertainment/recreation"; "region"; "territory/area"; "beautiful"; "to know" and "pretty" (Fig. 3).

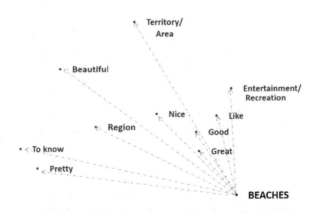

Fig. 3. EGO NETWORK of precession relationships: Lemma: BEACHES.
Legend: The lower the distance from the lemma to the lemma positioned in the centre, the greater the number of times this lemma appears before the central lemma.
Source: T-LAB 2018, questionnaire applied in July 2018.

Following the central idea of beaches, the festivals tend to frequently associate lemmas such as: "crowded"; "spectacular"; "meosudoeste" and "tranquil/tranquillity" (Fig. 4).

Fig. 4. EGO NETWORK of succession relationships: Lemma: BEACHES.
Legend: The smaller the distance from the lemma to the lemma positioned in the centre, the greater the number of times this lemma appears after the central lemma.
Source: T-LAB 2018, questionnaire applied in July 2018.

In summary, the results seem to show that the most positive image that festival-goers have come to have of the region after their participation in the festival is structured around the predominant idea of having good, beautiful and great beaches. They are also associated with shared images, although with less significance, with items such as animation/fun or meosudoeste that help to build an atmosphere/environment and holidays that the festivalgoers consider great and that they like.

The results obtained with T-LAB are corroborated by the analysis of thematic categorical content made using webQDA. The text units considered were coded based on the nine image dimensions defined by Beerli and Martin [8] (see Fig. 5), which allowed an understanding of which categories of image the festivalgoers emphasised most.

"Natural Resources" and "Atmosphere of the Place" have a greater number of references (42 and 71, respectively). "Beaches" was undoubtedly the impression/image that stood out most after their experience at the festival, leaving a positive image and being acknowledged as an element that was surprisingly positive (Ref14 "beaches and everything surprised me very positively"). However, there are also less positive impressions/images associated with beaches, such as the fact that there were a lot of people (Ref40 "I really liked the beaches, despite the number of people" and Ref24 "the beach is also very beautiful, although it was always very crowded"). As for the "Atmosphere of the Place" dimension, the impression/image of "animation and fun" was even more than initially thought (Ref21 "I thought it was different. It was very animated, I liked it a lot" and Ref29 "I thought it was different, it is very different from the coast and other festivals, we are in the countryside, in a very natural environment,

(MEO) Southwest Festival_Project
Issued by Ana Isabel

webQDA

Tree Codes

Name	Sources	Refs
Image after the festival_Question 7	0	0
Natural Resources	1	42
General Infrastructures	1	5
Touristic Infrastructures	1	12
Tourism, Leisure and Recreation	0	0
Culture, History and Art	1	10
Political and Economic Factors	1	3
Natural Environment	1	24
Social Environment	1	7
Atmosphere of the Place	1	71
Others	0	0

Fig. 5. Categorization of results of the image that festivalgoers came to have of the region after participation in the (MEO)Sudoeste festival.
Source: webQDA, questionnaire applied in July 2018.

the environment is great"), which reveals a differentiating element to promote the festival. In this category, it was possible to find text units expressing less positive ideas like Ref25 "some things I did not like; we could not eat in the cafés, they rejected us and ignored us", which shows some care is needed in the future.

Regarding the "Natural Environment" dimension, although the most referenced impressions are clearly those associated with "beaches", there are also references to the landscape as an element of the image that stands out (Ref9 "the landscape and the beaches are more beautiful than I thought"), denoting a positive surprise. Also worth mentioning is the "Tourism Infrastructures" dimension, highlighting high numbers of people and excess of people in certain places (Ref2 "cafes always full, difficult to find a place to eat") and references to some infrastructure needs such as the case of housing (Ref7 "could have more accommodation for young people"). It is also worth mentioning in the "General Infrastructures" dimension with references to the difficulty in getting to the place, due to lack of public transport and signs (Ref2 "It was not all good, reaching the southwest was difficult, there is no transport"). To conclude the analysis of the encodings of the remaining image dimensions, it is worth mentioning some positive impressions related to the hospitality of the local people and the great gastronomy associated with the idea of "you can eat cheaply".

5.2 Three Characteristics that Make the Alentejo Litoral/Southwest Region Unique and Different from All Other Places Visited in Portugal

The answer to RQ2: *What are the three characteristics that, according to the festivalgoers, make the region unique and different, in Portugal?* was obtained on the basis of analysis of the opinions shared in question Q8 (Table 1).

Table 1. Most identified lemmas/keywords in free elicitation on the three characteristics that distinguish the region

LEMMAS	Number of occurrences	LEMMAS	Number of occurrences
Beache_beaches	105	Beautiful	7
Nature	34	Zambujeira	7
Sunset	30	Different	5
Animation	27	Holidays	5
Good	24	Algarve	4
Sun	19	Watercourse	4
Food_gastronomy	16	Climate	4
Meosudoeste	14	Pretty	4
Fun	13	Sea	4
Environment	10	Music	4
Great	10	Landscape	4

Source: T-LAB 2019, questionnaire applied in July 2018.

Based on the words identified with T-LAB, cross-checked with the categorical analysis performed with webQDA (see Fig. 6), it was possible to identify that "Natural Resources" appears as the most representative image dimension, in which the most referenced image attributes are, in order of importance, "beaches" and "sunset".

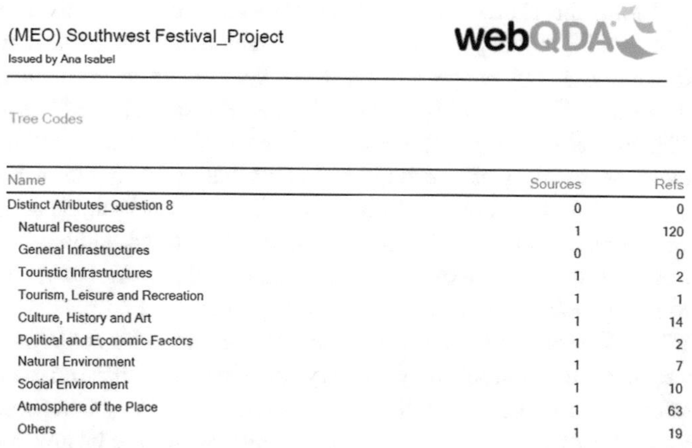

(MEO) Southwest Festival_Project
Issued by Ana Isabel

webQDA

Tree Codes

Name	Sources	Refs
Distinct Atributes_Question 8	0	0
Natural Resources	1	120
General Infrastructures	0	0
Touristic Infrastructures	1	2
Tourism, Leisure and Recreation	1	1
Culture, History and Art	1	14
Political and Economic Factors	1	2
Natural Environment	1	7
Social Environment	1	10
Atmosphere of the Place	1	63
Others	1	19

Fig. 6. Categorization of the results of free elicitation on the three characteristics that distinguish the region.
Source: webQDA, questionnaire applied in July 2018.

In a process of filtering, sorting and further data control that the use of webQDA allows, it can be seen that most of the references are restricted to the idea of "beautiful beaches" and "beautiful sunset". Next is the "Atmosphere of the Place" dimension, in which attributes such as "animation/entertainment" are highlighted, but very much associated with the experience of enjoyment of the beaches and not with animation that occurs in the destination/territory itself, when analysing the context of using this idea. In an analysis of the context in which the attribute is referenced, there are also some items in this dimension associated with "authentic environment", "hospitality of its people", and a certain differentiation from other destinations such as the Algarve (Ref4 "there are fewer people than in the Algarve" or "it is different from the Algarve"). It turns out that there are ideas and impressions that emerge about the destination which help, on the one hand, to better perceive the centrality given to the "beaches", such as references to "nature", to "sunset" and to "animation" and, on the other hand, to identify additional elements of the destination that can be enhanced, even if they are not substantial among the festival participants,. Take, for example, the experiences that can be created by accommodation around the "sunset", a simple natural element, but that can originate well-structured and creative tourist products; or in the case of animation, exploring ways of combining the work of accommodation and local tourist animation companies to further enhance the liveliness of the region.

6   Discussion of Research Findings

The experience of participation in the (MEO)Sudoeste festival in 2017 had an impact on the images formed of the Southwest/Alentejo Litoral region by festivalgoers who were in the territory for the first time: the vast majority (85.2%) had a more positive image.

When exploring the characteristics of the DI shared by the festivalgoers, and as Beerli and Martin's [8] model evidences, it can be seen that there are several categories of image. In this case, the image attributes are very much centred on the "natural resource" of "beaches" that qualify as "good, beautiful and/or great", that is, in the dimension "Natural Resources". The element "beaches" is clearly central to shared discourse, accompanied only at a second level of relevance by another natural attribute such as "the sunset". In terms of the "Atmosphere of the Place" dimension, ideas such as "entertainment/recreation" and "meosudoeste" are identified, although, less significant in shared narratives, they express the association they make between the festival and the surrounding animation. In addition, although there are references to "authentic environment", "hospitality of its people" and "a certain differentiation from other destinations such as the Algarve", they are very occasional and there are no objective or subjective references to other distinctive elements of the region (such as Cabo Sardão or Vila Nova de Milfontes), or even a stronger emphasis on the beauty of the natural coastal landscape. It can also be seen that there is no experience of visiting the territory, as answers are limited to what is an integral part of the festival site and/or associated with the contiguous locality of Zambujeira do Mar. The results point to DI being very influenced and conditioned by the very specific and sporadic relationship that festival owners have with the territory outside the festival grounds.

7 Conclusions

DI constitute a "combination" of impressions, sensations and opinions about a given destination/territory, including various dimensions such as "Natural Resources", "Culture, History and Art" or "Social Environment" and "Atmosphere of Place", that is, more objective and/or more affective attributes. In the formulation of DI, the distinctive characteristics of the destination are qualified and described as a result of the experiences developed by each person and/or shared with others, constituting holistic representations.

The DI shared by the festivalgoers who participated in the 2017 edition of the (MEO)Sudoeste festival are clearly distinguished as being positive; the image of the region is now different for the better because they have an almost one-dimensional characterization which very centred on the predominant idea of "beautiful beaches", accompanied only by another natural attribute such as "sunset" at a second level of relevance. This conclusion gives clues that call for reflection on the part of the various regional actors, among which the Municipality of Odemira stands out, on the real contribution of the festival in promoting and valorising the image of the region, in order, accordingly, to be able to adapt/tailor the respective strategies of promotion and dissemination of the destination.

Given the research questions and the type of material collected, of a quantitative and qualitative nature, the option of using a mixed methodology of analysis and combined use of qualitative analysis software made it possible to better explore and interpret the type of images described by the festivals. Given the recognition of the importance and impact that the festival has in the construction of the DI of the Alentejo Litoral/Southwest region in general and Odemira/Zambujeira do Mar in particular, it is considered appropriate to continue to explore and enrich this aspect of analysis. One suggestion is to cross the conclusions of this work with those extracted from the analysis made on the festival's promotional materials (videos, website, posters) and/or images (photographs, videos) and communication carried out on the social networks by the festivals (before, during and after an upcoming festival). Content analysis of this material will greatly enrich the understanding of the real effects of the experience of participation in the (MEO) Sudoeste festival on DI formed of the region.

References

1. Costa, A.P., Amado, J.: Content Analysis Supported by Software. Ludomedia, Oliveira de Azeméis (2018)
2. Hunt, J.D.: Image as a factor in tourism development. J. Travel Res. **13**, 1–7 (1975)
3. Gallarza, G., Saura, G., Garcia, H.: Destination image: towards a conceptual framework. Ann. Tourism Res. **29**(1), 56–78 (2002)
4. Gartner, W.C.: Temporal influence on image change. Ann. Tourism Res. **13**, 635–644 (1986)
5. Coshall, J.T.: Measurement of tourists' images: The repertory grid approach. J. Travel Res. **39**, 85–89 (2000)

6. Echtner, C., Ritchie, B.: The meaning and measurement of destination image. J. Tourism Stud. **2**(2), 2–12 (1991)
7. Jenkins, O.H.: Understanding and measuring tourist destination images. Int. J. Tourism Res. **1**, 1–15 (1999)
8. Beerli, A., Martín, J.D.: Factors influencing destination image. Ann. Tourism Res. **31**(3), 657–681 (2004)
9. Guerra, P.: Lembranças do último verão. Festivais de música, ritualizações e identidades na contemporaneidade portuguesa. https://www.researchgate.net/publication/300651289_Lembrancas_do_ultimo_verao_Festivais_de_musica_ritualizacoes_e_identidades_na_contemporaneidade_portuguesa. Accessed 18 June 2019
10. Saúde, S., Lopes, S., Borralho, C., Féria, I.: O Impacte Económico e Sociocultural do Festival MEO SUDOESTE no Concelho de Odemira. Sílabas & Desafios, Faro (2019)
11. Ferreira, P.: A utilização da metodologia de investigação-ação na intervenção social: Uma reflexão teórica. Intervenção Social **32**(34), 215–236 (2008)
12. Bardin, L.: Análise de conteúdo Edições 70. Lisboa, Portugal (2010)
13. Souza, de, F.N., Costa, A.P., Moreira, A.: webQDA - Qualitative data analysis (versão 3.0). Micro IO e Universidade de Aveiro, Aveiro (2016). www.webqda.net
14. Saldaña, J.: The Coding Manual for Qualitative Researchers. Sage, London (2009)

A Qualitative Study of Brazilian Scientific Production on the Methodologies of the Ergonomics of the Built Environment: Systematic Literature Review

Katia Alexandra de Godoi e Silva[(✉)] [ID]

Anhanguera Uniderp University, Campo Grande, MS, Brazil
katia.a.silva@educadores.net.br

Abstract. The Ergonomics of the Built Environment (EAC) is a scientific discipline whose study and implementation cover several areas of knowledge and deal directly he built space factors and their interactions with the user. This theme has been expanded gradually, in scientific circles, and with this research it is aimed to undertake an exploratory qualitative study of Brazilian scientific production in ergonomics in the context of the built environment, in the last four years (2015–2018), by means of a Systematic Review of the Literature with the use of the software webQDA. In this article, the concept of Systematic Review of the Literature and the research process from a protocol are presented. Digital resources, such as software for the organization and analysis of qualitative data, support the full process of conception and realization of a systematic review of the literature. In this sense, it is intended to make a contribution that encourages the scientific community to reflect and deepen unusual investigative procedures in this area of knowledge, because they contribute to the findings of the study.

Keywords: Systematic review of the literature ·
Ergonomics of the Built Environment · Software webQDA

1 Introduction - The Ergonomics of the Built Environment

The Ergonomics of the Built Environment (EAC), according to [1], is perhaps the most recent branch of Ergonomics, which encompasses the studies related to the influence of the physical environment in the development of the task by user and his or her relationships.

The elements considered by EAC are those related to environmental comfort, the environmental perception, adequacy of materials, colors and textures, accessibility, anthropometric measures and sustainability [2].

Upon considering these elements, there is a need for a comprehensive approach, to assess an environment under the ergonomic perspective, with the use of specific methodologies, involving the user's interaction with the environment, furniture, objects and the task, in this space.

It is known that EAC makes use of its own methodologies of ergonomics and other specific tools [1]. Under such perspective, in this article it is aimed to perform an

A. P. Costa et al. (Eds.): WCQR 2019, AISC 1068, pp. 253–261, 2020.
https://doi.org/10.1007/978-3-030-31787-4_21

exploratory qualitative study of Brazilian scientific production in ergonomics in the context of the built environment, in the last four years (2015–2018), by means of a Systematic Review of the Literature with the use of the software webQDA.

Based on the literature, the methodological path is introduced, from the protocol of a systematic review of the literature. Then, the process of analysis, and the results, constructed in the light of theoretical and methodological basis on the methodologies used in EAC investigation processes. In the last part, the considerations and perspectives of the study presented are found.

2   Methodologic Path - A Systematic Review of the Literature

A systematic review of the literature is a process of evidence-based research in the scientific literature, conducted by means of protocol. On this path, it is imperative that the research stages be recorded, "[…] not only so that this can be replicable by another researcher, but also to infer that the ongoing process follows a series of steps previously defined and absolutely respected in several steps." (Ramos, Faria and Faria [3, p. 23]).

For the cut in this study, the effect of the studies [4] and [5], some applicable steps are adopted in the context of this investigation: (i) objectives; (ii) equations of research by definition of Boolean operators; (iii) scope; (iv) inclusion criteria; (v) exclusion criteria; (vi) criteria for methodological validity; (vii) results; (viii) data treatment, as shown in Fig. 1.

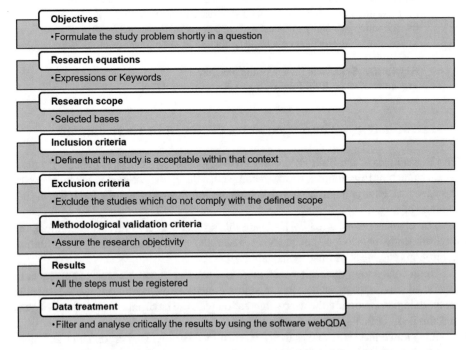

Fig. 1. Systematic literature review process. Adapted from [4, 5].

Following this process, for example, Table 1 was elaborated, in order to exemplify the systematic literature review process carried out in the course of this research.

Table 1. Example of the steps followed in the systematic literature review process.

Step	Explanation
Objective	Identify and understand the picture of the development of research in ergonomics in the context of the built environment
Research equations	Search expressions: Ergonomics; built environment
Research scope	Google Scholar Annals of major events in the area: International Congress of Applied Ergonomics (Conaerg) 2016; National Meeting of Ergonomics of the Built Environment (Eneac) 2018 and 2016; International Congress on Ergonomics and Usability of Humano-Technological Interfaces (Ergodesign) 2015 and 2017 Journals: journal Ação Ergonômica; journal Estudos em Design; journal Produção
Inclusion criteria	Articles published in scientific journals. Articles published in annals of events. Articles in the areas of Architecture and Urbanism and Industrial Design. Articles with free access
Exclusion criteria	Articles without abstract, articles published in other sources and outside the scope of ergonomics that are not in English and Portuguese
Methodological criteria of validity	Replication of the process by two researchers, for verification of the inclusion and exclusion criteria
Outcomes	Description of research and record of all the steps
Data treatment	Filter, critically analyze and describe the results with the use of the *software* webQDA

The research was held on 18th of February 2019, with the objective of identifying and understanding the overview of the research development in ergonomics in the context of the built environment, in scientific articles available in the databases of Google Scholar, in the annals of the events in the area in Brazil (Eneac 2018 and 2016; 16th Ergodesign 2015 and 2017) and in the journals Ação Ergonômica, Estudos em Design e Produção. After defined the thematic scope, it interested to map, on these data bases, the references relating to the term, according to certain criteria.

In view of the specificity of databases, the search process was initiated in order to circumscribe the object under study.

First, we selected, in the collection, articles written in Portuguese, published from 2015 until 2018, with the following keywords: "Ergonomy", "Ergonomic", and "built environment", with occurrence in the title of the article, without using any Boolean operator.

With the definition of this equation of research, the results for the keywords were found: 30 occurrences in journals and 158 occurrences in annals of events, totaling 188 articles, as shown in Table 2. Whereas on the data basis of Google Scholar, 13 occurrences were found of articles written in Portuguese; however, these articles were also identified in the databases consulted before. Then, the research process was replicated by a researcher who also used the above data bases.

Table 2. Journals, scientific events and number of articles found – 2015–2018.

Journals and scientific events	Number of articles found
Journal Ação Ergonômica	29
Journal Estudos em Design	1
Conaerg 2016	28
Eneac 2016	15
Eneac 2018	28
Ergodesign 2015	52
Ergodesign 2017	35
Total	188

After verification, the articles were saved and shared on webQDA software and Google Drive. The selection required particular filtering, because some were not suitable for the initial assumptions. In total, 42 occurrences were analyzed, as shown in Table 3.

Table 3. Journals, scientific events and number of articles validated – 2015–2018.

Journals and scientific events	Number of articles found
Journal Ação Ergonômica	3
Journal Estudos em Design	0
Conaerg 2016	5
Eneac 2016	8
Eneac 2018	11
Ergodesign 2015	9
Ergodesign 2017	6
Total	42

The work of analysis and contribution of webQDA software was conducted from this point and will be explained below.

3 Analysis Process - The Contribution of the *software* webQDA

To support the data analysis, we resorted to *webQDA software*, which is designed to support the qualitative research, especially in the phases of organization and the treatment of the collected data [6, 7].

These authors believe that, in spite of being "empty", the tool can be configured according to the researcher needs. The program is not directed to a specific type of design research, because its organization is based on the fundamentals of content analysis, more specifically in the content structure proposed by [8]: Organization of analysis (pre-analysis/material exploration, first inferences and interpretation); Coding (treatment of material to obtain the best representation of its contents); Categorization (data simplified representation); Inference (on which it is possible to relate to this kind of interpretation).

From this structure [8], it was important to understand the elements that organize the webQDA operation logic, to sort the data in this study, according to the three parties: "Sources", "Coding" and "Questioning".

For the cut of this article, "sources", first action of the researcher with the webQDA were used. This area can be organized according to the researcher needs; the types of documents; or the function of each [6, 7]. In this study, the sources used and organized for the constitution of material for analysis were the articles selected previously.

After using the "Sources" and even before starting working with the "Coding", it was chosen to use the "Questioning", on webQDA, in order to know, to delimit and create a theme for reflection during the study in question. It is, however, an initial and exploratory approach, that will continue to be the object of study in other moments of the research.

Thus, it was used the "Questioning" of webQDA, the tool "The most frequent words" (Fig. 2), from the following question: What are the 10 most frequent words contained in articles researched about the built environment ergonomics?

Fig. 2. Words cloud of articles researched generated by the *software* webQDA.

It was detected that the most frequent occurrences of the first ten words, in descending order, are: Environment (1,560), work (688), Ergonomics (520), space (497), users (490), analysis (464), activity (432), evaluation (374), comfort (373) and study (292). These words and the cloud announce a theme that can be analyzed in the corpus studied - Analysis of the Ergonomics Methodologies used in the Built Environment.

It is worth explaining that this theme was inserted into the "Coding", more specifically in the "Tree Codes", in the webQDA, and will be discussed and expanded in another time, because it requires a careful reading of the data extracts, with a view to create the dimensions, indicators, or categories, either descriptive or interpretive [6, 7]. By now, we will analyze under the perspective of the theme found in the corpus of scientific articles surveyed in a systematic review of the literature.

4   Results - Analysis of the Ergonomics Methodologies Used in the Built Environment

From the systematic review of literature and the use of the software webQDA, it was also resorted to the studies carried out by [2, 9, 10], in order to verify the most recurrent applications of the ergonomics methodologies of the built environment.

The studies [2] and [9] show that the methodologies used in Brazil are: post-occupancy evaluations [11]; environment/behavior relationship [11]; ergonomic intervention [9]; accompanied walk [12]; Ergonomic Work Analysis [13]; ergonomics in the relationship between man and the built space [14]; monitored displacement [15].

In this study, the use of the following methodologies was identified, during 2015–2018: post-occupancy evaluations [11]; Pre-project evaluation (APP) [16]; Ergonomic Work Analysis (AET) [13]; Ergonomics Intervention (EI) [9]; Method of Ergonomic Analysis of the Built Environment (Meac) [17]; Methodology for Construction Projects focusing on the User (MPCCU) [18], as shown in Fig. 3. The use of multi methodologies was also verified, i.e., the researcher uses different methodologies, in the same survey, in order to elucidate their issues.

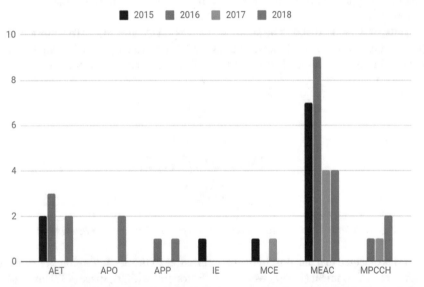

Fig. 3. Use of the methodologies of the built environment between 2015–2018. Research Data.

Meac [17] was the most widely used in the period 2015–2018. It was found that, from the analysis of the articles, a greater number of publications with the use of Meac, between 2015 and 2016. [1, p. 4] believe that the application of this methodology is connected "[…] to the group of researchers from UFPE [Federal University of Pernambuco], naturally due to the origin of the method and its close relationship between the author and the institution".

The MPCCH method (one of the most recent instruments, among the range of options for the researchers) shows moderate use, mainly by a small group of researchers, verified in recent editions of the congresses.

The survey also found that AET, APO, APP, IE and MCE appear discreetly, as methodologies used in EAC studies. In a relationship, specifically, the use of these methodologies, [1, p. 4] raise some predictions about the reasons that lead the authors to use more often this methodology, and also the discreet use of other methods.

Initially, on the preference of AET methodology [1, p. 4] explain that it is one of the alternatives taught "in the specialization courses in Ergonomics existing in Brazil, and that due to natural issues the authors of these instruments make use of such instruments to perform their studies." Another fact that justifies the use of this and other methodologies, is that the strictu senso post-graduation courses (Master's degree and Doctoral), "[…] use these methods in their studies and are sometimes taught for realization of studies - naturally each method of certain research group, uses the instrument that was developed at that institution".

The three most used methodologies in studies of Ergonomics of Built Environment - AET, Meac and MPCCH - will be briefly explained below.

The **work ergonomic analysis** has its origin in a French school, which mingles with the very origin of Ergonomics. According to [13] and [19], it is unfolded in five steps: analysis of demand, job analysis, analysis of activity, diagnosis and recommendations. Despite this unfolding, the main flow of this methodology includes two parts: the situational part and the analytical part. According to [1], the situational part refers to the analysis of the demand and the process ergonomic assessment, allowing some signs of improvement. The analytical part comprises the process of deepening the initial part, with a view to a work situation in context, from qualitative and quantitative data, which give reciprocal feedback.

The **ergonomics methodology in the relationship between man and the built space**, proposed by [17], has been structured in six stages, divided into two blocs. The first block deals with the environment physical analysis composed of three stages (global environment analysis; identification of the environmental configuration; assessment of the environment in use). The second block refers to cognitive analyzes (environmental perception; ergonomic diagnosis of the environment; ergonomic propositions for the environment).

[1] highlight that this method requires a global view of the researcher, and since the first phase until the last proposals for improvement to the situation assessed are offered.

The **methodology for projects construction focused on the human being** (MPCCH), presented by [18], is organized in four steps. The first refers to the design briefing, which aims to collect information about the environment. The second deals with the user profiles and adjusting group, which consists of the data collection about the environment users. The third one, the task analysis, allows to investigate the activities carried out in the building, which exhibits inadequacies, verifying the postures adopted by the users during these activities, in addition to encompassing measurements of environmental comfort. The fourth and last step, which relates to the users' needs, lies in the synthesis of all the problems encountered in building, highlighting and scoring the necessary adaptations and elaborating a set of ergonomic recommendations that meet the user specific needs.

5 Considerations

The objective of this study was to perform an exploratory qualitative study of Brazilian scientific production in ergonomics in the context of the built environment, in the last four years (2015–2018), by means of a Systematic Review of the Literature. Thus, it just had the intention to begin the process, since the definition of the research protocol by the exploratory analysis of data obtained from the use of the software webQDA.

Therefore, it was possible to observe, in the course of the research, the contents about the ergonomics methodologies of the built environment, in the period 2015–2018, on major national events, which are the exponents of studies conducted in Brazil: Conaerg, Eneac and Ergodesign. In addition to the national journals of influence in the scientific environment: journal Ação Ergonômica; journal Estudos em Design; journal Produção.

From the organization through the webQDA; validation; and the articles analysis, according to the criteria of the literature systematic review, it was observed initially a thematic analysis that meets the objective of the study - the analysis of the ergonomics methodologies used in the built environment. This theme helped us in finding that the methodologies most used in the ergonomic intervention in EAC in Brazil, were AET, Meac and MPCCH.

It is worth considering, on the one hand, that Meac, despite being the most widely used in the period 2015–2018, its analysis shows that there is a decline of articles publication between the years 2017–2018. On the other hand, MPCCH emerges as a finding of research, because it is a recent instrument, even with moderate use, but showing signs of growth of articles publications using this methodology, from 2018.

These findings, however, do not manage restricting the sources evaluation process to the proposal presented here; what is intended is to set strict criteria of scientific and methodological validity and that the result is the reflection of a mapping work and careful and explicit selection of bibliographical sources. As a consequence, it is intended to expand the systematic literature review for other data bases in the analysis.

References

1. Oliveira, G.R., Mont' Alvão, C.: Metodologias utilizadas nos estudos de ergonomia do ambiente construído e uma proposta de modelagem para projetos de design de interiores. Estudos em Design, Rio de Janeiro 23(3), 150–165 (2015)
2. Mont' Alvão, C., Villarouco, V.: Um novo olhar para o projeto. 2AB, Teresópolis (2011)
3. Ramos, A., Faria, P.M., Faria, A.: Revisão sistemática de literatura: contributo para a inovação na investigação em ciências da educação. Revista Diálogo Educacional, Curitiba 14(41), 17–36 (2014)
4. Gouch, D., Thomas, J., Oliver, S.: Clarifying differences between review designs and methods. Syst. Rev. 1(1), 28 (2012)
5. Saur-Amaral, I.: Revisão sistemática da literatura. Bubok, Lisboa (2010)
6. Souza, F.N., de Souza, D.N., Costa, A.P.: Asking questions in the qualitative research context. Qual. Rep. 21(13), 6–18 (2016)

7. Godoi e Silva, K.A., Almeida, M.E.B.: Combined use of software that supports research and qualitative data analysis potential applications for researches in education. In: Costa, A., Reis, L., de Neri Souza, F., Moreira, A., Lamas, D. (eds.) Computer Supported Qualitative Research Studies in Systems, Decision and Control, vol. 71. Springer, Cham (2017)
8. Bardin, L.: Análise de conteúdo. Edições 70, Lisboa (2004)
9. Moraes, A., Mont' Alvão, C.: Ergonomia: conceitos e aplicações, 3rd edn. 2AB, Rio de Janeiro (2007)
10. Araújo, M.C., Campos, F., Villarouco, V.: Cenário da produção científica brasileira sobre ergonomia do ambiente construído (2005-2015). Blucher Design Proc. **2**(7), 359–367 (2016)
11. Ornstein, S.W., Bruna, G., Romério, M.: Ambiente construído e comportamento: a avaliação pós-ocupação e a qualidade ambiental. Studio Nobel/ FAU-USP/ Fupam, São Paulo (1995)
12. Dischinger, M.: Designing for all senses: accessible spaces for visually impaired citizens. Department on Space and Process, School of Architecture, Chalmers University of Technology. Götebor, Suécia (2000)
13. Iida, I.: Ergonomia: projeto e produção. Edgard Blücher, São Paulo (2005)
14. Villarouco Santos, V.: Ergonomia do ambiente construído. ERGODESIGN. In: 2nd Congresso Internacional de Ergonomia e Usabilidade de Interfaces Humano – Tecnologia: Produtos, Programas, Informação, Ambiente Construído, Rio de Janeiro (2002)
15. Ribeiro, L.G., Mont' Alvão, C.R.: Ergonomia no ambiente construído: um estudo de caso em aeroportos. Dissertação (Mestrado) - Departamento de Artes e Design, Pontifícia Universidade Católica do Rio de Janeiro, Rio de Janeiro (2004)
16. Reis, T.C.: Contribuição da ergonomia nos processos de concepção de espaços de trabalho. Dissertação (Mestrado) - Departamento de Artes e Design, Pontifícia Universidade Católica do Rio de Janeiro, Rio de Janeiro (2003)
17. Villarouco Santos, V.: Construindo uma metodologia de avaliação ergonômica do ambiente – AVEA. In: 14th Congresso Brasileiro de Ergonomia, Porto Seguro (2008)
18. Attaianese, E., Duca, G.: Human factors and ergonomic principles in building design for life and work activities: an applied methodology. Theor. Issues Ergon. Sci. **13**(2), 187–202 (2012)
19. Iida, I., Guimarães, L.B.M.: Ergonomia: projeto e produção. Edgard Blücher, São Paulo (2016)

Sustainable Development in Primary Education – A Perspective from Official Portuguese Guiding Documents

Patrícia Sá[1,2(✉)] [iD], Patrícia João[1,2,3] [iD], and Ana V. Rodrigues[1,2] [iD]

[1] Department of Education and Psychology, University of Aveiro,
Aveiro, Portugal
patriciasa@ua.pt
[2] Research Centre on Didactics and Technology in the Education of Trainers,
University of Aveiro, Aveiro, Portugal
[3] Geosciences Centre, University of Coimbra, Coimbra, Portugal

Abstract. This study aimed to identify and characterize the perspectives of the concept sustainable development present in official Portuguese primary education guiding documents. Methodologically, this is a qualitative study framed by the interpretive paradigm and implemented through an exploratory and interpretative-descriptive strategy. In order to fulfil the defined objective, a documentary analysis was carried out, enabling the constitution and analysis of the corpus through content analysis. The content analysis instrument used resulted from the adaptation and revalidation of a previous published instrument. The webQDA software was used to categorize and analyse the data. The results show that the environmental dimension is privileged over the economic and socio-cultural dimensions. However, the three dimensions of analysis considered are present in the vast majority of analysed documents.

Keywords: Sustainable Development · Education · Primary level · Curricular guidelines

1 Introduction

The nature of the challenges posed to education and research in education by sustainability issues is such that, in recent years, there has been an explicit recognition by the academic community of the need to rethink and reorient education in order to equip students with the competencies to cope with the rapid changes, increasing complexity and uncertainty that characterise today's world [1, 2].

Research carried out in recent decades on Sustainable Development (SD) and its impact on the educational context has highlighted the need to reflect on new paradigms for education and educational research [3, 4], as well as on implications for perspectives on teaching and lifelong learning.

Concern with SD issues has also taken on a central role at the level of educational policies and national and international guidelines for education. Recently, at a national level, this concern is reflected in several of the official reference documents for educational practice. Given the diversity of documents and their importance for the

© Springer Nature Switzerland AG 2020
A. P. Costa et al. (Eds.): WCQR 2019, AISC 1068, pp. 262–273, 2020.
https://doi.org/10.1007/978-3-030-31787-4_22

regulation of educational practice, the authors consider it essential to comprehend what understanding of the concept of SD they convey. Thus, the present study focuses on Primary Education (PE) and aims to identify and characterise the concept of SD conveyed by official reference documents for teaching and learning at this level of education.

2 Methodology

As stated, this study aims to identify and characterize the understanding of the concept of SD present in the official guiding documents for teaching and learning in PE in Portugal.

Methodologically, this is a qualitative study framed in an interpretative paradigm [5, 6]. We opted for a descriptive-interpretative strategy of an exploratory nature [5], assuming, from the outset, a concern with describing patterns and characteristics of a certain area of interest (SD) emerging from the *corpus* for analysis (official guiding documents for PE).

In view of the research objective defined, a documentary compilation following a set of operations were carried out in order to identify and collect the documents for constitution of the *corpus* and its subsequent processing, operationalised through a content analysis [7].

Table 1. Official guiding documents for PE in Portugal.

Document organization	Documents
Curricular programme and aims	
1. Environmental Studies	Programme for Environmental Studies (1990) [8]
2. Portuguese Language	Programme and Curricular Aims for Portuguese Language in Primary Education (2015) [9]
3. Mathematics	Programme and Curricular Aims for Mathematics in Primary Education (2013) [10]
Essential Learning	
1. Environmental Studies	Essential Learning. Environmental Studies. First year. PE [11] Essential Learning. Environmental Studies. Second year. PE [11] Essential Learning. Environmental Studies. Third year. PE [11] Essential Learning. Environmental Studies. Fourth year. PE [11]
2. Portuguese Language	Essential Learning. Portuguese Language. First year. PE [12] Essential Learning. Portuguese Language. Second year. PE [12] Essential Learning. Portuguese Language. Third year. PE [12] Essential Learning. Portuguese Language. Fourth year. PE [12]
3. Mathematics	Essential Learning. Mathematics. First year. PE [13] Essential Learning. Mathematics. Second year. PE [13] Essential Learning. Mathematics. Third year. PE [13] Essential Learning. Mathematics. Fourth year. PE [13]
4. Citizenship and Development	Citizenship and Development – Primary and Secondary Education [14]
Student Profile	Student Profile at Completion of Compulsory Education (2017) [15]
Environmental Education for Sustainability	Referential Environmental Education for Sustainability [16]

The *corpus* for analysis consists of the official guiding documents for teaching and learning in PE, which are currently available on the website of the General-Directorate for Education (www.dge.mec.pt), in a total of twenty documents (Table 1).

The content analysis of these documents was carried out based on the adaptation and revalidation of the tool used for analysis of school textbooks for Environmental Studies in [17]. Considering the instrument's suitability for the aim of this study, the researchers opted to adapt it. However, considering the evolution of the theoretical reference framework and the diversity of documents constituting the *corpus* for analysis, it was deemed necessary to make changes to the original version. Thus, for the purposes of review and adaptation of the content analysis instrument, and in order to ensure its internal validity, the following were taken into account: (i) the aim of the study (considering the type and diversity of documents); (ii) the evolution of the reference literature with regard to the concept of SD; and (iii) the documents that would be subject to analysis, both in terms of their structure and content.

The adaptation of the content analysis instrument required the use of a set of systematic, interpretative and inferential procedures applied to the content of each document under analysis [7]. This procedure involved: (i) a free-floating reading of all documents in the *corpus* in order to identify trends and patterns related to the concept of SD and how it is present in each document; and (ii) operations of selection, encoding and reorganisation on the *corpus* with a view to its subsequent categorisation [7]. These procedures of text selection and division allowed for the reorganization and, in some cases, creation of categories and subcategories of analysis, in a dynamic and interactive process of construction of a categorisation matrix. Thus, in its final version, the content analysis instrument presents a mix of categories, some determined *a priori* and others inferred from the documentary *corpus*.

In addition, and aware that the act of encoding is an operation on meaning performed by the encoder and that it is not independent of the perspectives, expectations and hypotheses of the researcher [5], the revised and adapted instrument was subjected to a validation process. This validation was carried out by a panel of four senior researchers, three experts in Education for Sustainable Development and one in Social Science and Humanities Research Methodologies. These researchers commented on the structure and organisation of the instrument, as well as its suitability for the objectives of the study and the documents to be analysed.

The final version of the content analysis instrument used is structured in three dimensions of analysis: Environmental Dimension, Economic Dimension and Sociocultural Dimension. These Dimensions are understood as organising thematic axes and emerged from the main dimensions of the concept of SD identified in the reference literature [17–19]; among others). The Dimensions were organized into different categories and subcategories, as presented systematically in Table 2.

In this study, categories are understood as items with specific meanings according to which the content of the *corpus* was classified by researchers [20]. Thus, these are superordinate concepts that allow the researchers to situate their comprehension of the object of analysis to make it pertinent in relation to their stated objectives (Robert and Bouillaguet, 1997 *apud* [5], p. 28). The subcategories, by making explicit the broader meaning of the categories, facilitated identification of the recorded units and, consequently, the category analysis.

Table 2. Dimensions, categories and sub-categories of the analysis instrument.

Environmental
Planetary Limits
Natural resources - exploitation and consumption
World human population
Measures for protection, conservation and enhancement of environmental heritage
Risks and consequences of the use of natural resources
Ecosystems
Interactions and interdependencies in ecosystems
Imbalances in ecosystems
Abiotic factors
Biodiversity
Threats to biodiversity
Measures to preserve biodiversity
Measures to preserve ecosystem equilibrium
Science, Technology, Society
Use of Science and Technology for the benefit of Society and Environment
Risks of the use of Science and Technology for Society and Environment
Social and ethical issues in STSE
Economic
Consumption
Consumption in modern society
Sustainable consumption practices
Economic growth and environmental and social impact
Economic exploitation of natural resources
Environmental pressure caused by economic growth
Economic activities
Sociocultural
Ethics, citizenship and education
Recognition of the main problems/challenges for humanity and involvement in their resolution
Civil rights and duties
The role of education in training citizens
Ethical values
Cultural diversity
Knowledge and appreciation of cultural diversity
Homogenisation and its social, environmental, cultural and economic implications
Respect for cultural diversity and its traditions
Human Rights
Generations of Human Rights
Basic Needs
Asymmetries and their implications
Democratic values
Intolerance and conflicts
Valuing a culture of peace

The category analysis was performed based on the matrix defined using the *webQDA* software (www.webqda.net). An inferential analysis was then performed, which consisted of a process of logical deduction and reduction of meaning of all the information organised. The inferential analysis implied a reconfiguration of the data that, by being grouped according to a set of affinities, enabled the emergence of central topics that, in turn, were subjected to an interpretative analysis.

The result of the content analysis was also validated by one of the four senior researchers who had collaborated in the validation of the content analysis instrument.

3   Presentation and Discussion of Results

In order to make reading the results easier, and before moving on to their presentation and systematization, the codes assigned to each document integrating the corpus of analysis are presented (Table 3).

Table 3.   Codification of the documents analysed

Code	Documents	Characterization
1	Curriculum organization and PE Programme - Environmental Studies	Curricular document
2	Portuguese Language Programme for PE	Curricular document
3	Mathematics Programme for PE	Curricular document
4	Curricular Aims for Portuguese Language in PE	Curricular document
5	Curricular Aims for Mathematics in PE	Curricular document
6	Essential Learning. Environmental Studies. First year. PE	Guiding document
7	Essential Learning. Environmental Studies. Second year. PE	Guiding document
8	Essential Learning. Environmental Studies. Third year. PE	Guiding document
9	Essential Learning. Environmental Studies. Fourth year. PE	Guiding document
10	Essential Learning. Portuguese Language. First year. PE	Guiding document
11	Essential Learning. Portuguese Language. Second year. PE	Guiding document
12	Essential Learning. Portuguese Language. Third year. PE	Guiding document
13	Essential Learning. Portuguese Language. Fourth year. PE	Guiding document
14	Essential Learning. Mathematics. First year. PE	Guiding document
15	Essential Learning. Mathematics. Second year. PE	Guiding document
16	Essential Learning. Mathematics. Third year. PE	Guiding document
17	Essential Learning. Mathematics. Fourth year. PE	Guiding document
18	Essential Learning - Citizenship and Development	Transversal document
19	Student Profile at Completion of Compulsory Education	Transversal document
20	Referential Environmental Education for Sustainability	Transversal document

Given the diversity of the documents analysed, a decision was made to systematize in Table 4 the vertical and horizontal analyses carried out. Thus, the table summarizes the number of references that appear, by dimension, category and subcategory, in each of the twenty documents that make up the *corpus*. This form of presentation allows one to perceive the occurrence of each category of analysis in each the documents analysed and, simultaneously, to characterize each of the documents individually for all the categories of analysis.

Based on the information presented in Table 4, some of the most relevant results will be highlighted.

3.1   References by Size, Category and Sub-Category of Analysis

Based on the results of the horizontal analysis of all the documents in the *corpus*, the most privileged dimension was found to be the "Environmental" dimension, with a total of 226 references identified. In this dimension, the most visible category is "Ecosystems", with 92 references.

Of the three dimensions that make up the categorisation matrix, the "Economic" dimension is the least represented and only 22 references have been identified in the total number of documents.

Focusing the presentation of the results of the analysis by category, the category with the greatest number of references in the documents under study is "Ethics, citizenship and education" (Sociocultural dimension), with 106 occurrences. Analysis of this category also shows the importance given in the documents to "Role of education in training citizens", which is the most represented subcategory, with a total of 63 references.

As for subcategories, the subcategories "Environmental pressure caused by economic growth" and "Homogenisation and its social, environmental, cultural and economic implications", belonging respectively to the categories "Economic growth and environmental and social impacts" and "Cultural diversity", were never referred in the documents analysed.

3.2   References in Transversal Documents

Documents were considered transversal when presenting guidelines for all of the subject areas and years of schooling in PE, namely: (i) Profile of Students completing compulsory schooling; (ii) Environmental Education for Sustainability Reference; and (iii) Essential Learning - Citizenship and Development.

The analysis carried out on these documents showed that **Referential of Environmental Education for Sustainability** is the document containing most references, in absolute terms, with a total of 132. These are divided into the three dimensions of analysis, as follows: 88 in the "Environmental" Dimension, 12 in the "Economic" and 32 in the "Sociocultural". These results point to an overvaluation of the "Environmental" dimension in relation to the others, supported, for example, in the 50 references identified in the category "Planetary Limits". Aspects such as preservation of natural resources and adoption of conscious consumption practices are explicit and can

Table 4. Number of references found in the analysed documents, by subcategory.

Documents analysed	1	2	3	4	5	6	7	8	9	10	11	12	13	14	15	16	17	18	19	20
PLANETARY LIMITS	21	0	0	0	0	2	1	1	8	0	0	0	0	0	0	0	0	1	0	50
Natural resources - exploitation and consumption	10	0	0	0	0	0	0	1	4	0	0	0	0	0	0	0	0	0	0	10
World human population	0	0	0	0	0	0	0	0	1	0	0	0	0	0	0	0	0	0	0	1
Measures for protection, conservation and enhancement of environmental heritage	4	0	0	0	0	2	1	0	1	0	0	0	0	0	0	0	0	0	0	24
Risks and consequences of the use of natural resources	7	0	0	0	0	0	0	0	2	0	0	0	0	0	0	0	0	0	0	15
ECOSYSTEMS	33	0	0	0	0	2	11	7	9	0	0	0	0	0	0	0	0	1	0	29
Interactions and interdependencies in ecosystems	1	0	0	0	0	0	0	1	0	0	0	0	0	0	0	0	0	0	0	6
Imbalances in ecosystems	3	0	0	0	0	0	0	1	2	0	0	0	0	0	0	0	0	0	0	4
Abiotic factors	8	0	0	0	0	2	3	1	3	0	0	0	0	0	0	0	0	1	0	5
Biodiversity	13	0	0	0	0	2	4	1	3	0	0	0	0	0	0	0	0	0	0	3
Threats to biodiversity	4	0	0	0	0	0	1	1	1	0	0	0	0	0	0	0	0	0	0	4
Measures to preserve biodiversity	2	0	0	0	0	0	2	1	0	0	0	0	0	0	0	0	0	0	0	4
Measures to preserve ecosystem equilibrium	2	0	0	0	0	0	1	1	0	0	0	0	0	0	0	0	0	0	0	4
SCIENCE, TECHNOLOGY AND SOCIETY	8	0	0	0	0	5	3	5	4	0	0	0	0	3	3	3	3	0	5	9
Use of Science and Technology for the benefit of Society and Environment	5	0	0	0	0	5	2	4	4	0	0	0	0	3	3	3	3	0	4	6
Risks of the use of Science and Technology for Society and Environment	3	0	0	0	0	0	1	0	0	0	0	0	0	0	0	0	0	0	1	2
Social and ethical issues in STSE	0	0	0	0	0	0	0	1	0	0	0	0	0	0	0	0	0	0	0	1
CONSUMPTION	4	0	0	0	0	0	0	0	2	0	0	0	0	0	0	0	0	1	0	12
Consumption in the modern society	2	0	0	0	0	0	1	0	1	0	0	0	0	0	0	0	0	0	0	3
Sustainable consumption practices	2	0	0	0	0	0	0	0	1	0	0	0	0	0	0	0	0	1	0	9
ECONOMIC GROWTH AND ENVIRONMENTAL AND SOCIAL IMPACTS	1	0	0	0	0	0	0	0	0	0	0	0	0	0	0	0	0	0	1	0

(Vertical sidebar letters alongside rows spell **ENVIRONMENTAL** and **ECONO...**)

(continued)

Table 4. (continued)

M Economic exploitation of natural resources	1	0	0	0	0	0	0	0	0	0	0	0	0	0	0	0	0	0	0	0	0
I Environmental pressure caused by economic growth	0	0	0	0	0	0	0	0	0	0	0	0	0	0	0	0	0	0	0	0	0
C																					
Economic activities	0	0	0	0	0	0	0	0	0	0	0	0	0	0	0	0	0	0	0	1	0
ETHICS, CITIZENSHIP AND EDUCATION	*6*	*2*	*0*	*0*	*3*	*4*	*2*	*2*	*2*	*0*	*1*	*2*	*0*	*2*	*4*	*4*	*3*	*7*	*33*	*27*	
S Recognition of the main issues/challenges for humanity and involvement in their resolution	3	0	0	0	2	2	1	0	0	0	0	0	0	0	0	0	0	0	7	11	
o Civil rights and duties	0	0	0	0	1	0	0	0	0	0	0	0	0	0	0	0	4	4	2	6	
C The role of education in training citizens	2	2	0	3	0	1	0	1	2	0	2	4	4	4	4	4	3	3	24	8	
I Ethical values	1	0	0	0	1	0	0	0	0	0	0	0	0	0	0	0	0	0	0	2	
O *CULTURAL DIVERSITY*	*3*	*0*	*6*	*0*	*8*	*2*	*2*	*0*	*5*	*1*	*2*	*1*	*0*	*0*	*0*	*2*	*0*	*0*	*1*	*5*	*0*
Knowledge and appreciation of cultural diversity	2	0	0	0	1	1	2	0	2	1	2	1	0	0	0	0	0	1	2	0	
C Homogenisation and its social, environmental, cultural and economic implications	0	0	0	0	0	0	0	0	2	1	0	0	0	0	0	0	0	0	0	0	
U																					
L Respect for cultural diversity and its traditions	1	0	1	0	1	0	0	0	3	0	0	0	0	0	0	0	0	3	3	0	
T *HUMAN RIGHTS*	*15*	*0*	*0*	*0*	*0*	*3*	*4*	*1*	*5*	*0*	*1*	*0*	*0*	*2*	*0*	*0*	*0*	*13*	*32*	*5*	
U Generations of Human Rights	0	0	0	0	0	0	2	1	1	0	0	0	0	0	0	0	0	3	2	1	
R Basic Needs	7	0	0	0	0	0	0	0	0	0	0	0	0	0	0	0	0	0	1	1	
A Asymmetries and their implications	0	0	0	0	0	0	0	0	1	0	0	0	0	0	0	0	0	0	1	1	
L Democratic values	4	0	0	0	2	0	0	0	0	0	1	0	0	2	0	0	0	5	13	1	
Intolerance and conflicts	1	0	0	0	0	1	0	2	1	0	0	0	0	0	0	0	0	0	1	0	
Valuing a culture of peace	3	0	0	0	1	1	1	0	1	0	0	0	0	0	0	0	0	5	14	1	

be identified in several of the categorised recorded units, e.g. *"Adopt behaviour aimed at preserving natural resources in the present for future generations"* (p. 17).

In the **"Student Profile"**, 76 references were found, most of which belonging to the "Sociocultural" dimension, in a total of 70. In this dimension, most references can be found in the category "Ethics, citizenship and education", totalling 33. Also in this category, the 24 references found for the subcategory "Role of education in the training of citizens" show how the school is valued *"... as an environment fostering learning and development of skills, where students acquire the multiple literacies they need to put in practice, which must reconfigure itself to respond to the demands of these times of unpredictability and accelerated changes [...]"* (p. 7).

In the document **"Essential learning - Citizenship and Development"**, 23 references were identified, with most of the occurrences in the "Sociocultural" dimension, a total of 21. In this dimension, the category "Human Rights" stands out, with the subcategories "Democratic Values" and "Valuing a culture of peace" containing 5 references each. It is evident in this document that the humanist formation of students is valued, with teachers assuming *"the mission of preparing students for life, to be democratic, participative and humanist citizens, in an era of growing social and cultural diversity, in order to promote tolerance and non-discrimination, as well as suppress violent radicalism"* (p. 2).

3.3   References in Documents by Disciplinary Area

For this analytical perspective were considered the documents Curricular Aims of Mathematics and Portuguese Language, Programmes of Mathematics, Portuguese Language and Environmental Studies and Essential Learning for Mathematics, Portuguese Language and Environmental Studies.

In the documents regarding **Curricular Aims**, the document pertaining to the subject area of Portuguese Language stands out, with 11 occurrences, all in the "Sociocultural" dimension. Of these, 8 belong to the category "Cultural Diversity", subcategory "Knowledge and appreciation of cultural diversity".

Regarding the **Programmes**, Environmental Studies stands out, with a total of 91 references in the 3 dimensions: "Environmental" with 62 references, "Economic" with 5 and "Sociocultural" with 24. In the subcategories, two stand out with the greatest number of references, both from the "Environmental" dimension: subcategory "Biodiversity" with 13 references (considering themes related to living organisms in general), e.g. *"Recognize manifestations of plant and animal life (observe plants and animals in different phases of their lives)"* (p. 115)); and subcategory "Natural resources - exploration and consumption" with 10 references, e.g. *"Survey the most relevant mineral products of the region. Recognize mineral exploration as a source of raw materials (construction, industry...)"* (p. 127) and *"Recognize fishing as a source of raw materials (canned food, fish meal...)"* (p. 128).

In the documents related to identification of **Essential Learning**, the subject area of Environmental Studies is prominent with 91 references, followed by Mathematics with 29 references and Portuguese Language with 10 references. In the Environmental Studies, most references occur in the "Environmental" dimension, 58, particularly in the category "Ecosystems", with 29. However, the subcategory "Use of Science and

Technology for Society and Environment ", which belongs to the category "Science, Technology and Society", also in the "Environmental" dimension, presented the greatest number of references, a total of 15. As an example of essential learning for the first year, *"Recognize that technology responds to everyday needs and problems (electricity grid, water supply, telecommunications, etc.)"* (p. 7), and for the fourth year *"Recognize the importance of technological evolution for the evolution of society, relating objects, equipment and technological solutions with different everyday needs and problems"* (p. 9), two of the recorded units categorized in this dimension.

4 Conclusions

Methodologically, the exploration of the interactions between the various dimensions of analysis and the diversity of documents that made up the *corpus* was enhanced by the use of the *webQDA* software. This fostered a richer dynamic of analysis of the multiple viable matrices. It was also possible to cross dimensions, categories and subcategories with types of documents, years of schooling and subject areas, which has reinforced the validity and coherence of the study.

The use of this software facilitated both vertical and horizontal analyses, which are fundamental for the characterisation of the global (in the various documents) and particular (in each one) perspectives on how the concept of SD emerges in the *corpus*.

Regarding the analysis itself and the dimensions of analysis considered, the "Environmental" dimension is the most valued in the different documents analysed, with 226 references (out of a total of 471 for all documents). However, the "Socio-cultural" dimension is very close to these numbers, with 223 references. Of the three, the "Economic" dimension is the least represented, with only 22 references found in total.

Concerning the documents in the *corpus*, the analysis has shown that the different dimensions of the concept of SD are covered, to some extent, in the vast majority of the documents analysed. Only the document on Curricular Aims for Mathematics does not contain any reference in the categorisation matrix defined.

The document containing most references is the Referential for Environmental Education for Sustainability, with a total of 132 occurrences. Of these, 88 belong in the "Environmental" dimension, 12 in the "Economic" dimension and 32 in the "Socio-cultural" dimension.

The documents related to Essential Learning in Environmental Studies for the four school years of PE show a total of 91 references in the different dimensions considered, of which 58 are in the "Environmental" dimension.

It is also worth highlighting the Programme for Environmental Studies, with 91 references among the three dimensions of analysis considered: 62 in the "Environmental" dimension, 5 in the "Economic" dimension and 24 in the "Sociocultural" dimension.

In summary, the analysis carried out shows that, in Portugal, there are official documents that frame and guide the implementation of Education for Sustainable Development (ESD) from the first years of schooling. It is possible to observe that, in recent years, an effort has been made to integrate the international guidelines for ESD, considering SD from a multidimensional and transversal perspective. It is considered fundamental, however, that the initial and continuous education of teachers be rethought and made operational in order to enhance this orientation of educational practice. The implementation of ESD will depend on teachers' understanding of this guidance and what it entails in the reorganisation of their practices, so teacher education remains critical to achieving the SD goals by 2030.

Acknowledgments. This work is funded by: (i) National Funds through FCT - Fundação para a Ciência e a Tecnologia, I.P., under the UID/CED/00194/2019 project; (ii) Fundação para a Ciência e Tecnologia with the SFRH/BD/132272/2017 PhD grant, through the European Social Fund and Human Capital Operational Programme.

References

1. Bourke, R., Loveridge, J.: Exploring wicked problems and challenging status quo thinking through educational research. NZ J. Educ. Stud. **52**(1), 1–5 (2017)
2. Lönngren, J., Ingerman, A.: Avoid, control, succumb, or balance: engineering students' approaches to a wicked sustainability problem. Res. Sci. Educ. **47**, 805–831 (2017)
3. Pipere, A.: Envisioning complexity: towards a new conceptualization of educational research for sustainability. Discourse Commun. Sustain. Educ. **7**(2), 68–91 (2016)
4. Salite, I., Drelinga, E., Ilisko, D., Olehnovica, E., Zarina, S.: Sustainability from the transdisciplinary perspective: na action research strategy for continuing education program development. J. Teach. Educ. Sustain. **18**(2), 135–152 (2016)
5. Amado, J.: Manual de Investigação Qualitativa em Educação, 2nd edn. Imprensa da Universidade de Coimbra, Coimbra (2014)
6. Coutinho, C.: Metodologia de investigação em ciências sociais e humanas: teoria e prática, 2nd edn. Edições Almedia S. A, Coimbra (2018)
7. Bardin, L.: Análise de Conteúdo. Edições 70, Lisboa (2014)
8. ME-DEB: Organização Curricular e Programas. Ensino Básico - 1º Ciclo. Direção Geral da Educação, Lisboa (1990)
9. MEC: Programa e Metas Curriculares de Português para o Ensino Básico. Direção Geral da Educação, Lisboa (2015)
10. MEC: Programa e Metas Curriculares de Matemática para o Ensino Básico. Direção Geral da Educação, Lisboa (2013)
11. ME-DGE: Aprendizagens Essenciais - 1.º Ciclo do Ensino Básico|Estudo do Meio. Direção Geral da Educação, Lisboa (2018)
12. ME-DGE: Aprendizagens Essenciais - 1.º, 2.º e 3.º Ciclos do Ensino Básico|Português. Direção Geral da Educação, Lisboa (2018)
13. ME-DGE: Aprendizagens Essenciais - 1.º, 2.º e 3.º Ciclos do Ensino Básico|Matemática. Direção Geral da Educação, Lisboa (2018)
14. ME-DGE: Aprendizagens Essenciais – Ensino Básico e Secundário|Cidadania e Desenvolvimento. Direção Geral da Educação, Lisboa (2018)
15. ME-DGE: Perfil dos Alunos à Saída da Escolaridade Obrigatória. Ministério da Educação e Ciência, Lisboa (2017)
16. Pedroso, J.V. (Coord.): Referencial de Educação Ambiental para a Sustentabilidade para a Educação Pré-Escolar, o Ensino Básico e o Ensino Secundário. Ministério da Educação e Ciência, Lisboa (2018)

17. Sá, P.: Educação para o Desenvolvimento Sustentável no 1.°CEB: Contributos da formação de professores (Unpublished Doctoral Thesis). Universidade de Aveiro, Aveiro (2008)
18. Amador, F., Oliveira, C.B.P.: Integrating sustainability into the university past, present and future. In: Caeiro, S., Leal Filho, W., Jabbour, C., Azeiteiro, U. (eds.) Sustainability Assessment Tools in Higher Education Institutions, pp. 65–78. Springer, Switzerland (2013)
19. Sá, P., Lopes, J.B., Martins, I.P.: Sustentabilidade e Intercompreensão: Perspetivas e contributos de um centro de investigação em educação. Revista Lusófona de Educação **43**, 91–107 (2019)
20. Vala. J.: A análise de conteúdo. In: Silva e, A.S., Pinto, J.M. (Orgs) Metodología das Ciencias Sociais, 6th edn. Edições Afrontamento, Porto (1986)

Characteristics of the Pedagogical Supervisor in Context of a Constructive and Reflective Supervision

Susana Oliveira e Sá[1]([✉]) [iD] and Paulo Alexandre de Castro[2]([✉]) [iD]

[1] Institute of Higher Studies de Fafe (IESF)
and Interdiscipliniry Research Centre for Education and Development (CeIED),
Universidade Lusófona de Humanidades e Tecnologias (ULHT),
Lisbon, Portugal
susana.sa@iesfafe.pt
[2] Federal University of Goiás/Regional Catalão, Goiânia, Brazil
padecastro@gmail.com

Abstract. In the context of pedagogical supervision, it is imperative to clarify the essential characteristics of the pedagogical supervisor. It is pertinent that the process of pedagogical supervision regulation is acknowledged by the supervised, in order to substantiate the equity that everyone involved deserves. It is a qualitative study using the content analysis technique with webQDA® software. The sample contains 57 supervised, at initial training, in non-higher education schools in the North and Centre of Portugal, who answered a questionnaire with 4 open answers and were interviewed in a *focus group*. The conclusions point to mediation, leadership, collaboration and reflexivity, from a dialogic and innovative perspective, contributing to the professional development of peers. The regulation characteristic is not consensual among the supervised and, thus, there is no guarantee that the pedagogical supervision is taking place in a collaborative and formative way, as advocated in the regulations.

Keywords: Pedagogical Supervisor · Pedagogical Supervision · Qualitative research · webQDA®

1 Introduction

Portuguese public establishments of non-higher education that have a public education, specialized artistic education, private, cooperative, association contract or sponsorship (of these, all those whose income comes mainly from public funds), will be subjected to a third cycle of External Evaluation. The IGEC (General Inspection of Education and Science) document, which regulates such evaluation [1], defines, in its principles, among others, the "Promotion of the pedagogical practice supervision, namely in the classroom and in the school activities" [1, p. 2].

© Springer Nature Switzerland AG 2020
A. P. Costa et al. (Eds.): WCQR 2019, AISC 1068, pp. 274–287, 2020.
https://doi.org/10.1007/978-3-030-31787-4_23

However, this principle presupposes another, with moral and political connotation, that of equity. Equity, in this sense, is the same consideration for others or an action principle.

In this context, several questions concern us:

Is the importance/relevance of pedagogical supervision the same among all the IGEC assessors? Will all of these assessors adopt the same supervisor role in all schools?

How to assess an evaluation, which is meant to "(a) contribute to a better public understanding of the quality of the work in schools; (b) produce information to support decision making, in the context of the development of educational policies; (c) assess the effectiveness of the self-assessment practices of schools; (d) promote the quality of teaching, learning and the inclusion of all children and all students" [1, p. 2], if there may be, to begin with, no equality in the performance of the external evaluation carried out at schools, due to, not documents to fill, but different perspectives of pedagogical supervision (models, roles, functions, etc.), or the concept of pedagogical supervisor, among the teams of the IGEC itself, which may lead to a different look and opposite results in identical realities and contexts.

Once a literature review was carried out on the subject – pedagogical supervision – several sub-themes arise, such as: pedagogical supervision models, concept of pedagogical supervisor, the function of the pedagogical supervisor, the role of the pedagogical supervisor, pedagogical supervision scenarios, but almost nothing specific regarding the characteristics of the pedagogical supervisor.

1.1 Concept of Pedagogical Supervision

There are perspectives of different authors for the concept of supervision, resulting, therefore, in several models of Pedagogical Supervision, so one can say that the concept is clearly complex, but there is a convergence of the conceptualization in what concerns "interacting, informing, questioning, suggesting, encouraging, evaluating" [2, p. 33]. The professional dedicated to school supervision is available at the service of the community, as well as the students belonging to it, and the supervision is considered an assistance to the teaching activities so that the school is more efficient and reaches its purposes [3]. For Fuenlabrada and Weiss [4], pedagogical supervision is, for the school, a direct reference of technical, administrative and labour authority, and the same figure manages to interconnect the educational policies, as well as administrate the educational service and serve as pedagogical support and control of the teaching staff.

Some studies, carried out recently [5], indicate that collaborative cultures, among teachers, share characteristics of spontaneous, voluntary and development-oriented work. The construction of these cultures is guided by the operationalization and implementation of projects. Its efficiency is mainly observed when there is intrinsic motivation of the teachers to carry out projects, but also when the origin, being extrinsic, is adopted as significant.

1.2   Concept of Pedagogical Supervisor in a Constructive Way

According to Alarcão and Tavares [6], the concept of pedagogical supervisor is closely linked to the development and learning of the supervisor, the trainees (future supervisors) and the students (pupils) in the school context. According to Alarcão and Tavares [6, p. 47], this figure corresponds to:

> The act of supervising or guiding teaching, education, learning and development of the trainee or teacher-trainee is fundamentally part of the same structure underlying any teaching/learning process in which the development, education, teaching and learning emerge as inseparable elements.

If the focus is on the supervisor's activity as a teacher, then it results in the ability to help learn or unlearn to undertake in a different way [6]. If one associates the supervisor with the orientation of the pedagogical practice, the teaching-training process will be present, which echoes on the development of the trainees.

Initially, the supervisor figure was perceived by the teachers as "a process of hierarchy, impersonal and of inspection" that aims to control the education system, the curricula, the content, the teaching-assessment process, trying to reproduce more accurately what the system requires [7, p. 427].

The role played by the supervisor aims at the leadership of teachers and other people involved in the educational processes, with the purpose of producing an improvement in teaching and learning. The supervisor is considered as the figure who is at the service of all educational activities, assisting in the "planning, coordination and execution" [3, p. 28], considering the needs and aspirations of the pupils and the communities of the educational area, contributing to achieve several goals by "developing new perspectives, ideas, opinions, attitudes". The pedagogical supervisor carries out continuous work in the training of teachers, presenting effective alternatives adapted to the reality that surrounds them. In this way, the supervisor reinforces the self-monitoring in each teacher, activating the teacher's continuous productivity, since "the presence of the supervisor in the school imposes the re-dimensioning of the teachers' internal conduct" [8, p. 260].

According to Silva [9], this professional plays an important role in the transformation of the school, with strategic measures, and with a specific look that directs the focus to what is proposed and to the fact that decisions are made in the context experienced in the school's daily routine. In this way, Ferreira [10, p. 237] argues that:

> A new content is, therefore, imposed today for educational supervision, new relationships are established and new commitments challenge the education professionals [...] to another practice not only geared towards the quality of pedagogical work and its rigorous forms of realization, but also, and above all, a commitment to the construction of a new knowledge – emancipation knowledge - with public policies and the education administration in general.

This figure faces certain obstacles that concern the hierarchies to be respected and followed in the school system, in this highly diversified environment, it is necessary for the professional to develop strategies for valuing this diversity. It is up to supervisors to think about their practices, leading to significant transformations within the constructive school context [9]. The pedagogical supervisor establishes the dialectic of formation-development, providing the development of the members that compose the system, both personally and professionally.

Thus, it is imperative that the teacher who works with this method to prepare himself internally, knowing and analysing himself constantly, with the intention of extinguishing the ingrained difficulties, and by knowing them, controlling their non-exteriorization, thereby generating reflection [11, pp. 165–166].

The authors Páez and Bermúdez-Jaimes [12] have identified the competences that clearly describe the profile of a Montessory teacher and are composed of the following criteria: assertive communication, personal presentation, interpersonal relationships, commitment, mastery of the topics to be taught, awareness of the student's sensitive periods, thematic sequence, be a guide for students, appropriate handling of prepared environments and proactivity.

In a context of continuous training, the supervisor must have a collaborative relationship with the supervised, a relationship of mutual help, which creates possibilities for questioning, analysis, reflection so that collaboratively, critically, inquiringly, and consequently with a research spirit, where solutions are found, there is innovation, transformation and change. Supervisor and supervised are placed in a process of shared knowledge construction that, in order to be successful, must be in a healthy relationship of empathy [13].

1.3 The Concept of Reflection in Peer Supervision

According to Daresh [14], in the conceptualization of the reflection process, in order to achieve a clear, orderly, coherent and harmonious situation, one must modify a complex situation [15]. Still according to Daresh [14], this process of change consists, precisely, in a reflective process. This way, the act of thinking is based primarily on a problem that occurs, to which a solution is first imposed, and secondly, a search for the material needed to put it into practice. In this sense, three perspectives necessary for thinking are taken into account: open-mindedness, responsibility and dedication [15].

The reflective paradigm highlights the importance of teacher independence in the process of peer supervision. [16]. This arises the importance of teacher training and the development of the teachers' professional capacities in school, as a practicing space, not only in the classroom, but also in the good relations between the school and the community, which are conceived as fundamental for a critical reflection. Reflection is a process by which educational professionals develop a critical awareness of the social, cultural and ideological limits of education systems, which happens from reflection in practice and from practice itself [17–19].

Alarcão and Canha [20] establish that specific characteristics such as intelligence criteria, autonomy and responsibility are implicit in the reflective school. This is a school that plays an active role, promoting effective responses in the divergent and ambiguous situations of everyday life, and thus allows the perception of the different circumstances that result in different situations, in order to guarantee the progression process.

2 Methodology

The methodology is qualitative in nature using the technique of content analysis that has an "inferential function, in search of a meaning that is beyond what is immediately apprehensible and awaits the opportunity to be uncovered" [21, p. 303].

This inferential process obeyed rules in order to be reliable and allow the validation of the whole analysis process [22].

Thus, in this study one aimed to characterize a supervisor in pedagogical supervision context. The research question is which representation of the figure of the pedagogical supervisor is made by the supervised.

In order to answer this question, the following specific objectives were set out to be perceived by the supervised: (a) Identify the functioning of the pedagogical supervisor function in school context; (b) Know the meaning given to the figure of the pedagogical supervisor regarding their personal and professional competences; (c) Identify the pedagogical supervision practices in articulation with theoretical foundations and professional experience.

A questionnaire survey was used with 4 open questions, about the characteristics of the pedagogical supervisor, for the 57 supervised respondents.

2.1 Study Sample

The study was conducted in January 2019. The sample is composed of 57 supervised teachers (who voluntarily participated in the research), at the beginning of their careers, designated Training (or pedagogical practice) in pedagogical context (lasting one academic year), coming from two public higher education institutions in the North and Centre of the country. The sample contains 39 female teachers, 29 of whom are from the North of the country, and 18 male teachers, 7 of whom from the Centre of the country. Of the total number of supervised teachers from the North of the country, 35 are from the area of Exact Sciences and 5 from Humanities. From the Centre of the country, 11 are from Exact Sciences and 6 from Humanities area. The supervised ages range from 24 to 26 years old. The teachers were selected because the researcher and author of this work knew them personally.

The 57 early-career teachers were under pedagogical supervision with 14 supervisors (see Table 1), in groups of four people and a group of five in the North of the country.

Table 1. Characterization of the pedagogical supervisors

Pedagogical supervisors	North of the country	Centre of the country
Female	0	0
Male	10	4
Exact sciences area	9	3
Humanities area	1	1
Service time	25–32 years	28–35 years

(continued)

Table 1. (*continued*)

Pedagogical supervisors	North of the country	Centre of the country
Professional training	Master in education	Master in education
Academic degree	Masters	Masters
Ages	48–55	51–58
Project coordinators	3 coordinators	0
Publications	1 who publishes	0
Professional situation in the school	School board	School board

The pedagogical supervisors form a more or less uniform sample in terms of the macro-social aspect: (a) they are male; (b) they have a stable professional career, as they belong to the school board or the school grouping; (c) they have the same academic and professional qualifications; (d) they belong to the same age group; (e) they have the same service time, which allows them to be placed at the same level of career progression.

The sample only differs at the micro-social level, that is: there are three supervisors who are coordinators of National Projects and one inserted in a research centre, recognized by the Foundation for Science and Technology (FCT).

2.2   Data Collection

The data collection procedure was done, in writing, by invitation for research addressed to the Director of the School Institution or School Grouping, to which the supervisors belonged. In total: 4 Schools and 10 School Groupings. The Schools and 7 School Groupings were located in the North of the country and 3 School Groupings in the Centre.

Each supervised teacher was asked, at the beginning of the month, to fill out a questionnaire survey with four open questions about the perception of the most relevant characteristics of their supervisor, that favoured his/her practice in context of pedagogical supervision.

The questionnaire was anonymous, enclosed in an envelope (duly sealed and addressed), submitted for completion and sent to the researcher and author of the article.

For ethical purposes, anonymity was maintained: (1) of the supervised; (2) of the supervisor (it was expressly requested in the survey not to write the name of the Supervisor) and (3) of the School or School grouping where the Pedagogical Supervision was taking place.

At the end of the month, focus interviews were conducted on the 14 independently supervised groups.

The semi-structured interview followed the fluency of the conversation, in order to question the functioning of the training group in a school context, taking into account the relationships with the supervisor.

These interviews were carried out by the researching author, audio-recorded with the express consent of the participants and carried out in each school or school grouping, in the room dedicated to the training group. Each interview lasted on average 1 h.

In order to ensure anonymity, the questionnaire surveys were identified by the abbreviations Q1 to Q57, and the focus interviews were identified by the abbreviations E1 to E14.

2.3   Categories and Indicators

Taking the objectives of the study into account, one sought to provide a detailed and rigorous description, in order to guarantee the validity or credibility of the qualitative study [21]. Some authors [23, p. 9] refer to "the need to establish some strategies. Among them, one highlights the triangulation of the various collected sources, that is, one looks at the same phenomenon from different angles"; one also chose to focus on triangulation of data – a modality that proves whether the information collected is confirmed by another (theoretical), and one turned to the transparency of the whole process that guarantees the reader the merit, credibility and reliability of the research [24, p. 151].

Table 2. Dimension, categories and subcategories of the Characteristics of the Pedagogical Supervisor in school context

Dimension	Categories	Subcategories	Definitions
Pedagogical supervisor	Mediator	Learning	Promotes the learning of the supervised
		Reflection	Promotes the reflection of the supervised
		Context	Considers the supervised as the participant of the process
		Mentor	Is the mentor of the well-being of the supervised
	Leader	Peers	Works with the peers of the supervisor
		Teach	Teaches the supervised
		Regulation	Promotes the regulation of the supervised
		Learn	Teaches the supervised to learn
		Help	Helps individual and collective progression of the supervised
	Collaborative	Interaction	Interacts with the supervised in school context
		Share	Shares the different knowledge of the supervised for the collective and individual progression
		Knowledge	Promotes the exchange of knowledge within the group and with the supervised
		Opinion	Listens to the opinion of the supervised
		Exchange	Provides the sharing of their knowledge with the supervised
		Practice	Shares their pedagogical and scientific practice with the supervised
	Reflective	Reflexivity	Works in an environment of reflexivity with the supervised
		Observation	Observes the supervised constructively
	Regulator	Guide	Is the guide of the knowledge and the rules of the supervised in school context and in training
		Training	Considers that training is a process

One did not have pre-defined categories and decided to opt for emerging categories (empirical, inferential categories). For this process, one questioned the data corpus inserted in the internal sources of the webQDA® [25], which is a content analysis software in qualitative research. The 200 most frequent words were questioned, conditioned to a minimum of 3 letters. The most mentioned words were: Peers (112), Supervisor (64), Pedagogical (48), Supervised (15), Training (15), Context (12), Reflection (12), Learning (10), Teaching (10), Reflexivity (8), Observation (8), Knowledge (7), Interaction (6), Mentor (5), Guide (5), Regulation (3), Learn (3), Practice (3), Teach (3), Help (3), Share (3) and Exchange (3).

In this study, one presents the dimension, the categories and indicators of the characteristics of the pedagogical supervisor in school context, represented in Table 2.

3 Data Analysis

In the present research, one proceeded to the free-floating reading of the 57 questionnaires and 14 focus interviews. When the content analysis was carried out [26], the ideas were cut out into reference units, words or phrases, texts contained in the information material produced, which corresponded to clear, objective and meaningful ideas in the context of the research. Subsequently, after deep reading, the reference units were grouped into indicators, which later allowed us to clarify the definition of each of the categories. As frequency unit, one took the reference unit, which was counted as many times as present in the discourse. The data analysis and treatment was carried out with the support of content analysis software in qualitative research, webQDA®, through open procedures, corresponding to a permanent process of progressive creation, in which the reflection and analysis of data are rigorously and constantly triangulated, which makes the methodological process reliable [22].

One proceeded, in an early stage, to the analysis of the results of the supervised teachers from each educational institution. Afterwards, in order to validate categorization, to guarantee a greater reliability of the study, one requested the opinion of 2 specialized researchers, through online collaborative work, supported by the software webQDA®. These validated the emerging categories.

Subsequently, one proceeded to a comparative analysis between the results of the questionnaire responses of the supervised from the different institutions and the focus interviews.

3.1 Analysis Matrix

One was aware of the characteristics of each community and the absence of theoretical referential. Therefore, it was important to create a homogeneous analysis matrix, which contained three objectives: (a) not lose sight of the research questions; (b) allow a triangulation and comparison between the various data corpora; (c) allow a comparison between contexts. The matrix created facilitated these objectives and is presented in Table 3:

Table 3. Internal coherence of the research for the dimension "Characteristics of the Supervisor in Pedagogical Context"

Research question	Research objectives perceived by the supervised	Analysis *Corpus*	Analysis type	Observations and expectations
Which representation of the figure of the pedagogical supervisor is made by the supervised?	– Identify the functioning of the pedagogical supervisor function in school context; – Know the meaning given to the figure of the pedagogical supervisor regarding their personal and professional competences; – Identify the pedagogical supervision practices in articulation with theoretical foundations and professional experience	Questionnaire surveys Interview *focus group*	Content analysis	Perspective of non-competition, acquiring that of collaboration Innovations in the professional development of the supervised Supervisor's role allowing knowledge sharing, teaching to learn and contextualizing the supervised at school, promoting regulation and reflection

One began with the "free-floating reading" [26], in order to establish an initial contact with the documents. This was followed by a further reading due to the wealth and extension of the *analysed* corpus. After this stage, the larger categories began to emerge inductively, in accordance with the pre-established objectives and the results which ensued from the reading of the questionnaires and interviews. Since these were replicated in the 10 institutions, one found 1 dimension: Pedagogical Supervisor. The remaining were introduced as categories.

4 Analysis of Results

Once the previous steps were completed, we started to question the data to answer the research question, obtaining the respective matrix of questioning facilitated by the software webQDA®. From the questioning of the data collected, one verified that an important characteristic of the Pedagogical Supervisor is being a mediator: (a) of the learning of the supervised; (b) of the refection of the supervised [12]; (c) of the intervention on the conflict context. This mediation was performed turning to the well-being (27 reference units), the managing of the exchange of knowledge in different stages within the group (10 reference units), the constructive observation of classes "(21 reference units) and the listening to the opinion of the group (19 reference units). Here are the examples of comments: "What was important to me was that whenever there was a problem in the group work, the supervisor would intervene straight away, to immediately talk to all the colleagues" (Q1), "Collaborative work was essential in order to clarify the logic of some of the planning, promoting reflection, but with no arguments" (E3). "Class observation was performed naturally, without stress, because we knew that the goal was to learn and not to evaluate" (Q51).

Being a mediator, that is, being the mentor of the well-being, is being the one who avoids conflict and enables the supervised to a healthy environment where discussion is healthy and well accepted. This characteristic is not referred in any study, but nowadays it arises, due to the state of tension in which each school lives and the training contexts, which is also translated into a classification, due to the importance of the final classification, as it might launch them to higher rankings (ahead of their colleagues) in the public recruitment processes. If one directs this situation to the teacher evaluation, the Pedagogical Supervisor must bear in mind that training in school mediation will be necessary, to prevent conflicts to achieve harmony and teacher interaction.

From the questioning of the data collected, one verified that an important characteristic of the Pedagogical Supervisor is being a leader, promoting: (a) peer work (13 reference units); (b) the teaching of the supervised (10 reference units); (c) regulation (2 reference units); (d) teaching to learn (19 reference units). This leadership is performed turning to the training of the supervised as a process. Considering the supervised as a participant in the process and the supervisor a peer in the context of supervision. Here are examples of comments: "I have always felt at ease within the group to ask any kind of questions, pedagogical or scientific" (E5), "My supervisor always listened to me and helped me with my scientific doubts" (Q32); "I have learned that in supervision having doubts is not a bad thing, what is bad is stay in doubt or not having doubts at all" (Q44).

Being a leader, that is, working with peers. It is not being a leader in the narrow interpretation of an immediate superior, it is a peer among peers. There are few studies that point this way. One, for example, from Sá [16], highlights that this work doesn't mean in pairs (of two), but in groups of people whose theme work is the same. Thus, if the supervised have to plan a unit, the supervisor should lead them to plan it with their peers (teaching colleagues) from the same school group, who are teaching that same unit to the same school year.

From the questioning of the data collected, one verified that an important characteristic of the Pedagogical Supervisor is being collaborative, as most studies advocate [6, 12, 13, 17, 18], that is, it helps to share the different knowledge or stages of knowledge in the group for individual and collective progression, promoting: (a) interaction with the group and the school context (12 reference units); (b) the sharing of different knowledge at different levels among the members of the group, so that their collective or individual progression is possible and at the own rhythm (28 reference units). The collaborative work is an advantage for a productive and fast work, and is evidenced in the exchange of knowledge in the group (9 reference units); in listening to the opinion of the supervision group (12 reference units) and sharing and practising the experience of its pedagogical and scientific practice (28 reference units). Here are some examples of comments: "The supervisor always gave his personal example, both scientific and pedagogical for us to understand" (Q57), "My supervisor deconstructed his scientific reasoning to teach me and clear my doubts" (Q32); "In group, my colleagues and I always counted on our supervisor for everything. Falling into a school is horrible and he always integrated us, never neglecting the planning work" (E14).

The work of the collaborative supervisor most evidenced in the studies [5–7, 12, 20], stands out in the sharing of his/her pedagogical and scientific practice and in the sharing of different knowledge and stages of knowledge, existing in the supervision group, for the collective and individual progression. Here are some examples of comments: "If I didn't understand, my supervisor would give other examples and tell the group to do others things, giving me his full attention until I understood" (Q1), "Sometimes, I lacked a click and he explained with an experienced example and it was enough" (E7); "Sometimes, when it was complicated for us to understand, he would ask one to play the teacher and the rest to play the students… and then it always… do you know what I mean?" (Q17).

From the questioning of the data collected, one verified that an important characteristic of the Pedagogical Supervisor is being reflective (in promoting the reflection of the supervised), which is in accordance with what some studies refer [2–4, 6, 11, 13, 17, 18], promoting: (a) reflexivity (17 reference units); (b) constructive observation (11 reference units). The studies do not refer to constructive observation, they only refer to grids, without explaining how they are approached with the supervised. Reflective work is evidenced in thinking before acting and the reason for the action. Here are some examples of comments: "If I opted for a strategy, the supervisor would always ask why I had chosen that strategy and not the one suggested by the manual" (Q13), "It was interesting to see that there was a great distance between theory and practice, and when we put it into action that distance was gigantic" (E7).

From the questioning of the data collected, one verified that an important characteristic of the Pedagogical Supervisor is being a regulator, although no study refers to this characteristic. Being a regulator means establishing and demonstrating the regulatory documents of the Training, that is, a controller within the regulations [8]. Regulation is a methodology adjusted within the norms accepted by the regulations of either the School or the School Grouping, as well as of the several Higher Education Institutions from which the supervised originate, as well as of the MEC (Ministry of Education and Science), promoting: (a) a guide of knowledge and rules of the supervised in training (without references); (b) supervision as a process (without references).

Although regulatory work may seem a disadvantage, it makes supervision more equitable if the rules are common to all 57 supervised. This view is not consensual in all supervision groups, since not all see the supervisor as a guide of knowledge and rules and promoter of regulation, visible in the process of supervision and in peer work. Here are examples of comments: "I have never understood the school rules" (E9), "My supervisor never explained, much less showed the standards of university supervision" (E11); "I do not know the rules of the school and supervision from my supervisor, I know them because I researched them" (E11).

5   Final Considerations

In order to answer the question of the study, the supervised identified five characteristics in the supervisor: mediator, leader, collaborative, reflective and regulator.

The characteristic of mediating supervisor was identified in the role of pedagogical supervisor in school context. This was valued by the supervised in the sense of empowering them with tools to avoid conflict.

The meaning attributed to the figure of pedagogical supervisor was understood, regarding their personal competences, as well as the articulation between the practice of pedagogical supervision and professional experience, explained in the characteristics of being collaborative and reflective, already explained in the literature [6, 12, 13]. The studies of Alarcão and Tavares [6] corroborate it, in the sense that the change of the paradigm of teacher training is based, therefore, on the transmutation of "being subjected" to "being the subject" of this training, becoming a "privileged actor of changes", in a school setting of cooperative and reflective formation.

In order to know the meaning attributed to the figure of the pedagogical supervisor regarding their personal and professional competences, that characteristic of leader emerged, which contradicts Montessori [11], in the sense that it is only necessary to enable teachers (supervisors) for observation and experimentation.

In short, the supervised consider that they learn by reflecting on their actions, since knowledge is also acquired through practice, in collaborative work among peers and through their professional experiences, and that the role of the pedagogical supervisor must take into account a conduct of egalitarian dynamic interaction and without overrating hierarchies. The pedagogical supervisor acts differently, that is, the supervisor must give a meaning to their integrated action within a certain context.

In this study, there was no coherence of opinions in the regulator supervisor characteristic. It should be noted that the sample only contained 57 supervised.

This situation brings us to the concern expressed initially: the fruitful and urgent need for equity.

If the regulations are not known, how will all supervised be assured that they are having equal pedagogical supervision? Or in a collaborative and formative logic, as the regulations generally advocate, which the supervised affirm, defend and intend.

In attempting to answer the first question, the supervision carried out by [1] is questionable, which may lead to an absence of equity.

Thus, what will be in question will be an innovative dialogic [18] of regulation and of mediation contributing to the professional development of the peers. These new ideas are added to their self-critical position, enabling improvement, equity and transparency of the methods currently implemented in supervision.

References

1. Inspeção-Geral de Educação e Ciência (IGEC): Terceiro Ciclo de Avaliação Externa das Escolas (2019)
2. Vieira, F.: Supervisão – uma prática reflexiva de formação de professores. Edições Asa, Rio Tinto (1993)
3. Nérici, I.: Introducción a la Orientación Escolar. Kapelusz, Buenos Aires (1990)
4. Fuenlabrada, I., Weiss, E.: Prácticas escolares y docentes en las escuelas primarias multigrado. Consejo Nacional de Fomento Educativo, México (2006)
5. Formosinho, J., Machado, J., Mesquita, E.: Formação, Trabalho e Aprendizagem –Tradição e Inovação nas práticas docente. Edições Sílabo, Lisboa (2015)
6. Alarcão, I., Tavares, J.: Supervisão da prática pedagógica. Uma perspectiva de desenvolvimento e aprendizagem. Edições Almedina, Coimbra (2010)
7. Smyth, J.: Teacher as collaborators in clinical supervision: cooperative learning about teaching. Teach. Educ. 24, 60–68 (1984)
8. Leal, A.B., Henning, P.C.: Do exame da supervisão ao autoexame dos professores: estratégias de regulação do trabalho docente na supervisão escolar. Currículo sem Front. 9, 251–266 (2009)
9. Silva, A.M.C.e: A supervisão escolar e as intervenções do supervisor no processo de ensino e aprendizagem (2013)
10. Ferreira, N.S.C.: Supervisão educacional para um escola de qualidade: da formação à ação. In: Supervisão educacional para um escola de qualidade: da formação à ação. Cortez, São Paulo, pp. 235–254 (2010)
11. Montessori, M.: A Descoberta da Criança. Pedagogia Científica. Kírion, Campinas, SP (2017)
12. Bermúdez-Jaimes, M.E., Mendoza-Páez, A.M.: Teacher evaluation in Montessori Education: a proposed tool. Educ. y Educ. 11, 227–252 (2008)
13. Rudduck, J., Flutter, J.: How pupils want to learn. In: Pollard, A. (ed.) Readings for Reflective Teaching in Schools, pp. 73–74. Bloomsbury, London (2015)
14. Daresh, J.C.: Leading and Supervising Instruction. Corwin Press, California (2006)
15. Alarcão, I.: Formação reflexiva de professores. Estratégias de supervisão. Editora Porto, Porto (1996)
16. Sá, S.: A interação entre pares: que lugar na avaliação do desempenho docente? Rev. Lusófona Educ. 37, 27–43 (2017)
17. Zeichner, K.M.: A formação reflexiva de professores: ideias e práticas (1993)
18. Vieira, F.: A experiência educativa como espaço de (trans) formação profissional. Lingvarvmarena 2, 9–25 (2011)
19. Moreira, M.A.: A avaliação do (des)empenho docente: perspectivas de supervisão pedagógica. In: Pedagogia para a Autonomia – Reconstruir a esperança na educação. In: Atas do 4.º encontro do GT-PA. Universidade do Minho. Centro de Investigação em Educação (CIEd), Braga, Portugal, pp. 241–258 (2010)
20. Alarcão, I., Canha, B.: Supervisão e colaboração: uma relação para o desenvolvimento. Porto Editora, Porto (2013)

21. Amado, J., Costa, A.P., Crusoé, N.: A técnica de análise de conteúdo. In: Manual de Investigação qualitativa em educação. Coimbra Universty Press, Coimbra, pp. 301–355 (2017)
22. Costa, A.P., Amado, J.: Análise de Conteúdo suportada por software. Ludomedia, Oliveira de Azeméis – Aveiro (2018)
23. Sá, S.O., Costa, A.P.: Critérios de Qualidade de um Estudo Qualitativo (Carta Editorial). Rev. Eixo. 5, 9–12 (2016)
24. de Souza, D.N., Costa, A.P., de Souza, F.N.: Desafio e inovação do estudo de caso com apoio das tecnologias. In: Investigação Qualitativa: Inovação, Dilemas e Desafios. Ludomedia, Oliveira de Azeméis, Aveiro, pp. 143–162 (2015)
25. Costa, A.P., Moreira, A., Souza, F.N.: webQDA – Qualitative Data Analysis (2019). https://www.webqda.net/
26. Bardin, L.: Análise de conteúdo. Edições 70, São Paulo (2013)

Interprofessional Education and Collaborative Work in Health: Implementing is Needed!

Maria Vieira[1] (ID), Rafaela Rosário[2,3,4] (ID), Ana Paula Macedo[2,3(✉)] (ID), and Graça Carvalho[4] (ID)

[1] School of Medicine of the Federal University of Alagoas,
Maceió 57072-900, Brazil
vieiramlf@uol.com.br
[2] School of Nursing – ESE, University of Minho, 4710-057 Braga, Portugal
amacedo@ese.uminho.pt
[3] Health Sciences Research Unit: Nursing,
Nursing School of Coimbra, 4710-057 Braga, Portugal
[4] Research Center on Child Studies, Institute of Education,
University of Minho, 4710-057 Braga, Portugal

Abstract. Proactive actions to implement good collaborative practices that result in improved health care is critical since the health professionals' training. This study aims to understand the process of education and interprofessional collaborative work in training courses in the health area of one Portuguese university. Professional trainers of Nursing, Psychology, Sociology and Medicine courses participated in Focus Groups. Content analysis of the transcripts was carried out with the use of the NVivo software version 12.0. From the data obtained five categories emerged: conceptual clarity; barriers/challenges; potentialities; strategies; and suggestions. Also, subcategories were defined, such as lack of communication, a students' profile, and difficulty in operationalizing InterProfessional Education (IPE) and collaborative Work (CW). Weaknesses and threats surpassed strengths and potentialities. It is hoped that this study can contribute to a reflection that helps strengthens the curricular changes necessary to make work education viable.

Keywords: Interprofessional Education · Interprofessional collaboration · Collaborative practice in health · Professional identity · Teamwork

1 Introduction

The InterProfessional Education (IPE) in health is studied and encouraged since the decade of sixty, but only in 2010, the World Health Organization published a study with conceptual pillars and recommendation for its implementation and improving people's health care. This document called "Framework for Action on Interprofessional Education & Collaborative Practice" [1] considers IPE "occurs when students from two or more professions learn about, from and with each other, to enable effective collaboration and improve health outcomes".[1, p. 10] This concept reinforces pioneers such as Baar et al. [2]; McNair et al. [3] and is corroborated by others, including Baar

Baar and Low [4], Frenk et al. [5]; Reeves et al. [6]. Reeves et al. [6] introduce to the IPE the concept of "learning interactively together, for the explicit purpose of improving the health or wellbeing of patients/customers".[6, p. 657]

Concerning the Collaborative Work (CW) in the health field, the most relevant recommendations highlighting the competencies to be developed by students and health professionals are expressed in publications of the expert groups of the Centre for the Advancement of Interprofessional Education - CAIPE [7] of England, the Canadian Interprofessional Health Collaborative - CIHC [8] and the Interprofessional Education Collaborative - IPEC [9] of the United States.

The Collaborative Work (CW) in Health is a need in today's world that demands comprehensive and quality care for health service users. To develop CW naturally, the professional identity of health professionals must be built based on IPE, at health professionals' undergraduate/graduate studies or other initial training courses on health. [10] When referring to CW, there is an emphasis on teamwork, prioritizing the competencies common to all health professionals, those specific to each profession, and the vital collaborative competencies to work together. [11] Such competencies, as well as abilities, values and common sense, are fundamental for working with other health professionals. [8]

For health professionals to develop teamwork, it is important to remember the WHO motto: "Learning together to work together". [12] Therefore, it is crucial to invest in teacher training and in the strategy of IPE and practice [13], which are widely documented in the United Kingdom, Canada, the United States and Australia [14]. This idea is intensified by the worldwide movement of reforms of undergraduate/graduate courses in health, in which the reflective processes of teaching practice, specifically interprofessionalism, are part of the discussion, and it is increasingly necessary to include this topic in the training curricula of future health professionals. In this sense, the present study intends to answer the question: how is interprofessional education (IPE) and collaborative work (CW) taking place in health training courses? Thus, this study was developed to understand IPE and CW, from the perspective of professional trainers of health and sociology courses of the University of Minho, in the north of Portugal.

2 Methodology

This is an exploratory study, with a qualitative approach, based on the concept of IPE. [1] To avoid confusion between IPE or CW at health initial training (undergraduate students) and in-service activity (health professionals), in this study it was decided to analyse only the process of initial training courses on health, focusing on the learning related to the health services users.

The study scenarios were the courses of medicine, nursing, psychology and sociology of the University of Minho (UMinho). The selection of subjects took into account the leaders responsible for these courses and the invitation to participate was made by e-mail or in-person with a letter of presentation from the researcher. At least two professional trainers from each course were invited to participate in a Focus Group. The nine participants were professional trainers acting in the health area of these courses. This data collection tool was chosen to identify subjective aspects related to the theme studied in a qualitative approach. [15]

The focus groups interviews were rather difficult to set up due to problems of reconciling the participants' agendas and the last minute unforeseen events. Despite all the difficulties, the place was provided by the School of Medicine in a private room with a table and mobile chairs that facilitated the organisation of a circle of interviewees and enabled their speeches more easily. The moderator and observer in the focus groups were not professional trainers, nor university employees, to avoid conflicts of interest and constraints on the part of the participants.

The Consent Form was read collectively, and signed, individually, after the researcher's assurance about the confidentiality of the information, the volunteering of the answers, and other ethical information before starting the audio recording of the speeches. After signing the Consent Form, the members of the focus group were asked to choose a pseudonym to ensure anonymity and facilitate the subsequent transcription process. The pseudonyms chosen in the first focus group were: Van Gogh (P, for psychology), Rui (M1, for medicine), Ricardo (M2), Graça (N1, for nurse-1), Sofia (N2) and Kia (N3). The second focus group was smaller and no pseudonym was used to identify the participants. It consisted of professional trainers of sociology (S), nursing (N4) and nurse/psychology (N/P).

The first focus group took 58 min, while the 2nd group 85 min. The semi-structured script for the Focus Groups was done according to Amado [16] and consisted of the following questions that were simple, clear and open: How do you perceive interprofessional education in the area of health? What about collaborative work? In your course, is there interprofessional education and collaborative work in health? What are the facilitating factors for interprofessional education and collaborative work in health? What are the barriers, constraints, or difficulties with effective interprofessional education and collaborative work in health? What are the strategies used for the development of interprofessional education in your course? Describe your experience with interprofessional education and collaborative work.

The data were transcribed and reviewed, constituting the corpus of this study that was imported into the NVivo Software Version 12.0, where the excerpts of interest were selected and stored in the respective nodes or categories and subcategories. This analysis of qualitative data content was based on Amado who refers that content analysis is an investigative method that encompasses diverse processes such as the development of concepts and the interpretation of results. [16] The process of analyzing the content takes six steps: *(i)* definition of the problem and the objectives of the work; *(ii)* explanation of a theoretical reference framework; *(iii)* constitution of a corpus of documents; *(iv)* attentive and active reading; *(v)* formulation of hypotheses; and *(vi)* categorization. [16]

This study was approved by the Subcommittee on Ethics for Social Sciences and Hu-manas of the University of Minho - SECSH 036/2018.

3   Results and Discussion

The following words stand out in the word cloud of both focus group interviews as referring to IPE and CW (Fig. 1): we/us (*nós*), work (*trabalho*), areas (*áreas*), being (*são*), medicine (*medicina*), professionals (*profissionais*), training (*formação*) thinking (*pensar*), health (*Saúde*) and collaboration (*colaboração*). These results reflect the teachers' concern with the training of health professionals to work in the 21st century, in line with the literature [1, 2, 4]. Nursing (*enfermagem*) is the most frequent profession referred, which is in agreement with the literature. [17]

Fig. 1. Word cloud from focus Groups interviews about IPE and CW in Health, UMinho.

From the teachers' statements, four categories and subcategories emerged: *(i)* conceptual clarity; *(ii)* barriers and challenges regarding the subcategories: teaching in interprofessionality, inadequate teaching plans and curricula, XXI century student profile, teamwork versus in-group, operationalisation difficulty and inadequate communication and language; *(iii)* Potentialities; and *(iv)* strategies and suggestions for IPE and CW in health. These categories and subcategories are presented below.

3.1   Conceptual Clarification About IPE and CW

There is a variety of concepts used by students, health professionals and professional trainers, along with lots of terminologies to designate collaborative practices or collaborative work in health. Silva [18] reports this lack of conceptual clarity and further say that often people make confusion between the collaborative work at initial professional training (undergraduate courses) and the in-service (health professionals) collaborative work. In addition to this conceptual confusion, there are also diverse interpretations resulting from questionnaires application. [7, 19, 20] To make it clear, it has been assumed that the interprofissional education (IPE) is a special case of professional education in which best practices of teaching and learning are applied to different contexts. [21]

Initially, the professional trainer participants' dialogues about the concepts of IPE and CW were confusing as, for example, the nursing trainer N3 said:

> *"IPE in health, I understand it as an objective work, a developed work...I do not know... In a multidisciplinary team, even in the way of sharing responsibility, but... that responsibility must be shared by decision making, attending to the health mechanisms of the patient."*[N3]

On the other hand, the medicine trainer M2 was very clear when he said:

> *"If we allow an educational environment to the students, where they start to deal with all the professions in the health field, they will be more prepared to deal with the practical work, and capable of working with those different professions. This improves the understanding of the perspectives that each* [profession] *has about the patient and the person, but also the perspective that each profession has about themselves and the others."*[M2]

Moreover, and regarding clarity, he added:

> *"More than having a doctor or a nurse talking with Medicine students about the work of Nursing, Medicine or Psychology, is putting them doing things together, is creating conditions for working together, understanding their role and the others' role in the team."*[M2]

Therefore, there is a need to achieve a conceptual consensus among professional trainers, to develop activities, courses, collaborative practices, active methodologies that are motivational and that allow an effective IPE and CW.

3.2 Barriers and Challenges

There are many barriers and challenges for the implementation of IPE and CW. Therefore, there is a need to reinforce the training of health professionals to be able to face the difficulties. [18]

The concern with the training and human resources has led to a high amount of texts related to the barriers and challenges for the exercise of IPE and CW in health. These difficulties in implementing IPE and CW have also been observed in other fields of research, such as language, communication, leadership, management of interpersonal and interprofessional conflicts. [18, 19, 22–25]

The barriers/challenges found in the present study are described below.

3.2.1 Interprofessionality Training

The first challenge immediately perceived by the interviewed professional trainers was the feeling of being unprepared for IPE or CW as they had difficulties in giving examples of other professions to work with, lack of communication, student profile, demotivation and lack of institutional incentive, among others.

The professional trainers are aware of their resistance to the interprofessional training process. The sociology trainer [S] said:

> *"I think the main difficulty is the implementation... eh... I think it is ourselves, really, professional trainers because we are... probably crystallized in what has been so far, the profession, the limits and the transmission of what I know."*[S]

The need for having more professional trainers involved with the IPE was also debated, as well as the initiative to develop IPE and collaborative practices, as the medicine trainer M1 said:

> "...need to educate other people, in short, colleagues who were not so involved in this process [...] is being a bit creative to try, to experiment. There is no one formula that works for everyone, it depends a lot on the context... it depends a lot on the kind of professionals that we want to contact, the type of work [...] the first step is for those who have responsibilities in recognizing the importance of the subject and to understand why IPE has to be a reality. Then we must have this meeting of wills."[M1]

The way how to seek for the implementation of IPE and CW is in line with what Meleis [24] proposes: to be creative, to establish plans, to define the best practices, to develop accreditation mechanisms, to educate for teamwork, to use innovative approaches, draw curriculum based on competencies and, most importantly, to recognize the urgent need to invest time, energy, intelligence and the resources to enable the implementation of IPE and CW in health.

Researches on professional trainers' training has been the focus of their professional identity and professionalization since the 2000s; they focus on the professional trainer, his/her opinions, representations, knowledge and practices as they seek to give voice to the professional trainer and better know his/her way of teaching, and so to find ways to obtain quality in education. This fact supposes a collaborative work. [26]

3.2.2   Teaching Plans and Curricula Are Inadequate

This study showed that professional trainers want a shift in the paradigm of the professional training in the health area, expressing the urgent need to adjust teaching plans and the curriculum itself, as the nursing trainer N3 said:

> "... this change is urgent! ... thinking about the training plans of the institutions that train either doctors or nurses... because there are here, I would say, some cultural issues that makes this collaboration not beneficial."[N3]

Professional trainers perceive that there is an inadequacy in teaching plans and curricula to operationalize IPE, however, they are still motivated to go further. In this sense, the nursing trainer N3 mentioned the will to do it in the future:

> "We will develop cases together and the teams will be mixed: students of Medicine and Nursing. So, [at the moment] we don't have any results... let's say, of what will happen."[N3]

In contrast, the nursing trainer N1 said that they do it already:

> "In our reality, we already have collaboration with other areas in the curriculum of our course [in nursing] this collaboration already exists! [...] in the case of Nursing, already exists and is a reality."[N1]

This statement shows that the IPE is already in the progress. This phase supposes the need for professional trainer's development and meetings for planning activities that can be incorporated into the curricula of health courses.

Formal interprofessional education is easier to be accepted than interprofessional practice, which development depends on political will and professional trainers and students' compliance. [14]

3.2.3   Profile of the 21st Century Higher Education Student

In the 21st century, the students of the health courses present themselves with both positive and negative idiosyncrasies that need to be addressed. Positive experiences cited by the professional trainers were the facility that their students from different courses have to make joint activities, as reported by the nursing and psychology trainer [N/P] and the sociology trainer [S]:

> "Very quickly, they were together and with a great will! Indeed, several times, I believe that the problems are in our head!"[N/P]

> "Other positive dot in student profile: It is a big challenge that collaborative leadership! Good!"[S]

Professional trainers further referred other negative and challenging experiences, namely the fact that the students have a lack of general culture, lack of interest in class, difficulty in reasoning and decision-making, demotivation, lack of commitment, irresponsibility with tasks, difficulty in writing etc., as can be seen in the examples below expressed by the sociology trainer [S] and the medicine trainer [M1]:

> "Things are very bad, that is, it is very difficult!!! They do not put the assignments, they do not share the work, they do cheating to the end and leave."[S]

> "... when they have to make decisions together,... they are not used to it... in fact, collaborative work is critical."[M1]

It was also revealed the student resistance to the process of IPE and CW, as expressed by the sociology trainer [S]:

> "How difficult it is for students to work in groups! [...] we also assume that... they do not want to be with other colleagues from other disciplinary areas."[S]

From these statements, it becomes clear that it is important to make collaborative work effective.

3.2.4   Differences Between Team Work and Group Work

In the Health Services daily work, it can be seen a mismatch between the training process of the health professions and the focus on teamwork. It is important to clarify the differences between teamwork and group work. In a teamwork, the goal, performance and responsibility are shared and the members' abilities are complementary; in the case of group work, the aim is to share information, but the responsibility is individual and personal and skills are diverse. [27]

Ideally, one would expect communication and dialogue between the students of the various health professions since the beginning of their undergraduate studies, so that after graduation they would be able to develop collaborative work in their professional service, making activities in teamwork easily. [10, 18]

Studies on group work adopt the terms collaboration and cooperation to characterise it, but collaborative work is a well-established and recognized term worldwide. "The verb to collaborate derives from *laborare* - to work, to produce, to develop activities for a particular purpose".[28, p. 214] Indeed, the medicine trainer M1 said:

> "... whether the team works synchronously or whether there is some dissonance of a simple piece of information, that is shared... we have to work and then we have the trainer who will meet them with the perspective of the interaction difficulties they had."[M1]

3.2.5   Difficulties on Operationalizing the IPE and Teamworking

Because of the teachers' feeling of unpreparedness and demotivation, the confusion of professional identity roles, the need for conflict resolution, teachers and students' resistance to IPE and CW, it becomes very difficult to put IPE into operation. Additionally, there are "institutional, faculty, student, curricular, and corporatism resistances". [25]

The medicine trainer M2 pointed out structural disorganization, demotivation and resistance as barriers to IPE and CW in health:

> "There are barriers of the organization of the different structures and [lack of] motivation of the different professionals... Not everyone is sufficiently motivated, not everyone knows the virtue of this type of actions... they have some resistance..."[M2]

The medicine trainer M1 demystified the myth about being complicated to join people from different courses in the same activity because they think differently. He said:

> "The first step is to see that they [other professionals] are exactly like us and have no difference at all from the point of view of the brain [laughs], or anything."[M1]

In addition to structural and personal issues, there is the problem of funding. This dimension was recalled by the nursing and psychology trainer [N/P]:

> "It is very expensive to assure the thing. Anyway, we must study and learn together to work together in the future."[N/P]

In this sense, the medicine trainer M1 said:

> "If people are used to working together, it's a lot easier: Ah, I know! I got it! Ready!"[M1]

3.2.6   Difficulties in Communication and Inadequate Language

In this study, the lack of students' communication between them and of trainers' communication between them appeared very strongly, as well as the use of inadequate language which make IPE and CW difficult, as seen in these examples by the sociology trainer [S] and the medicine trainer [M1]:

> "...we need to adjust our language and, of course, look for more crosswise examples."[S]

> "Two [students] were using, apparently, different language, but they were both saying the same thing. The diagnosis was the same. [However] collaborative work only works if, in fact, there is this effort, I think. [...] it really takes a language approach."[M1]

This communication barrier to IPE, observed in collaborative practices, can be overcome by increasing the frequency of interprofessional activities. Interdisciplinarity/ interprofessional education and professionalism should be the way of setting up health teams focused on the integrality of care at all levels. [13]

3.3 Potentialities for Implementing IPE and CW

Several potentialities were observed for favouring IPE and CW in health. The one that stood out most was the teacher motivation and awareness. In this sense the medicine trainer [M1] and the nursing trainer [N2] said:

> "... it is to introduce a change... and some teachers have some sensitivity to the issue [...] it is possible!... you just must have the will to develop interrofessional activities and to try with methodologies that ensure attending a large number of students... and that are well accepted by students."[M1]
> "We have great conditions because we are the only university [in Portugal] that has both courses: Medicine and Nursing."[N2]

Another potentiality is the development of skills to work as a team, and more supporters are needed. Indeed, previous studies have pointed points out CW potentialities among teachers: "... collaborative work between trainers has the potential to enrich their way of thinking, acting and solving problems, creating possibilities for success in the difficult pedagogical task."[28, p. 218]

3.4 IPE and CW Strategies and Suggestions

In this study, the characteristics of CW were observed by the professional trainers who described several experiences, involving health courses, as well as Sociology and Law courses linked to health subjects. Indeed, in collaborative work, the members of a group establish horizontal relationships, shared leadership, with confidence and responsibility in the development of actions. [28] Such CW can occur among trainers, being an excellent learning space, socializing knowledge, identifying weaknesses and strengths, as well as opportunities for changes in pedagogical practices. Furthermore, collaborative work among students facilitates socialization, meaningful learning and development of communication skills, autonomy, leadership and decision making. [28]

The interviewed trainers mentioned some specific activities, such as simulation of clinical cases with medicine and nursing students together. Students accepted very well these practical activities and the trainers were pleased and motivated to move forward, as the nurse N1 said:

> "... to put together the students, in mixed groups, to solve the practical cases where the collaboration of the two areas is necessary to respond to the needs of the simulated moment [...] in the sense of developing this interprofessionalism... there are initiatives of the students, for example, joint debates on subjects... I remember an activity... [of nursing students] with law students on issues of vaccinations legislation".[N1]

Only a few suggestions to promote IPE and CW were given. However, the nursing trainer [N1] proposed:

> "Maybe I would create a School of Health, a School that would have Medicine and Nursing together, maybe... that would have all the areas involved and that both courses would be thought in this logic [IPE and CW], maybe this is the ideal."[N1]

4 Conclusions

In this study, there were other barriers and difficulties to be faced than the potentiating factors in the construction of IPE and CW in health, indicating how hard it is to implement this way of working. After all, the traditional uniprofessional education has been rooted for many years which keeps the power of the isolated professions. On the other hand, the professional trainers were very receptive and motivated to build together a training course more adequate to the health demands of the population. Therefore, it is necessary to learn to work together by joint practices of the different professions, to get the availability and respect to work with other professions.

This required change involves professional trainers' motivation and skills development, institutional encouragement and support, joint planning, students' sensitization to "learn together, work together" and other strategies. For this to happen, one must believe in the benefits to all people involved; it requires the willingness to act, commitment and involvement to achieve the proposed goals and to include them in the curriculum of the schools of the health professions.

A metodologia utilizada neste estudo deu conta de compreender a subjetividade das questões envolvidas no processo de implantação da IPE e das práticas colaborativas em saúde em uma universidade do norte de Portugal. Enfim, esse estudo poderá subsidiar ações no processo de implementação e/ou manutenção da IPE e trabalho colaborativo em saúde nas Instituições de Ensino Superior que tenham interesse em melhoria da qualidade da formação dos profissionais que cuidam da saúde dos seres humanos.

The methodology used in this study made it possible to understand the subjectivity of the issues involved in the implementation process of IPE and CW practices in health at the University of Minho. Although being a case study, this work may support actions in the process of IPE and CW setting up and/or improving health in higher education institutions that are interested in improving the quality of the training of professionals who are involved in the health care of people and communities.

Acknowledgements. This work was partially funded by Portuguese national funds through the FCT (Foundation for Science and Technology) within the framework of the CIEC (Research Center for Child Studies of the University of Minho) project under the reference UID/CED/00317/2019.

References

1. WHO – World Health Organization: Framework for action on interprofessional education & collaborative practice. Geneva, World Health Organization. Disponível em: (2010). http://whqlibdoc.who.int/hq/2010/WHO_HRH_HPN_10.3_eng.pdf
2. Baar, H., Hammick, M., Freeth, D., Koppel, I.: Effective Interprofessional Education. Blackwell Publishing Ltd., London (2005)
3. McNair, R., Stone, N., Sims, J., Curtis, C.: Australian evidence for interprofessional education contributing to effective team work preparation and interest in rural practice. J. Interprof. Care **19**, 579–594 (2005)

4. Baar, H., Low, H.: Introducing Interprofessional Education, Centre for the Advancement o Interprofessional Education – CAIPE, London, UK (2013)

5. Frenk, J., et al.: Health professionals for a new century: transforming education to strengthen health systems in an interdependent world. Lancet Commissions 376(9756), 1923–1958 (2010)

6. Reeves, S., Fletcher, S., Baar, H., Birch, I., Boet, S., Davies, N., Kitto, S.: A BEME systematic review of the effects of interprofessional education: BEME Guide no. 39. Med. Teacher 38(7), 656–668 (2016)

7. CIHC - Canadian Interprofessional Health Collaborative: Interprofessional Education & Core Competencies (2007). http://www.cihc.ca

8. IPEC – Interprofessional Education Collaborative: Core competencies for interprofessional collaborative practice: 2016 update. Washington, DC: Interprofessional Education Collaborative (2016)

9. Peduzzi, M.: Educação Interprofissional para o Desenvolvimento de Competências Colaborativas em Saúde *[Interprofessional education for the development of collaborative skills in health]*. In: Toassi, (Org.). Interprofissionalidade e formação na saúde: onde estamos?. São Paulo, Brasil, pp. 40–48 (2017)

10. Rossit, R.A.S., Freitas, M.A.O., Batista, S.H.S.S., Batista, N.A.: Construção da identidade profissional na Educação Interprofissional em Saúde: percepção dos egressos [*Professional identity construction in Interprofessional Health Education: perception of graduates]*. Interface comunicação, saúde e educação 22(Supl.1), 1399–1410 (2018)

11. Rossit, R.A.S., Batista, N.A., Batista, S.H.S S.: Formação para a integralidade no cuidado: potencialidades de um projeto interprofissional *[Training for integral care: the potential of an interprofessional project]*. Revista Internacional de Humanidades Médicas. Disponível em, vol. 3, no. 1 (2014). http://tecnociencia-sociedad.com/revistas/

12. WHO – World Health Organization: Learning together to work together for health. Report of a WHO study group on multiprofessional education of health personnel: the team approach. In: World Health Organization Technical Report Series. Geneva: World Health Organization, vol. 769, pp. 1–72 (1988)

13. Batista, N.A., Batista, S.H.S.S.: Docência em saúde: temas e experiências [Health training: subjects and experiences], 2ª edn. Editora SENAC, São Paulo, Brasil (2014)

14. Ceccim, R.B.: Conexões e fronteiras da interprofissionalidade: forma e formação *[Connections and frontiers of interprofessionality: form and training]*. Interface comunicação, saúde e educação 22(Supl.2), 1739–1749 (2018)

15. Dias, C.A.: Focus Group: Tech. Collect. Data Qualitative Res. 10(2), 1–12 (2000)

16. Amado, J.: Manual de Investigação Qualitativa em Educação [*Handbook of Qualitative Research in Education]*, 2ª edn. Imprensa da Universidade de Coimbra, Coimbra (2014)

17. Peduzzi, M., Norman, I., Coster, S., Meireles, E.: Adaptação transcultural e validação da readiness for interprofessional learning scale no Brasil *[transcultural adaptation and validation of the readiness for interprofessional learning scale in brazil]*. Revista da Escola de Enfermagem da USP 49(Esp2), 7–15 (2015)

18. Silva, R.H.A.: Interprofessional education in health graduation: evaluation aspects of its implementation in Marília Medical School (famema). Educar em Revista, Curitiba: Editora da UFPR 39, 159–175 (2011)

19. McPherson, K., Headrick, L., Moss, F.: Working and learning together: good quality care depends on it, but how can we achieve it? Quality in Health Care, 10, suppl II, pp. ii46–ii53 (2001)

20. Oliveira, C.M., Batista, N.A., Batista, S.H.S.S., Uchoa-Figueiredo, L.R.: A escrita de narrativas e o desenvolvimento de práticas colaborativas para o trabalho em equipe *[The writing of narratives and the development of collaborative practices for teamwork]*. Interface, Comunicação, Saúde **20**(59), 1005–1014 (2018)

21. Freeth, D.: Interprofessional Education. in Swandick, T.: (Org.), Understanding Medical Education: Evidence, Theory and Practices, pp. 81–96. Wiley, Ltd., London, UK (2014)

22. Thadani, K.: Critical review: Traditional versus interprofessional education: Impact on attitudes of health care professionals. Copyright (2008)

23. Paradis, E., Whitehead, C.: Louder than words: power and conflict in interprofessional education articles, 1954-2013. Med. Educ. **49**, 399–407 (2015)

24. Meleis, A.I.: Interprofessional education: a summary of reports and barriers to recommendations. J. Nurs. Scholarsh. **48**(1), 106–112 (2016)

25. Nuto, S.A.S., Lima Júnior, F.C.M., Camara, A.M.C.S., Gonçalves, C.B.C.: Avaliação da Disponibilidade para Aprendizagem Interprofissional de Estudantes de Ciências da Saúde [Assessment of the Availability for Interprofessional Learning of Health Sciences Students]. Unifor, Fortaleza (2016)

26. André, M.: Formação de professores: a constituição de um campo de Estudos *[teacher training: the constitution of a field of studies]*. Educação, Porto Alegre **33**(3), 174–181 (2010)

27. Peduzzi, M.: Multiprofessional healthcare team: concept and typology. Rev. Saúde Pública **35**(1), 103–109 (2001)

28. Damiani, M.F.: Understanding collaborative work in education end its benefits. Curitiba, Educar **31**, 213–230 (2008)

Author Index

© Springer Nature Switzerland AG 2020
A. P. Costa et al. (Eds.): WCQR 2019, AISC 1068, pp. 301–302, 2020.
https://doi.org/10.1007/978-3-030-31787-4

Printed in the United States
By Bookmasters